Proceedings of the 13th Reinventing Space Conference

Scott Hatton
Editor

Proceedings of the 13th Reinventing Space Conference

Springer

Editor
Scott Hatton
The British Interplanetary Society
London, United Kingdom

ISBN 978-3-319-32816-4 ISBN 978-3-319-32817-1 (eBook)
https://doi.org/10.1007/978-3-319-32817-1

Library of Congress Control Number: 2017960954

Printed on acid-free paper

This Springer imprint is published by the registered company Springer International Publishing AG part of Springer Nature.
The registered company address is: Gewerbestrasse 11, 6330 Cham, Switzerland

Contents

1 **Exploring our Solar System with cubesats and nanosats** ... 1
Anthony Freeman, Charles Norton

2 **Cubesat-scale high-speed laser downlinks** ... 7
Richard Welle, Siegfried Janson, Darren Rowen, Todd Rose

3 **The worlds first commercial SAR and optical 16-satellite constellation** ... 19
Amar Vora, Liam Sills, Andrew Cawthorne, Caroline Slim, Rachel Bird, George Tyc, Wade Larson

4 **Resource Prospector: A cost effective lunar resource pathfinder** ... 29
D. Andrews

5 **Firefly – new generation of low cost, small launch vehicles designed to serve the rapidly growing small satellite market** ... 39
Tom Markusic, Shey Sabripour, PJ King, Andy Bradford

6 **HeL1o : The first cubesat to L1?** ... 49
Cyrille Tourneur, Matthew Stuttard, Markos Trichas, Jonathan Eastwood, John Bellardo

7 **Innovative small launcher** ... 59
A.J.P. van Kleef, B.A. Oving, C.J. Verberne, B. Haemmerli, M. Kuhn, I. Müller, I. Petkov

8 **The Google Lunar XPRIZE - past, present and future** ... 75
Andrew Barton, Nathan Wong, Derek Webber

9 **SEAHAWK: A nanosatellite mission for sustained ocean observation** ... 83
Hazel Jeffrey, Hessel Gorter, Alasdair Gow, Craig Clark, Alan Holmes, Craig Herrin, Linda Sasaki, John Morrison

10 **Cost disruptive reflector surface for large deployable antennas** ... 95
Mike Lawton, Juan R Reveles, Zhong You, Ashley Dove-Jay, Amjad Khan, Vincent Fraux

11 **An analytical, low-cost deployment strategy for satellite constellations** ... 107
Ciara McGrath, Emma Kerr, Malcolm Macdonald

12 **A new era in space flight: The COTS model of commercial partnerships at NASA** ... 117
Dennis Stone

13 **IODISPLay: Capturing European needs and capabilities for in-orbit demonstration of space technologies** ... 125
Giovanni Binet, Gabriele Novelli, Celestino Gomez Cid, Marco Bolchi

14 **Optimal (not opportunity) orbits for rideshare payloads** ... 137
Chris Pearson, Marissa Stender, Chris Loghry, Joe Maly

15 Gateway Earth - low cost access to interplanetary space 149
Derek Webber

16 New superlight class launch vehicles From Yuzhnoye 171
Olexander Degtyarev, Mykola Lytvyn, Oleh Ventskovsky

17 Small satellite launch vehicle from a balloon platform 177
Kieran Hayward, Jose Mariano Lopez Urdiales

18 Reinventing constellations: the effectiveness of rideshare approaches for constellation deployment 193
Joseph W. Gangestad, James R. Wilson, Kristin L. Gates, John V. Langer

19 MicroLaunch: The electric rocket 207
A. Bacon

20 Design criteria of remote sensing constellations of small satellites with low power electric propulsion and distributed payloads 229
Salvo Marcuccio, Stefan Gregucci, Pierpaolo Pergola

21 Feasibility study of LTA launch system for micro and smaller 239
Andrea Testore

22 Could reusable air-launch break the space access paradigm? 249
David J. Salt

23 First steps towards the Kingston space shot: low altitude test vehicle 265
Adam M Baker, Malcolm Claus

24 Enabling solutions for small satellite space access 273
Andreas Jonsson, Magnus Engström

25 Development of low cost propulsion systems for launch and in space applications 281
Peter H. Weuta, Neil Jaschinski

26 Electromagnetic launch to space 323
Ian R. McNab, Timothy R. Wolfe

27 The prospect for orbital airliners 331
David Ashford

28 Dynamically supported launch infrastructure 333
John Knapman

29 Fostering technology innovation in space through national activities: The Swiss example 343
Grégoire Bourban, Volker Gass, Johann Richard

Exploring our Solar System with cubesats and nanosats

Anthony Freeman
Jet Propulsion Laboratory, California Institute of Technology,
MS 301-335, 4800 Oak Grove Drive, Pasadena CA 91109, USA;
Tel: (818) 354 1887
e-mail: anthony.freeman@jpl.nasa.gov

Charles Norton
Jet Propulsion Laboratory, California Institute of Technology,
MS 301-165, 4800 Oak Grove Drive, Pasadena CA 91109, USA;
Tel: (818) 393-3920
e-mail: charles.d.norton@jpl.nasa.gov

ABSTRACT

The Jet Propulsion Laboratory (JPL) is NASA's lead center for robotic exploration of our solar system. We are known for our large, flagship missions, such as Voyager, which gave humanity its first close look at Jupiter and Saturn; and the Mars Rovers, which have excited millions worldwide with their daring landing exploits. Less familiar to those outside NASA may be our role in developing the Kepler mission, which has discovered more than 2000 planets around other stars; or the recently launched Soil Moisture Active Passive (SMAP) mission, one of many JPL Earth Science missions.

A recent JPL initiative has emphasized low cost missions that use rapidly evolving technology developed for cubesats and nanosats to explore our solar system. Costs are significantly lower (by one or two orders of magnitude) than for conventional JPL missions, and development time is also significantly shorter. At present 21 such cubesat flight projects are under way at the laboratory with various partners: some in flight, some in development, some in advanced formulation. Four are planned as deep space missions. To succeed in exploring deep space cubesat/nanosat missions have to address several challenges: the more severe radiation environment, communications and navigation at a distance, propulsion, and packaging of instruments that can return valuable science into a compact volume/mass envelope. Instrument technologies, including cameras, magnetometers, spectrometers, radiometers, and even radars are undergoing miniaturization to fit on these smaller platforms. Other key technologies are being matured for smallsats and nanosats in deep space, including micro-electric propulsion, compact radio (and optical) communications, and onboard data reduction. This paper will describe missions that utilize these developments including the first two deep space cubesats (INSPIRE), planned for launch in 2017; the first pair of cubesats to be sent to another planet (MARCO), manifested with the InSight Mars lander launch in March of 2016; a helicopter "drone" on Mars to extend the reach of future rovers; plans for a Lunar Flashlight mission to shine a light on the permanently shadowed craters of the Moon's poles; a Near Earth Asteroid cubesat mission; and a cubesat constellation to demonstrate time series measurements of storm systems on Earth.

From these beginnings, the potential for cubesats and nanosats to add to our knowledge of the solar system could easily grow exponentially. Imagine if every deep space mission carried one or more cubesats that could operate independently (even for a brief period) on arrival at their target body. At only incremental additional cost, such spacecraft could go closer, probe deeper, and provide science measurements that we would not risk with the host spacecraft. This paper will describe examples including a nanosat to probe the composition of Venus' atmosphere, impactors and close flybys of Europa, lunar probes, and soft landers for the moons of Mars. Low cost access to deep space also offers the potential for independent cubesat/nanosat missions – allowing us to characterize the population of near Earth asteroids for example, deploy a constellation around Venus, or take closer looks at the asteroid belt.

KEYWORDS: NASA Deep Space Missions, cubesats, nanosats

S. Hatton (ed.), *Proceedings of the 13th Reinventing Space Conference*,
https://doi.org/10.1007/978-3-319-32817-1_1

INTRODUCTION

In early 2013, a small group at JPL formed what became known as the 'cubesat kitchen cabinet', with the aim of creating an environment in which cubesats and nanosats for science could prosper at the laboratory. Science in this case of JPL means observations for Earth Science to study our planet's environment and climate, Astrophysics observations to explore the origins of the Universe and how stars form, Heliophysics measurements to study the interaction of the Earth and the Sun, and Planetary Science to study our neighbors in the solar system. At the time the group was formed, we had just two active cubesat projects in development – IPEX[1] and CHARM[1] (later re-named RACE), and had successfully flown just one cubesat, MCubed/COVE-2[1] (a re-flight of a prior mission from 2011 that experienced an on-orbit anomaly shortly after deployment). IPEX and MCubed/COVE-2 were technology demonstration missions, while RACE, lost during the Antares Orb-3 launch explosion, would have been the first radiometer science mission to measure liquid water path and precipitable water vapor. The potential for cubesat/nanosat missions to enable science return was evident from a number of internal studies we conducted, and NASA's Innovative Advanced Concept (NIAC) program had funded a groundbreaking study of deep space cubesats led by Rob Staehle of JPL[2]. Usually NIAC studies have a time horizon decades out but it was clear that the pace of change in the world of cubesats was picking up speed, bringing what was thought to be a distant future in much closer. Surveying the cubesat community of the time, it seemed obvious that cubesats and nanosats in Low Earth Orbit were primed for exponential growth and that the potential was there for deep space cubesats/nanosats to take off in similar fashion.

The cubesat kitchen cabinet is largely an adhoc group, led out of JPL's Innovation Foundry. Membership comprises managers at the lab who are of course enthusiastic about cubesats and nanosats, but also knowledgeable about missions in a NASA context, and in a position to make decisions (or strongly influence them) about the laboratory's investments, promote good ideas, fund studies, and steer people with good ideas towards the right funding opportunity or partnership for that opportunity. As a measure of our success, at the time of writing, JPL has twenty-one 'live' cubesat projects, at different stages of development with our partners, but all funded. Four of these projects are deep space cubesat missions.

MISSIONS

Figure 1 illustrates exploration of our solar system using cubesats/nanosats. The inner solar system is considered accessible to both free-flying and mother-daughter configuration cubesats/nanosats, while the outer solar system is currently compatible with just the mother-daughter configuration. Deep Space missions are challenging, whatever their size, and there are lots of problems to solve for deep space cubesats/nanosats, including: propulsion; communications at large distances; surviving the radiation environment; power management; attitude determination

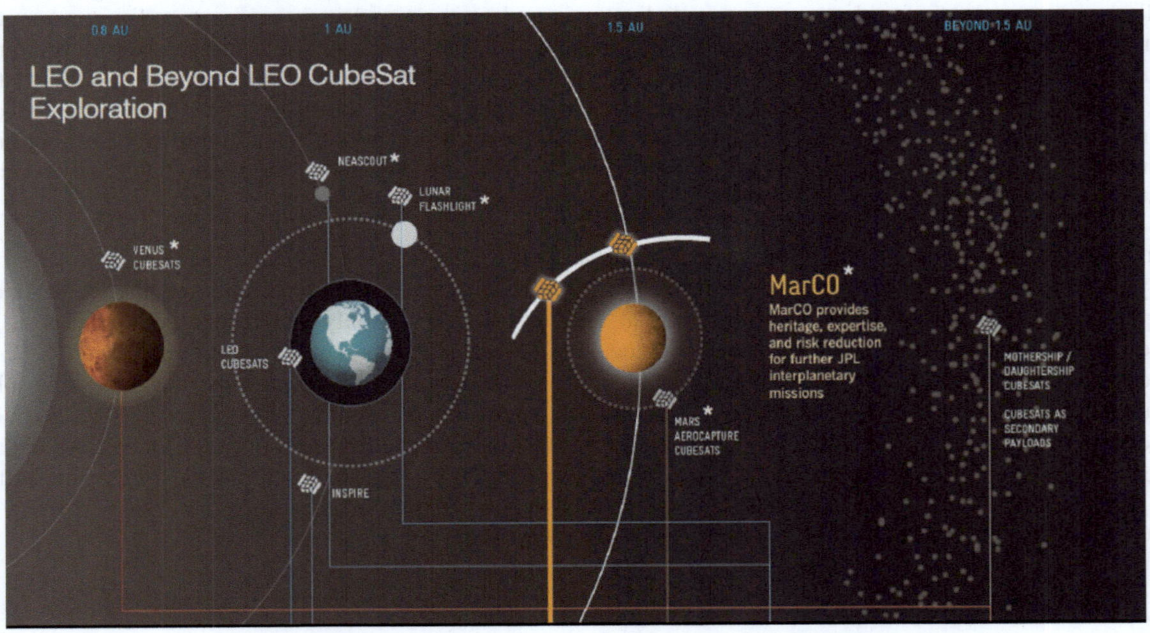

Figure 1: Exploring our solar system with cubesats

and control, thermal balancing; energy storage; proximity operations; autonomy; mission assurance and reliability; multi-mission ground operations; planetary protection; hazard avoidance; and flight software standards. With INSPIRE and MarCO, the first true deep space cubesats, JPL is tackling these problems, as pathfinders for the space science community, whose interest in such missions is building rapidly[3]. The INSPIRE (Interplanetary NanoSpacecraft Pathfinder In Relevant Environment) spacecraft are already assembled, integrated and flight-qualified, awaiting only a ride on an Earth escape trajectory. INSPIRE (Figure 2) will flight prove key technologies for deep space cubesats, and demonstrate operations, communications and navigation of such missions. Each of the two spacecraft host a compact magnetometer, to characterize the Sun's magnetic field. The MarCO (Mars Cube One) pair of spacecraft are scheduled to launch on the same vehicle as NASA/JPL's InSight mission, but will make their way independently to Mars. These first interplanetary cubesats will execute a flyby of the red planet, during which time they will relay engineering telemetry from InSight as it lands, out of direct line of sight from Earth.

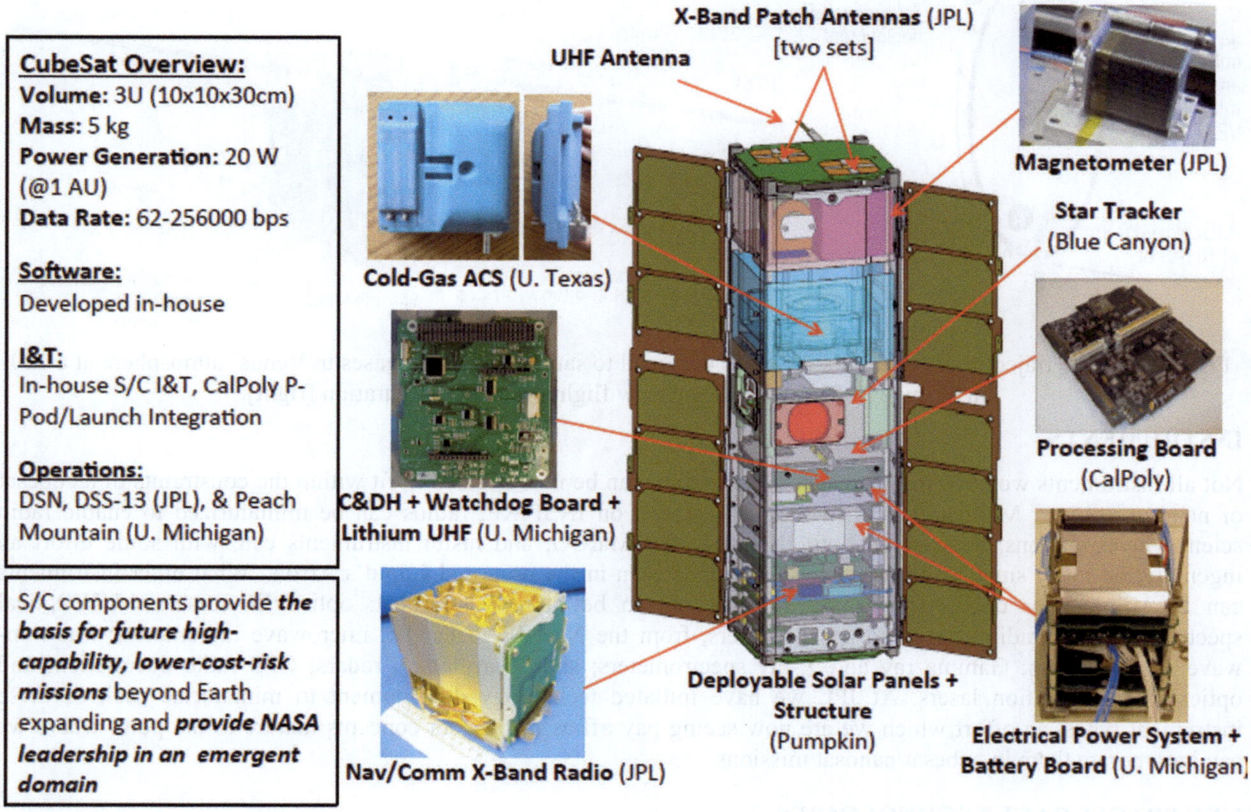

Figure 2: The InSight Spacecraft will flight prove critical technologies for deep space cubesats

Other Deep Space cubesat projects that are currently active at JPL include Lunar Flashlight[1], a mission that will utilize lasers to illuminate the permanently darkened craters of the Moon's poles, to probe the composition of the regolith there; and NEAScout[1], which will deploy a solar sail to achieve a trajectory that puts it on a path to rendezvous with a Near-Earth Asteroid.

JPL recently proposed seven cubesat/nanosat missions as technology demonstrations to augment larger missions proposed to NASA's Discovery program[4]. Each cubesat or nanosat was carried in a mother-daughter configuration to its destination in the solar system, which ranged from Venus to main belt asteroids, a Jovian comet, and Phobos[5]. Each provided a unique capability that augmented the science of the companion (mother) Discovery mission. Some were flybys of the target body to offer a closer look than could be risked with the main spacecraft, others provided insitu measurements.

In particular, the "Cupid's Arrow" nanosat, proposed as an optional additional payload on the VERITAS Venus orbiter, enables high payoff science at a fractional additional cost. Released by the mother spacecraft after Venus orbit insertion, the nanosat uses a new, ultra-compact Quadrupole Ion Trap Mass Spectrometer (QITMS) with unrivaled sensitivity to determine atmospheric noble gas abundances and isotope ratios.

As proposed, the VERITAS mother ship carries Cupid's Arrow to Venus' orbit (Figure 3a) and it is deployed during aerobraking. The nanosat requires a relatively small nudge of 1.25 m/s of ΔV in the along-track direction to send it on its path, dipping below Venus' homopause, at less than 120 km altitude. The compact mass spectrometer then pops open a cap and ingests a sample for analysis. The nanosat exits the atmosphere and relays its measurement data via a UHF communication link to VERITAS from a range of ~1000 km. Cupid's Arrow leverages mature, flight-qualified Avionics and other subsystems developed for MarCO and INSPIRE. The total mass for Cupid's Arrow is less than 25 kg.

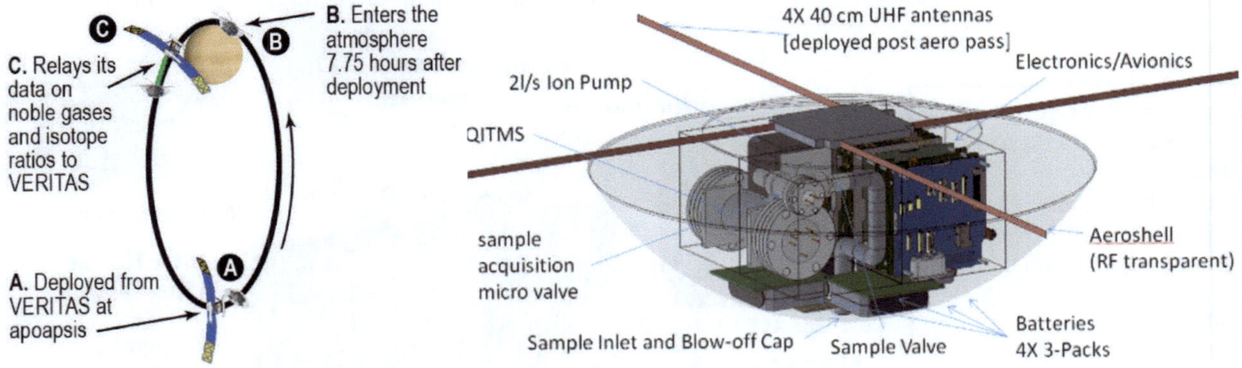

Figure 3: a) The trajectory for Cupid's Arrow is designed to sample the noble gases in Venus' atmosphere at a 120 km altitude [left]; b) Cupid's Arrow flight system configuration [right].

INSTRUMENTS

Not all instruments we need for deep space exploration can be miniaturized to fit within the constraints of a cubesat or nanosat volume. Magnetometers fit the bill, as seen on INSPIRE, radios can be miniaturized to enable radio science investigations, as seen on both INSPIRE and MarCO, and insitu instruments can, with some effort an ingenuity, be made small and low-power enough, as seen in the proposed Cupid's Arrow. What other instruments can be tailored for cubesats/nanosats? It turns out to be quite a long list: optical/IR cameras; UV/Optical spectrometers; IR radiometers and spectrometers, from the Near-IR to Far-IR; microwave radiometers; Sub-mm-wave spectrometers; Gamma ray and X-ray spectrometers; short wavelength radars; GPS radio occultation; and optical communication lasers. At JPL we have initiated technology development to miniaturize each of these instrument types, an effort which we are now seeing pay off as instrument concepts mature to the point where we can incorporate them in cubesat/nanosat missions.

KEY SPACECRAFT TECHNOLOGIES

JPL has invested in some critical spacecraft technologies (Figure 4)needed for deep space cubesats/nanosats, including: a low mass radio transponder; reflectarrays for X-band and Ka-band telecom; a compact, deployable

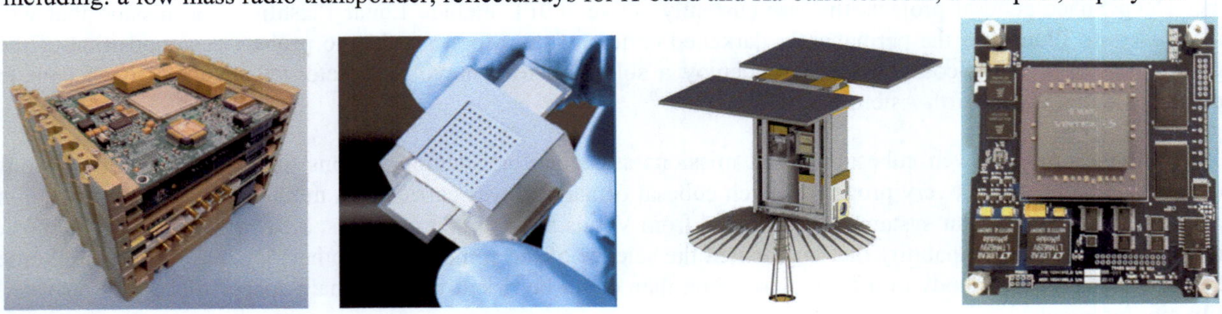

Figure 4: Examples of JPL spacecraft technology development for cubesats/nanosats, and the corresponding mission they will be demonstrated on. From left to right – the deep space transponder (INSPIRE and MarCO), micro-electric spray propulsion (TBD), compact, deployable 0.5 m diameter reflector (RainCube), and onboard data reduction board (M-Cubed/COVE-2)

Ka-band 0.5 m diameter reflector antenna; Micro-Electric Propulsion (MEP) that can provide up to 1 km/s of Delta-V; the design of a Deep Space P-POD to deploy cubesats on mother-daughter configuration missions; and onboard data reduction and data handling to significantly reduce science data volumes.

INFRASTRUCTURE

To support the current and projected surge in deep space cubesats/nanosats, JPL has also invested in improvements to our infrastructure as illustrated in Figure 5.

Figure 5: From left to right: Communication and Navigation protocols using the Deep Space Network; Team Xc for fast formulation of cubesat/nanosat mission/spacecraft concepts such as MarCO and Cupid's Arrow; the Cubesat Development Lab for cubesat development, integration and test.

PROJECTING FORWARD

The "cubesat kitchen cabinet" at JPL made a projection three years ago of exponential growth in deep space cubesat/nanosat missions. So far, as seen in Figure 6, based on the current projection of planned missions from NASA, ESA, and commercial entities, our prediction appears to be on track. Of course, these plans are fluid – as

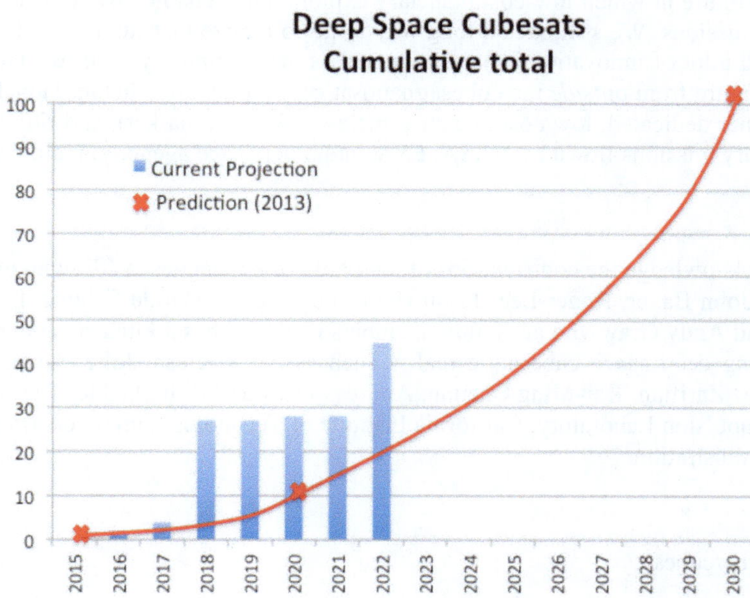

Figure 6: Projection of Exponential Growth in Deep Space Cubesats and currently planned Deep Space Cubesat Missions out through 2022.

anyone involved in the space business for an extended period knows: launch manifests can change overnight. So, having seen the pace of developments in deep space cubesats/nanosats accelerate over the last few years, what is the secret to ensuring this projection of exponential growth becomes reality? We would argue that, to adapt a famous saying from NASA's past by former Administrator Dan Goldin, they need to be executed Faster, *Smarter* and Cheaper than conventional space missions. What we mean by that is summarized in Table 1 in a list of do's and don'ts for cubesat and nanosat missions of the future that explore our solar system.

Table 1: What Faster, Smarter, Cheaper does and does not mean in the context of cubesat/nanosat missions.

Short (1-2 year) development cycles	✔	Lengthy (8-15 year) development cycles with extended Phase A	✖
Take advantage of today's incredible advances in instrument miniaturization	✔	Only use heritage instruments	✖
Take advantage of today's incredible advances in software	✔	Don't fly current-generation microprocessors. Insist all science processing be done on ground	✖
Cubesat form factor encourages use of modular subsystems	✔	Design everything to meet a specific set of mission requirements	✖
Keep launch costs low	✔	Dedicated launches on rockets designed primarily for 1000s of kg GEO comsats	✖
Create ride-along opportunities on <u>all</u> planetary missions	✔	Everything has to be decadal survey-approved science	✖

SUMMARY

We have described a bright future in which low-cost planetary exploration is enabled by compact, but capable, deep space cubesat and nanosat missions. We still have a long way to go to realize that future but the pace of change in this area is accelerating, and a lot of innovation is happening across the community. The two factors that will have the most influence on this future from *outside* the cubesat/nanosat community are whether launch costs can be kept low (and in particular whether dedicated, low cost launch vehicles make it to market), and ride-along opportunities can be created on <u>all</u> planetary missions flown by NASA, ESA, and other space agencies, including the UK.

ACKNOWLEDGMENTS

The authors would like to acknowledge the contributions of many of our colleagues at JPL who make up our cubesat community, but especially John Baker, Roger Lee, Jason Hyon, Peter Kahn, Harald Schone, Les Deutsch, Pamela Walker, Pat Beauchamp, and Andy Gray who are fellow members of the 'cubesat kitchen cabinet'; and some of the practitioners who are making deep space cubesats a reality such as Andy Klesh, Julie Castillo-Rogez, Courtney Duncan, Lauren Halatek, Pez Zarifian, Kar-Ming Cheung, Andreas Frick and Rob Staehle. The work described here was performed at the Jet Propulsion Laboratory, California Institute of Technology, under contract with the National Aeronautics and Space Administration.

REFERENCES

1. http://www.jpl.nasa.gov/cubesat/
2. Staehle, R., et al, 2012. "Interplanetary CubeSats: Opening the Solar System to a Broad Community at Lower Cost," Phase I NIAC Study Report
3. Small Satellites: A Revolution in Space Science, KISS Institute for Space Studies Report, July 2014.
4. Discovery 2014 Announcement of Opportunity , NASA Solicitation: NNH14ZDA014O, Nov 2014.
5. Sherwood, B., et al, 2015. "Planetary cubesats come of age", Proceedings of the 66th International Astronautical Congress, Jerusalem, Israel, Oct 2015.

Reinventing Space Conference
BIS-RS-2015-70

CubeSat-Scale High-Speed Laser Downlinks

Richard Welle
Siegfried Janson
Darren Rowen
Todd Rose

The Aerospace Corporation
PO Box 92957
Los Angeles, CA, 90009

13th Reinventing Space Conference
9-12 November 2015
Oxford, UK

S. Hatton (ed.), *Proceedings of the 13th Reinventing Space Conference*,
https://doi.org/10.1007/978-3-319-32817-1_2

CubeSat-Scale High-Speed Laser Downlinks

Richard Welle, Siegfried Janson, Darren Rowen, Todd Rose
The Aerospace Corporation
PO Box 92957, Los Angeles, CA, 90009
welle@aero.org

ABSTRACT

The recent surge in the development of new commercial services hosted by satellites in low Earth orbit will lead to rapid increases in the demand for data downlinking, presenting a challenge for conventional radio-frequency communication systems. Optical communication systems offer a significant but so far unrealized potential for ultra-high-volume downlinking. A number of demonstration laser communication missions have flown in space, but no optical systems are in routine operational use. Existing optical communication systems are typically too large for use in next-generation small commercial satellite systems.

The Optical Communication and Sensor Demonstration (OCSD) is a three-spacecraft program that will provide a technology demonstration of the first laser communication system designed for use in CubeSats. Simplicity and low mass is achieved by dispensing with the gimbal system and using the spacecraft ACS to point the laser beam. The engineering model of the OCSD, launched in October 2015, was intended for risk-reduction for the two flight units to follow in spring 2016. A software anomaly cut short the testing of the ACS, but the spacecraft will still be used to test many other new flight systems. The anomaly will not delay the launch of the two flight units, which are expected to achieve downlink rates of up to 622 Mb/s. The OCSD mission demonstrates the approach to satellite development using the CubeSat paradigm, making use of frequent and inexpensive access to space to shorten development cycles, accept higher risks, and test by flying.

KEYWORDS: **[CubeSats, Laser Communication, Flight Test]**

INTRODUCTION

The development of a competitive market in space launch systems, the continuing miniaturization of electronics and sensors, and the development of new satellite standards such as the CubeSat standard that encourage rapid development cycles, have combined to create a surge in the development of new commercial services hosted by satellites in low Earth orbit (LEO). Some of these satellites have already reached orbit, but many more are yet to come. The volume of data that will be generated in LEO, with the consequent downlinking requirements, will tax the capacity of conventional radio-frequency (RF) communication systems. An alternative that has been investigated and demonstrated to varying degrees is the use of optical systems for downlinking. Although a number of systems have been demonstrated, none have been developed for the small satellites typical of next-generation space systems, and are far too large for CubeSats. At the time of this writing, no operational satellites are known to depend on optical communication links for primary download of data. Impediments to broad use of optical systems for data download include system complexity, pointing requirements, concerns with atmospheric propagation, and the lack of extensive flight heritage.

A number of demonstration laser communication missions have flown in space. The TESAT Laser Communications Terminal (LCT) has flown on the Terra-SAR-X and NFIRE satellites. This terminal has demonstrated optical communication at rates up to 5.6 Gb/s at ranges out to 8000 km. However, the terminal as flown has a mass of 35 kg and a volume of 150 l, and requires 120 W of power to operate. The next generation of this terminal is expected to see the mass increase to 53 kg, the volume to 250 l, and the power requirement to 160 W. This will allow the range to extend to greater than 45,000 km, enabling optical communication to geosynchronous Earth orbit (GEO) distances.[1]

The Lunar Laser Communication Demonstration (LLCD) experiment flew with the Lunar Atmosphere and Dust Environment Explorer (LADEE) spacecraft in late 2013. This system successfully demonstrated bidirectional optical communication between a spacecraft in lunar orbit and a terminal on the Earth's surface at data rates up to 622 Mb/s.[2] The spacecraft terminal on this mission included a coarse pointing gimbal, a fine steering mechanism, and a telescope.[3] The modem alone, independent of the output optics, has a volume of 15 l, and a mass of about 11 kg.[4]

A third demonstration mission, specifically designed for LEO-to-ground transmission is the Optical Payload for Lasercomm Science (OPALS). This mission consists of a space terminal mounted on the International Space Station (ISS) that transmits to the JPL Optical Communications Telescope Laboratory (OCTL). The space terminal in this case has a mass above 150 kg and a volume above 600 l.[5] (This large size was in part because some of the components were not vacuum rated, so had to be enclosed within a pressurized volume.)

A common feature of each of these demonstrations is that they make use of a space terminal that provides a low-to-moderate-power laser beam with a very narrow divergence angle, and a beam steering capability built into the terminal. These features drive the size, mass, and power of the terminal to values that are unsupportable on a broad range of next-generation space systems that use very small satellites. In particular, each of the terminals is far larger than a typical CubeSat, and so cannot be considered for application on these types of satellites.

The performance of a laser communication system is ultimately constrained by the ability to collect photons from the laser at the receiving station. The number of collected photons can be increased by increasing the irradiance (power per unit area) at the receiver, or by increasing the size of the collector, which will allow collection of more light from a lower-irradiance beam. However, increasing the collector size quickly leads to significant cost increases, so the preference is to increase the irradiance at the receiver. This can be accomplished by increasing the power of the laser, or decreasing the divergence (cone angle) of the laser beam. Increasing the power of the laser presents technical challenges, in laser design, laser fabrication, power-system design, and dealing with waste heat. Decreasing the laser beam divergence is, in principle, relatively easy, but transfers the technology challenge to the pointing system required to get the laser beam on the receiver. It is important to note, in addition, that increasing the power of the laser provides a linear increase in the irradiance, while decreasing the beam divergence reduces the spot area as the square of the divergence angle. Thus, doubling the laser power will result in a doubling of the irradiance at the target, while reducing the beam divergence by a factor of two will result in a factor of four increase in the irradiance. Provided, of course, that the beam still hits the target, which depends on the precision of the beam pointing system. Since increases in laser power provide only a linear benefit in irradiance, and lead to corollary issues with power supplies and thermal management, while pointing improvements lead to quadratic improvements in irradiance, it is easy to understand why developers of high-end laser communication systems concentrate efforts on beam pointing systems.

THE OPTICAL COMMUNICATION AND SENSOR DEMONSTRATION

The Aerospace Corporation, as part of its continuing series of technology-demonstration CubeSats (called AeroCubes), initiated a new project three years ago to investigate the utility of laser communication systems for application in CubeSats. The project is called the Optical Communication and Sensor Demonstration (OCSD).[6] There are two aspects of CubeSats that will affect the development of a compatible laser communication system. The first, and most obvious, is the severe limitations on size, mass, and power. As noted above, in a typical CubeSat, the mass and volume of the entire CubeSat is well less than the mass or volume of any of the laser terminals used in prior flight demonstrations. The second aspect of CubeSats may be, at first, less obvious but it is no less significant. Most CubeSats are developed and built using processes that fall within what is known as the CubeSat paradigm. In contrast to typical space systems, the relatively low cost of most CubeSats, combined with easy access to space, allows for significantly greater risk and significantly shorter development times. The low cost of launch allows developers to fly prototype models for risk reduction, followed quickly by the next generation of the spacecraft. The consequence of this for the development of a laser communication system is that it can be an iterative process, with frequent flight testing; potentially with multiple generations of development flying each year. This is the process used throughout the Aerospace CubeSat program, and is evidenced in the development of the OCSD.

The mass and volume of a CubeSat are constrained by a standardized launch-vehicle interface. Although there are some variations, most CubeSats conform to a version of the CubeSat design specification first developed by California Polytechnic State University in San Luis Obispo, California, and maintained by the CubeSat Project.[7] A standard, one-unit (1U) CubeSat is nominally a cube with dimensions of 10 cm on each side. Larger CubeSats are defined by stacking cubes (or units) such that a 3U CubeSat is 10 by 10 by 34 cm (a little extra volume is available when only one satellite is deployed from a launch tube). CubeSats as large as 18 units have been proposed, but none larger than six units has yet flown. Within the size constraints, particularly of the smaller CubeSats (3U or less), a laser terminal with all the capabilities of those flown on larger satellites would present a serious engineering challenge. However, there is still significant opportunity for less-capable (but still very capable) laser communication systems in CubeSats.

The most challenging aspects of the space-based laser terminals that have flown to date are the beam-shaping optics and the beam-pointing mechanisms. However, the requirements for complex beam shaping and pointing subsystems are driven by the application. In the three cases described above, the goals included one or more of; ultra-high data rates, very long ranges, and/or the ability to mount on a space vehicle that is not capable of pointing the beam unassisted. CubeSats, in contrast, are typically dedicated satellites with one, or at most a few, flight objectives. They will typically have data requirements that may tax CubeSat-scale RF systems, but are very modest compared to high-end optical communication capabilities. In addition, the spacecraft itself can be designed to accommodate the requirements of the laser communication system, including beam pointing. These factors allowed the OCSD to compromise in the requirements for beam pointing, which allows significant reduction in system complexity relative to the optical communication demonstrations that have taken place to date.

The AeroCube Optical Communication and Sensor Demonstration (OCSD) mission has as its primary goal to explore and demonstrate the potential utility of optical communication for downlinking of data from CubeSats or CubeSat-scale spacecraft in Low Earth Orbit (LEO). These spacecraft have been developed with the support of NASA's Small Spacecraft Technology Program (SSTP) under the Space Technology Mission Directorate. The OCSD mission consists of three spacecraft. In keeping with the CubeSat interface control requirements, the laser system and its concept of operation (CONOPS) were designed to fit within the volume, mass, and power limitations of a 1.5U CubeSat (10x10x17 cm, 2.4 kg). In keeping with the CubeSat paradigm, the three satellites were developed in two models, with OCSD-A being the engineering model, and OCSD-B & -C being refined versions that will be the flight models. The engineering model, shown in figure 1, was launched in October 2015, and the flight models are expected to be launched in early 2016.

Figure 1. Photograph of OCSD-A

The design trade that allowed for a laser terminal small enough to fit in a 1.5U CubeSat was the decision to eliminate the gimbal system, hard mount the laser to the spacecraft body, and to depend on the spacecraft attitude control system (ACS) to point the laser transmitter at the receiver. The consequence of this was that we had to accept the pointing capability of the ACS as the controlling parameter in setting the beam divergence of the laser. This divergence, along with the range and the laser power, define the laser irradiance at the receiver and, consequently, the maximum data rate of the system.

When the design effort started there were several uncertainties, including the expected performance of the ACS and the maximum laser output power that we could expect to generate within the mass and volume constraints of the CubeSat. Most of the development effort in this program has been focused on improving the performance of the ACS, and on developing a laser transmitter compatible with a CubeSat.

ACS

The attitude control system uses several sensors to determine the spacecraft attitude, and sets of reaction wheels and magnetic torque rods to control the attitude. The sensors include several to provide coarse (approximately 1 degree) attitude knowledge, two additional independent sensor systems for fine attitude knowledge, and two different microelectromechanical (MEMS) gyroscopes to measure spacecraft rotation rates. The coarse sensors include magnetometers, infrared Earth-nadir and Earth-horizon sensors, and sun sensors. Of these, the sun sensors, Earth-nadir sensors, and magnetometers have flown on several previous AeroCube spacecraft, and their performance is well-understood.[8] The Earth-horizon sensors have flown previously on AeroCube-6, however this is a spin-stabilized spacecraft so, while the sensors have been confirmed to operate in space, they have not been demonstrated in the same operational mode (3-axis stabilized) that will be required on OCSD. Based on experience with previous flights, this suite of coarse attitude sensors is expected to provide attitude knowledge to within one degree without difficulty.[8]

The fine attitude sensors include star cameras and an uplink laser beacon detector, both of which were developed specifically for this mission. The star cameras, which have been described in detail elsewhere,[8] are part of a larger camera system developed for this mission (see figure 2). Several prior AeroCubes flew with simple VGA cameras that were integrated using hardware and processes developed several years ago and first flew on AeroCube 2.[9] These cameras produced relatively-low-resolution photographs that were compatible with the limited downlink capability (data rates less than 500 kb/s) of earlier AeroCubes. The camera system in OCSD includes a single camera control board that operates five cameras. One of these is a high-definition video/still camera with the primary purpose of collecting enough data to fill the downlink capacity of the laser system. A second camera provides a wide angle view from the nadir face of the spacecraft and is intended primarily for obtaining contextual images during laser downlinking; essentially a photographic verification of the laser pointing direction. The third camera is used in imaging a partner spacecraft during proximity operations (the second mission objective of the OCSD). The last two cameras are star cameras, optimized for imaging of the starfield to provide fine attitude knowledge during laser downlink

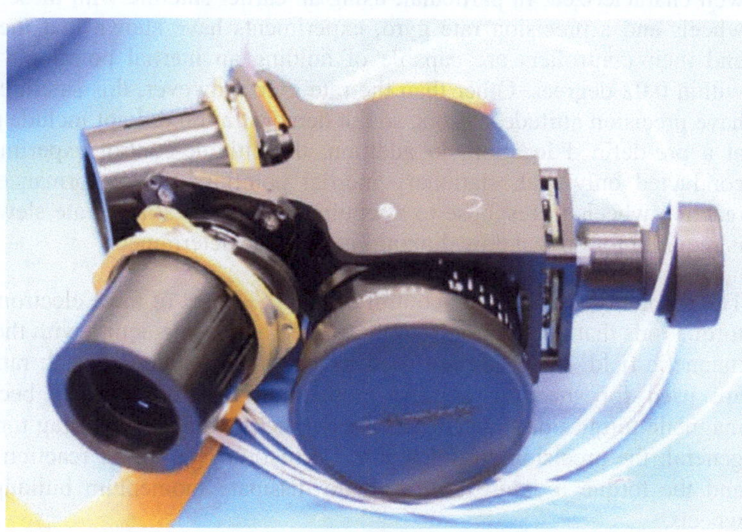

Figure 2. Photograph of the OCSD camera assembly. The two star cameras are on the left.

The star camera system includes the two cameras and two microprocessors for analyzing the images to obtain pointing information. The first of these processors works on the raw image by executing various filtering steps to extract the pixel (or sub-pixel) locations of stars in the image. These locations are then transferred to a second processor that compares the locations to a star catalog to determine a pointing direction. Because the coarse orientation of the satellite is known, the star locations can be compared to a small subset of the whole-sky star catalog. The system has been tested on the ground and demonstrated to be able to provide pointing knowledge to approximately 0.02 degrees, with updates at a rate of at least once per second.

The second fine attitude sensor is an uplink beacon detector. This sensor consists of a quad photodiode and appropriate optics to focus the light from a laser beacon onto the quad diode. The laser beacon will be mounted on the optical ground station to provide a signal that the spacecraft can use for alignment. The alignment knowledge provided by the quad photodiode is expected to be close to the 0.02 degrees provided by the star camera system.

The OCSD is intended as an experiment/demonstration mission, with the goal of determining what is possible and what works well in a CubeSat form factor. The two fine attitude sensors are designed to operate independently of one another so that either is capable of providing the attitude knowledge necessary to point the laser at the ground station. On the spacecraft side, the quad photodiode is simpler than the star cameras, in design complexity, in operations, and in total mass and volume. However, the star camera system has the advantage of operating without the need to provide an uplink beacon from the ground station. In the end, the performance of the two systems will be characterized and compared to inform designs of future systems that use laser downlinking of data from LEO.

To augment the attitude sensors, the spacecraft are equipped with a pair of MEMS rate gyros to measure spacecraft rotation rates. The attitude sensors, particularly the star cameras, provide intermittent data on spacecraft attitude. During laser downlinking, the laser beam (and therefore the spacecraft attitude) must track a fixed location on the ground. From LEO, this tracking may require slew rates up to about 1 degree per second (depending on orbit altitude). To maintain tracking between star camera fixes, the ACS will use data from the rate gyros. The two gyros are of different types and operate independently; one is a high precision rate gyro while the other is of lower precision. The high-precision rate gyro is substantially larger and requires more power than the low-precision gyro. However, it is unclear whether the low-precision rate gyro can provide sufficient precision to keep the laser on target between fixes from the star cameras. Again, testing of the two systems will provide performance data and inform future designs.

Attitude control is provided mainly by a set of reaction wheels, as shown in figure 3. The design of these wheels is the same as has flown on seven previous satellites in the AeroCube series, and their performance on orbit is well characterized. In particular, using an earlier satellite with these reaction wheels and a precision rate gyro, experiments have shown that the wheels and their controllers are capable of holding an inertial pointing vector to within 0.02 degrees. Other than the rate gyro, however, this satellite did not have precision attitude sensors, so the demonstration did not include pointing at a pre-defined location. In addition, the attitude-control experiment was conducted only with stationary inertial pointing; the performance of the reaction wheels in response to a requirement for a variable-rate slew (which is required for optical downlinking) is yet to be determined.

Figure 3. CubeSat-scale reaction wheels. The total height of the assembly is about 3 cm.

The second type of attitude-control actuators is a set of three electromagnetic torque rods that apply torques to the spacecraft by interacting with the Earth's magnetic field. These are not used for precision attitude control; rather they are used for shedding angular momentum that accumulates because of magnetic torques on the vehicle, or asymmetric atmospheric drag torques. In general, the precision attitude-control actuation falls to the reaction wheels, and the torque rods are used only to dissipate momentum buildup in the wheels.

Laser transmitter

The original plan, when OCSD was proposed, was to use a simple diode laser, with a relatively wide beam divergence, and data rates limited to about 5 Mb/s. While very modest compared to other laser communication systems, this would still represent an order-of-magnitude increase in data rates relative even to the most sophisticated CubeSat-scale RF communication systems flying at that time. During the preliminary design phase, we realized we had an opportunity to go for a stretch goal of 50 to 200 Mb/s by improving the attitude control of the spacecraft, and developing a CubeSat-scale and flight-qualified version of a fiber-amplified diode laser that had been developed for other applications. In doing the trade study, we considered the expected pointing capability of the satellite, the expected laser output power, and the collector area of the available ground receiver.

In order to maximize the potential data rate, the decision was made to develop a fiber-amplified laser with two-stages of amplification, which would allow the laser to reach 10 W of optical output power. This power level, combined with an anticipated 0.15 degrees of pointing precision, and the 30-cm-diameter ground receiver, would be sufficient to demonstrate a 100 Mb/s downlink rate. Details of the design of this laser are presented in reference 10. The challenges with this laser were fitting it into the available volume (about 250 ml), providing the power required to operate it (about 60 W), and dealing with the waste heat (about 50 W). To achieve the volume constraint, the various components (seed laser, two pump lasers, and filters) were mounted in various locations on a single plate, and about 20 meters of gain fiber were spooled into various loops embedded in grooves on the plate. The plate also acted as the heat sink for the laser. A photograph of the flight laser is shown in figure 4.

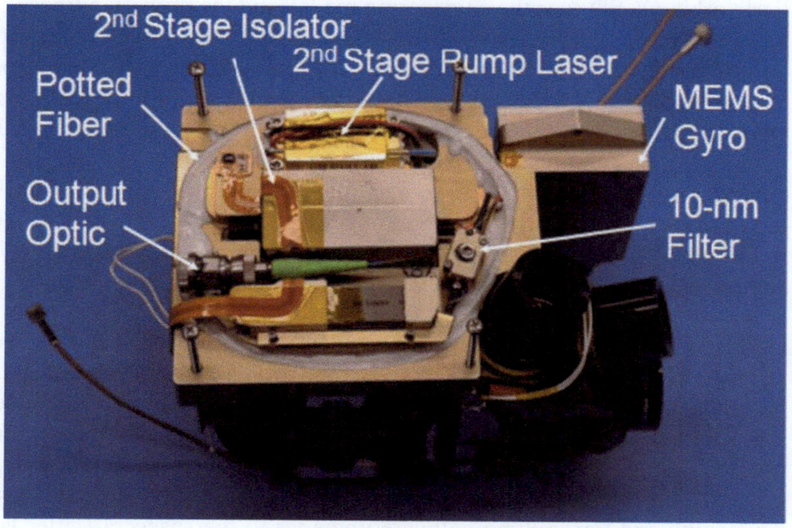

Figure 4. Photograph of the OCSD-A communications laser.

As we were completing the laser build and integrating the engineering model of the satellite, it became apparent that the need to have each of the optical components pre-assembled with correct optical fiber lengths to match the component placement on the plate, as well as the locations of the fiber grooves, presented a serious challenge both in building the laser and integrating it into the available volume in the spacecraft. At the same time, testing of the various ACS components, and modeling of the ACS as a whole, indicated that it would not be unreasonable to expect ACS performance significantly better than 0.15 degrees. Since the irradiance on the target (and therefore the available data rate) scales with the square of the pointing precision, but only linearly with laser power, we chose to modify the design of the flight units by simplifying the laser to include only a single amplification stage. The second-generation flight laser is shown in figure 5.

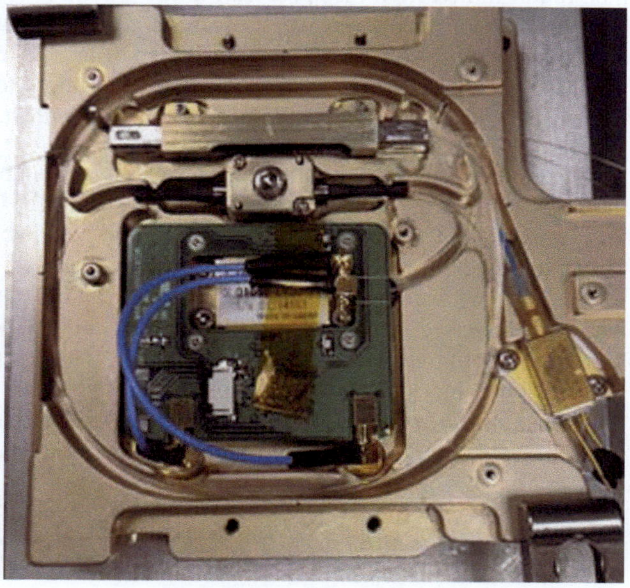
Figure 5. Photograph of the OCSD-B laser.

The elimination of the second amplification stage was expected to reduce the laser output power by about a factor of five. To compensate for this and achieve the same data rate would require an improvement in the pointing by a factor of 2.2. Ground testing of the ACS sensors, specifically the star cameras, as noted above, indicated that attitude knowledge should be possible to a precision of 0.02 degrees. Flight testing of the reaction wheels, using an on-board MEMS rate gyro for comparison, indicated that the reaction wheels should be able to provide control authority to the same 0.02 degree precision. As such, we anticipated using a smaller beam divergence on the flight units to enable higher data rates than were expected with the engineering model, even with the reduction in laser output power. The intention was to use the engineering model to get on-orbit flight experience with the ACS, and to use the results to inform a decision about the final beam divergence for the flight units.

FLIGHT EXPERIENCE

The OCSD mission consists of three spacecraft. The first, OCSD-A, was launched on October 8, 2015, into a low Earth orbit. OCSD-A is the engineering model, and the primary purpose of this spacecraft was buy down risk for the two flight models, which are scheduled to follow it into orbit in the second quarter of 2016. Risk is reduced when lessons are learned, and lessons can be learned in the development process, in the build process, and in on-orbit operations. With OCSD, a number of lessons were learned in the development process; to the extent possible, these were incorporated into the design of both the engineering model and the flight units. Additional lessons were learned in the final integration of the engineering model. While too late for changes in the engineering model, these integration lessons can be incorporated into design modifications in the flight units. A key example of this was the realization of the difficulty of building a two-stage laser in a package small enough for integration into the spacecraft. The design change - moving to a single-stage laser - could not be implemented in time for the engineering model flight, but will be incorporated in the two flight units.

CubeSats fly almost exclusively as secondary payloads on launch vehicles that are otherwise carrying a primary mission to space. The launch provider and primary customer agree on a launch schedule, and any secondary payloads have to conform to that schedule or they do not ride. In the case of OCSD-A, the spacecraft hardware was ready in time to meet the delivery deadline, but the software was not yet in its final form. Since the entire AeroCube program is focused on technology-development, the software that operates the spacecraft (as well as the ground systems) is actually evolving continuously as lessons are learned. As such, we have developed processes for developing and transmitting software upgrades to the spacecraft after launch. These processes involve developing and testing new software in ground simulator units to demonstrate new or improved functionality. Once the software is deemed ready for transmission to the spacecraft on orbit, it is first transmitted to a ground test unit identical to the orbiting spacecraft for verification and validation. If it passes this test, the software is uploaded to the spacecraft, then retransmitted from the spacecraft to the ground so that the integrity of the upload can be confirmed. Once the integrity is confirmed, then the software is committed to memory on the spacecraft. These processes have been demonstrated with hundreds of uploads to ten different CubeSats prior to the launch of OCSD-A, and were thought to be robust. As such, we chose to take the opportunity to fly the engineering model of OCSD, even knowing that extensive software uploads would be required after launch.

To ensure communications reliability, new software is uploaded to the spacecraft in discrete blocks. The extent of the modifications and upgrades to the ACS software for OCSD-A required uploading of many blocks of code and this required more communications time than was available in a single pass over a ground station, so it was spread over two passes. In the interval between the two passes, the satellite underwent a reboot, and the ACS processor rebooted on software that was partially old code and partially new code. Unfortunately, this was a configuration that had not been tested on the ground, and incompatibilities between the two partial versions of the code resulted in the processor booting into a non-functional state. The architecture of the spacecraft requires that each individual processor be able to install its own software upgrades. Since the ACS processor was now booting into a non-functional state, it was no longer able to install additional code, nor is it possible for it to be reset to the original code. As such, the ACS in this spacecraft is permanently lost.

While a costly lesson, the anomaly is not a problem inherent in the design of the spacecraft, and it will not affect the delivery and launch of the next two spacecraft in the series. In addition, the software architecture will be modified before the next delivery to include a non-changeable boot-loader section of code in each processor. This code will ensure that the processors will always boot into a "safe" mode before executing any modifiable code, and will ensure that the processors will always be capable of installing code modifications if code errors are discovered on orbit.

Beyond the modification to the code upload procedure, the primary impact of this anomaly on the flight units is that it will not be possible to test the performance of the ACS before their launch. As such, it will be necessary to set the beam divergence of the two lasers based on estimates of ACS performance. Although a final decision has yet to be made, it is likely that the two satellites will be operated with different beam divergences, one with a wide divergence and low data rate to ensure that minimum goals are met, and the other with a narrower divergence and higher data rate, which, if the ACS performs as expected, will allow successful attainment of a stretch goal.

APPLICATIONS OF OPTICAL COMMUNICATION

The OCSD-A ACS software anomaly notwithstanding, the potential value of optical communication in CubeSats and other very small satellites is still very high. The principal advantage of optical systems is that they are fundamentally capable of substantially higher data rates than comparable RF systems. A survey conducted in 2014 on the state of the art in CubeSat communications systems[11] showed that, of 144 CubeSats flying at that time, 121 had communication systems with data rates below 10 kb/s, 16 had data rates between 10 kb/s and 1 Mb/s, and only seven had data rates above 1 Mb/s. Since then, Planet Labs has launched several sets of CubeSats with X-band transmitters having data rates up to 120 Mb/s.[12] Although the 120 Mb/s data rate represents a major advance in RF communication for CubeSats, it is not clear how quickly or how much higher this rate can go, so there is certainly potential for optical systems in CubeSat applications requiring substantial downlink capacity. Although laser systems could, in principle, encounter bandwidth limitations similar to those encountered by RF systems, the reality is that such limits will not be encountered until data throughput of optical systems reaches many orders of magnitude higher than RF systems. Two factors drive this. First, a laser operating in the 1550 nm band has an effective carrier-wave frequency of $2x10^{14}$ Hz, or 200 THz. The data capacity of a carrier wave at this frequency is exceptionally large. By multiplexing a set of optical bands into one transmitter/receiver combination, data rates well into the Tb/s range are theoretically possible. The second factor is that the laser beams are both very narrow and directional. As such, the communications bands can be shared among many satellites simply by pointing the beams at different receivers, and multiple receivers that are located at the same site can operate simultaneously by looking at different satellites.

Optical communication systems have other advantages beyond data rates. In general, they can be expected to have lower mass and power requirements than RF systems with comparable data rates. The optical systems also present an interesting security advantage. In principle, any wireless communication signal can be intercepted by a receiver that is in the beam pattern of the transmitter. Even very narrow RF beam patterns tend to be fairly broad, covering a large ground area, and a clandestine RF ground station a few km away from the intended receiver will be able to collect the signal. In contrast, an optical system can have a beam pattern with a spot size on the ground measuring as small as a few meters. With such a system, it is relatively easy to control all locations on the ground where an interception might take place, simply by placing the receiver within a secure environment larger than the spot size. Of course, CubeSat-scale optical systems as described here will not have such small spot sizes. However, a beam with a divergence of 0.05 degrees downloading data from 1000 km will have a spot size on the ground of slightly under 1 km, so providing enough land to locate a secure downlink receiver is not particularly challenging.

Finally, there is the regulatory environment. Operation of an RF communication system requires regulatory licensing, and the available bandwidth is limited; such licenses are getting harder to obtain. In contrast, as of this writing, no license is required for operation of a laser communications system in space. Operators of space-based laser systems (or terrestrial lasers directed into space) that are funded, developed, or operated by various U.S. government agencies including DoD, NSF, and NASA, work with the Laser Clearinghouse, an office within the DoD that is charged with ensuring that such lasers do not pose a risk to any space assets. However, there is, as yet, no formal requirement for non-government laser systems to be operated in coordination with the Laser Clearinghouse. In the long run, it may become necessary to establish automated processes for ensuring that laser communication systems do not interfere with each other, or with other space assets.

Although laser communications systems have many potential advantages relative to RF systems, there are also two key implementation issues. The first, as discussed above, is that the satellite must be able to direct the beam onto the receiver with a high degree of precision. In CubeSats, this will impose requirements on the spacecraft ACS that will translate into mass and cost. However, for many missions, including Earth imaging missions, the spacecraft may already have pointing requirements that are more than adequate for the laser communication system. In this case, the pointing requirements of the laser communication system will not add to the mass and cost of the spacecraft.

The second key implementation issue is that laser communication is possible only with a clear line of sight between the source and detector; clouds will completely obstruct the communication channel. As such, a laser communication system must be operated on an as-available basis. Whether this is a problem or not depends on the mission. Some missions, a weather observatory for example, will produce data that have a very short shelf life. If the data is not downloaded and made available to forecasters within a few hours it will be of little value. On the other hand, a mission observing phenomena associated with climate change, for example, will have little need to provide immediate download of data. Climate modelers work with large data sets spanning many years; if the latest data is a few days old rather than a few hours old, it will matter little.

For those mission requiring prompt downloads, there are two options to reducing data latency with optical downlinking systems. The first is to establish an extensive network of optical ground stations, most or all located in regions with historically low average cloud cover. In this case, a satellite will collect data and store it until an opportunity comes to download to a clear station. The number and locations of stations, as well as the orbit of the spacecraft, will determine the statistical likelihood of getting data down within some specified interval. By increasing the number of ground stations, this likelihood can be made large, but never unity.

An alternate method is to establish a network of relay satellites in orbit that can provide optical crosslinking. By this method, if a satellite with data for download is not in range of a ground station with clear skies, the data can be transmitted optically to another satellite, or even through a series of satellites, until it reaches a satellite that is within view of a clear ground station. Again, this method does not provide a certainty of a clear channel to the ground, but the statistical likelihood of a clear channel can be increased to nearly unity by providing enough ground stations and enough relay satellites. If a relay node can be designed to fit within the form factor of a CubeSat, then the low cost and frequent access to space provided by the CubeSat design specification will make it possible to build such a network of relay nodes at a very modest cost compared to a similar system using larger satellites.

SUMMARY

The AeroCube OCSD program has been developed to demonstrate laser communication in a CubeSat. The program consists of three spacecraft, an engineering model and two flight models. The new technologies developed for this program include an upgraded attitude-control system, and a CubeSat-compatible communications laser. The OCSD-A, an engineering unit, has already flown, but encountered a serious anomaly during software updates in the ACS. Although this has disabled the ACS, limiting the value of the engineering model as a risk-reduction pathfinder, there is still value, both in the lessons learned in preparing the engineering model for flight, and in flight testing of the subsystems not disabled by the ACS anomaly, including the newly-developed star cameras and software-defined radio.

Laser communication has potential as a valuable tool for enabling missions in CubeSats and CubeSat-scale spacecraft that require downlink rates that are significantly larger than those available with conventional RF communication systems. The small scale and rapid development cycle of CubeSats offers the potential for rapid growth in this field, leading to a LEO network of optical relay satellites that can provide downlinking of data with high bandwidth and near-zero latency.

REFERENCES

1. Björn Gütlich, Rolf Meyer, Sabine Philipp-May, Michael Lutzer, Ines Richter, Anke Pagels-Kerp, and Markus Bibus, "German Roadmap on Optical Communication in Space," Proc. Int. Conf. on Space Optical Systems and Applications (ICSOS) 2014, S3-4, Kobe, Japan, May 7-9, 2014.

2. Don M. Boroson, "Overview of the Lunar Laser Communication Demonstration," Proc. Int. Conf. on Space Optical Systems and Applications (ICSOS) 2014, S3-4, Kobe, Japan, May 7-9, 2014.

3. Jamie W. Burnside, Stephen D. Conrad, Allen D. Pillsbury, and Catherine E. DeVoe, "Design of an Inertially-Stabilized Telescope for the LLCD," Proc. SPIE 7923, Free-Space Laser Communication Technologies XXII, March 2011.

4. Steven Constantine, Laura E. Elgin, Mark L. Stevens, Joseph A. Greco, Kenneth Aquino, Daniel D. Alves, and Bryan S. Robinson, "Design of a High-Speed Space Modem for the Lunar Laser Communications Demonstration," Proc. of the SPIE, 7923, Free-Space Laser Communications Technologies XXIII, February, 2011.

5. Bogdan Oaida, "OPALS: Optical Payload for Lasercomm Science," 1st Annual ISS Research and Development Conference, Denver Colorado, June, 2012. URL: http://www.astronautical.org/sites/default/files/issrdc/2012/issrdc_2012-06-27-0815_oaida.pdf

6. Siegfried W. Janson and Richard P. Welle, "The NASA Optical Communication and Sensor Demonstration Program: An Update," SSC14-VI-1, 28th Annual AIAA/USU Conference on Small Satellites, Logan Utah, August 11-14, 2014.

7. California Polytechnic State University at San Luis-Obispo, "CubeSat Design Specification, Ver. 13," 2015, URL: http://www.cubesat.org/images/developers/cds_rev13_final2.pdf

8. Rowen, Darren W., Utter, Alexander C., Dolphus, Richard M., and Eddson, Alcid M., "A Miniature, Low-Power, Star Tracker for Precision Pointing Nanosatellites," paper AAS-15-045, 38th Annual AAS Guidance and Control Conference, Breckenridge, CO, Jan. 30 to Feb. 4, 2015.

9. David A. Hinkley, "Picosatellites at The Aerospace Corporation," Chapter 20 in Small Satellites: Past, Present, and Future, AIAA and The Aerospace Press, 2009.

10. T.S. Rose, S.W. Janson, S. LaLumondiere, N. Werner, D.A. Hinkley, D.W. Rowen, R.A. Fields, and R.P. Welle, "LEO to Ground Optical Communications from a Small Satellite Platform," Proc. of the SPIE 9354, Free-Space Laser Communication and Atmospheric Propagation XXVII, San Francisco, CA, Feb 7, 2015.

11. Bryan Klofas, "CubeSat Radios: From Kilobits to Megabits," Ground Systems Architectures Workshop, Los Angeles, CA, Feb. 2014. URL: http://gsaw.org/wp-content/uploads/2014/03/2014s09klofas.pdf

12. http://spaceflight101.com/flock/

BIS-RS-2015-57

The World's First Commercial SAR and Optical 16-Satellite Constellation

Amar Vora, Liam Sills, Andrew Cawthorne, Caroline Slim, Rachel Bird
Surrey Satellite Technology Ltd.
Tycho House, Surrey Research Park, Guildford, Surrey; 01483 803803
a.vora@sstl.co.uk

George Tyc, Wade Larson
UrtheCast
#33 – 1055 Canada Place, Vancouver, BC, Canada; +1 604 669-1788
gtyc@urthecast.com

ABSTRACT

UrtheCast plans to build, launch and operate the world's first fully-integrated, multispectral Optical and Synthetic Aperture Radar (SAR) commercial constellation of Earth observation satellites. These will be deployed over multiple launches in 2019 and 2020. Known as the Constellation, it will comprise of 8 Optical and 8 SAR satellites flying in two orbital planes, with each plane consisting of four satellite pairs. Each pair of satellites will consist of a dual-mode, high resolution Optical satellite (video and pushbroom) and a dual-band high resolution SAR satellite (X-band and L-band) flying in tandem.

The Constellation will provide an unmatched space-imaging capability, including high collection capacity, Optical and SAR data fusion, weather-independent high resolution imaging using the SAR, target revisit, and imaging latency. By flying the satellites in tightly-paired SAR and Optical tandem formations, the Constellation is expected to offer a number of innovative capabilities, including on-board real-time processing, cross-cueing between the satellites, and real-time cloud imaging on the leading SAR satellites that enables cloud avoidance in the trailing Optical satellites. By employing two orbital planes, the Constellation will allow for maximum revisit rates in the mid-latitudes, while providing global coverage extending to the poles.

This paper will describe how the envisaged constellation will create new opportunities for both businesses and government with an altogether new and responsive way to addressing applications.

Surrey Satellite Technology Ltd. (SSTL) is the strategic implementation partner for the satellite design and build and will use its considerable experience in designing spacecraft constellations to tackle this new challenge. This paper will provide some insight into the mission engineering approach that goes into a constellation of this complexity and performance. It will also provide an overview of the benefits of this strategic partnership between UrtheCast and SSTL.

KEYWORDS: UrtheCast; SSTL; Constellation; SAR; Optical; Earth Observation; Tandem; Cross-Cueing; Cloud-Free;

INTRODUCTION

UrtheCast, although a relatively new name in the industry, has already emerged as a key player in the downstream space industry. The company's novel vision to democratise Earth observation has, in part, been achieved by establishing several sovereign space capabilities in a relatively short period. This includes the Generation 1 cameras that were docked onto the ISS in 2013 that has been producing ultra high definition videos (via the Iris instrument) and medium resolution imagery (via the Theia instrument). Without the usual power, mass and thermal constraints of a standalone spacecraft, the use of the ISS dramatically changes the economics of Earth observation from space. In addition to this, the newly acquired Deimos Imaging by the company comes with the added benefit of the use of the fully operational medium resolution Deimos-1 and high resolution Deimos-2 satellites. The final piece of the company's grand vision is the recently announced Constellation mission; a 16-satellite state-of-the-art constellation of SAR and Optical satellites, built and tested in partnership with SSTL, UK. The combination of SAR and Optical satellites forms a constellation whose performance and functionality is unparalleled in the Earth observation domain.

S. Hatton (ed.), *Proceedings of the 13th Reinventing Space Conference*,
https://doi.org/10.1007/978-3-319-32817-1_3

SSTL has been building satellites for Earth observation since 1985. The Disaster Monitoring Constellation (DMC) satellites proved that it is possible to acquire humanitarian, political and technological value from Earth imaging using cost effective small satellites and, since then, SSTL has continued to push the boundaries of low cost satellite capability by pioneering and advancing the small satellite design approach. The UrtheCast Constellation mission is proof of this, with challenging requirements designed to provide novel applications amidst the 'New Space' era in the ever-growing Earth observation industry. As such, SSTL's experience and approach perfectly places them to be the strategic implementation partner of the UrtheCast Constellation.

THE CONSTELLATION

Constellation Design

A few of the many unique features of the Constellation is the satellites it is comprised of and the orbits they will be flying in. For typical LEO altitudes (600-800 km), the laws of physics dictate that the image resolution achievable through a relatively small aperture is in the mid-to-low resolution region. However, very high resolution data can be collected from a relatively small aperture at very low altitudes – a region typically unused by commercial operators.

The Constellation is formed of 8 Optical satellites and 8 SAR satellites at an altitude of 450 km split across two orbital planes: a sun synchronous plane and a medium inclination plane. The satellites are equally distributed in each plane, with the SAR satellite leading the Optical satellite by a few minutes to enable effective cross-cueing operations, as shown in Figure 1.

The combination of both planes enables an optimum revisit in the mid-latitude regions to be achieved, while providing global coverage that extends to the poles.

The revisit, combined with the tandem formation of the satellites enables a fusion of different datasets with very low latency, creating and enhancing various applications.

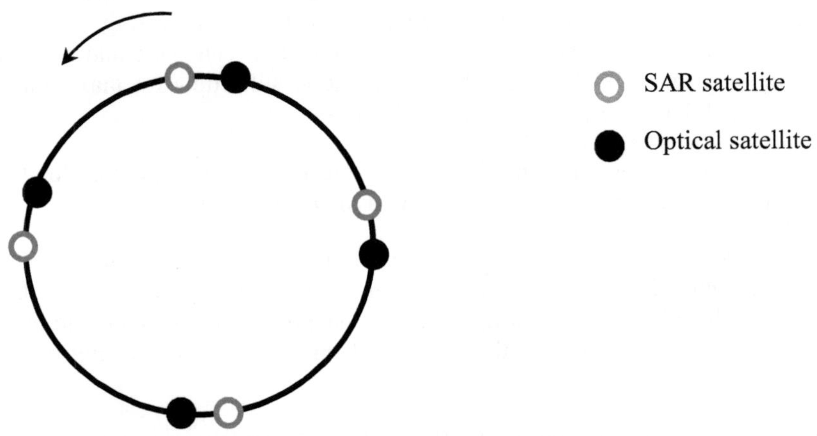

Figure 1. Constellation Design.

This Isn't Just Another Constellation!

The 'New Space' phase has been at an all-time high. The past few years has seen a significant rise in announcements of "game changing" constellations in the Earth observation, science and telecommunications domain. So is this mission truly unique or is it just another cool idea? The UrtheCast Constellation is the world's first combined SAR and Optical commercial constellation. The goal of the mission is to collect data that can be processed into very high resolution, 0.5 m-class still and video Optical imagery, and high resolution L-Band and X-Band SAR products. The SAR sensor incorporates a patented technology that gives it the ability to image simultaneously in quad-polarisation L-band and single-polarisation X-band from the same sensor. This is a feat never achieved before.

These products will serve a variety of end users and markets such as traditional Earth observation commercial and civil applications, data analytics and 'big data' applications, and the nascent social media and consumer applications. The individual performance of each satellite, as described in the coming sections, speaks for itself, however, the other aspect that differentiates this constellation from the rest is its innovative concept of operations. This will also be detailed in the coming sections.

Dual-Mode Optical Camera

The Optical satellites within the constellation include of a pair of sensor suites – the dual-mode camera and the meteorological camera. The dual-mode camera can be operated either in pushbroom or video mode.

The pushbroom sensor uses a 64-stage Time Delayed Integration (TDI) architecture, digitised to 14 bits. This sensor yields a nominal 12.29 km swath (at nadir for 450 km altitude) and is comprised of a panchromatic channel giving 0.5 m-class imagery and six multispectral channels giving 2 m-class imagery: blue, green, yellow, red, red-edge and near-infrared (NIR). The video sensor uses a 20 MPixel CMOS detector that yields a nominal 2.5 km by 1.9 km footprint (at nadir for 450 km altitude) at up to 30 FPS, digitised to 12 bits. This detector uses a Bayer filter that provides three spectral channels (red, blue and green) giving 0.5 m-class imagery.

Augmenting the dual-mode Optical camera is a meteorological camera (MetCam), providing additional spectral channels, albeit at a lower resolution, designed to measure the impact of the atmosphere on the imagery and enable its correction during ground image processing. The data from the MetCam is not included in the distributed product though.

Dual-Band Synthetic Aperture RADAR

The SAR satellites within the constellation include a range of sensor suites – the dual-band SAR, AIS receivers and the cloud camera. The dual-band, L- and X-band, SAR can be operated in one of three modes: SpotLight, StripMap and ScanSAR.

The SpotLight mode is able to acquire 1 m-class (X-band) and 5 m-class (L-band) imagery with a nominal size of 5 km by 5 km.

The StripMap mode is able to acquire 2 m-class (X-band) and 10 m-class (L-band) imagery with a nominal swath width of 10 km.

The ScanSAR mode is able to acquire 10 m-class (X-band) and 30 m-class (L-band) imagery with a nominal swath width of 25 km when both bands are operated together. When operating L-band alone, the ScanSAR mode is able to acquire 30 m-class imagery with a swath width of up to 100 km.

The L-band SAR supports the full complement of polarisation options, including single, dual, quad, linear-compact and circular compact pole. The X-band SAR supports VV polarisation only.

The SAR data can also be used to generate interferometric products.

Complementing the SAR payload are the AIS receivers which, when combined with SAR data on-board, provide useful information on potential targets of interest in the maritime regions, for the trailing Optical satellite to then investigate further.

Adding to the SAR payload and AIS receivers is a cloud camera (CloudCam), providing continuous cloud coverage to assist the trailing Optical satellite with its image acquisition campaign. The rationale for this is described in the following section.

Novel Concept of Operations

Traditionally, the power hungry nature of SAR sensors combined with their day and night imaging ability have driven SAR satellites to fly in dawn-dusk orbits to maximise power generation and the payload duty cycle. This, however, hasn't been the case for Optical satellites due to the less than optimal ground illumination conditions in such orbits. For the Constellation, the tandem formation of both types of spacecraft drives the local solar time of the sun synchronous plane to that more suited for an Optical satellite. However, it is this combination of SAR and Optical satellites that forms a constellation whose performance and functionality is unparalleled in the Earth observation domain. This is a result of each SAR-Optical satellite pair in the constellation being able to uniquely interact with each other in real-time to optimise and enhance the data acquired by the Constellation.

Studies have shown that approximately 67% of the Earth's surface is typically covered by clouds, with only 30% of land usually cloud-free[1]. With Optical satellites frequently being susceptible to acquiring cloudy images, it is important to manage on-board resources effectively, to maximise the amount of useful imagery acquired and increase revenue potential. As clouds can't be moved and it is both difficult and expensive to provide the satellites with real-time cloud coverage from the ground, why not just avoid them autonomously? As detailed in the previous section, each of the leading SAR satellites will have the ability to take real-time cloud imagery via the CloudCam. A continuous stream of cloud maps will be sent over to the trailing Optical satellites, which will then process the data on-board to determine which cloud-free areas of interest it can image to optimise its image acquisition campaign.

The cross-cueing capability of the Constellation provides another completely unique level of service, where the SAR satellite can operate in a wide-area surveillance mode (using ScanSAR) and be used in combination with the on-board AIS sensors and cloud camera to determine if a target of interest (TOI) appears that can be imaged by the trailing Optical satellite. If so, the SAR satellite can immediately send the position of the TOI to the Optical satellite which can then re-task itself to manoeuvre to take a very high resolution image of the target within minutes of the detection.

These are just a few of the unique concept of operations that illustrates the rationale of this tandem formation and the potential of the mission.

What You Get From the Constellation

A combination of the state-of-the-art sensors on-board each satellite in the Constellation and the innovative concept of operations generates unique and useful datasets of which a wide range of information can be extracted from.

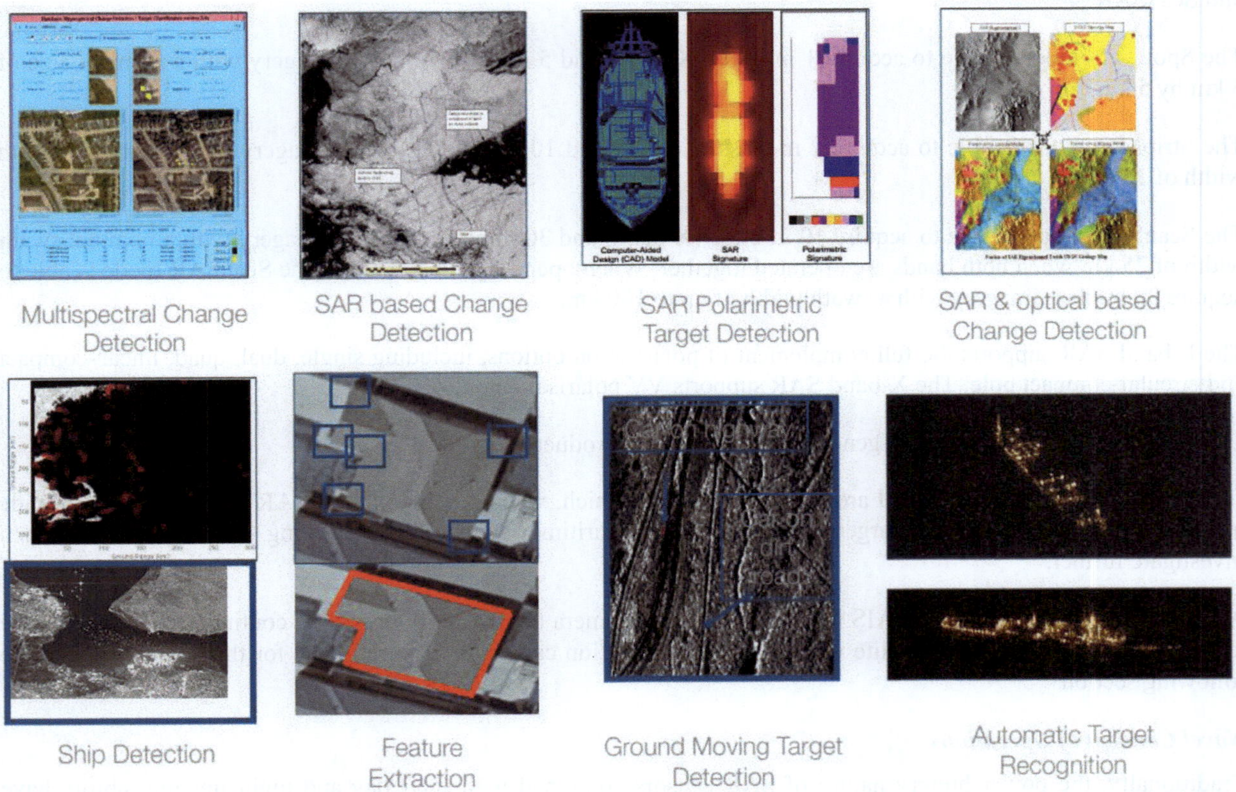

Figure 2. SAR and Optical Information Product Types.

The Constellation will generate an industry standard set of Rapid Positioning Capability (RPC) Model and Ortho Model (OM) products and videos. These products will include the imagery (all Optical sensor spectral channels and all SAR sensor polarisations) and the associated metadata for both sensors.

One of the main advantages of the Constellation is the ability to fuse SAR and Optical data. The SAR and Optical data are highly complementary in terms of the information that can be extracted from each source.

The polarisation and dielectric measurement provided by SAR data can support determination of material classification, wetness, structure, texture and roughness information about the scene that the Optical data often cannot. For example, agriculture and forestry applications based on Optical data need to include a correction for soil moisture. And the SAR data can also assist in differentiating plant and tree types based on their polarisation information. The SAR interferometric products can also be used to measure minute variations in the Earth's surface.

The spectral measurement provided by Optical data can support determination of signature classification information about the scene that SAR data often cannot. For example, the spectral signatures of different types of man-made objects, vegetation and geologic features are all well characterised in Optical data, and less so in SAR data.

The addition of a time-series of images acquired by the video sensor over the arc of acquisition geometries yields an even deeper understanding of the scene due to the 3D surface model and motion vector information that it provides.

Consequently, the fusion of Optical imagery, SAR imagery and interferometric information, 3D surface model and motion vector products yields a suite of products where there is significantly more information that can be extracted from any individual data source alone.

With co-incident Optical and SAR imagery, 3D surface model and motion vector information, the accuracy and range of possible applications becomes even more interesting because it eliminates a variety of unaccountable sources of error, typical of most fusion products resulting from the misregistration due to weather conditions, solar illuminations, temporal scene changes, viewing geometries, etc.

For example, in forestry biomass estimation, the X-band SAR sensor is used to locate the tops of the trees and the L-band SAR sensor is used to locate the bottom of the tree, thus yielding an accurate stand height. The pushbroom sensor is used to perform spectral classification to determine tree species and stand density. The video sensor is used to construct a 3D surface model of the scene and correct for any errors in the intermediate results. The data is then fused, in conjunction with the appropriate forestry models, in order to estimate the biomass.

The processing of high resolution SAR data will benefit greatly from being fused with accurate 3D surface models and motion vector information, thus yielding accurately focused imagery. SAR data has been traditionally processed assuming a smooth Earth or a coarse DEM, and therefore tends to be somewhat out of focus and suffer from layover and shadow artefacts. Since SAR data relies on the Doppler phase history, objects that are moving are therefore mis-located.

The cyclic nature of the solar illumination variations from the medium inclined orbit will allow construction of shadow-free 3D image models of cities by fusing both Optical and SAR data acquired over several orbits, giving both colour and structural information for all surfaces.

The following images illustrate a simple example of multi-sensor data fusion, combining X-band SAR, L-band SAR and multispectral Optical imagery, resulting in a very content-rich information product.

Figure 3. Samples of X-Band SAR Image (Top Left), L-Band SAR Image (Top Right), Multispectral Optical Image (Bottom Left), Fused Image (Bottom Right).

At the End of the Day, it's About the Applications!

The standard and fusion products discussed earlier will feed into applications and services such as data analytics, site monitoring and wide-area reconnaissance.

The unique combination of the multispectral Optical data and the X-band and full quad-pole L-band SAR data provides for many unique applications. As described in the previous section, the L-band SAR data has the unique ability to penetrate through the forest canopy to measure biomass, for example, and detect objects under trees, while the X-band SAR reflect from the top of the canopy to support the biomass estimates and provides higher resolution. When combined with the multispectral Optical data, this provides for information rich fusion products.

As another example, in the rapidly growing area of 'big data' analytics, the dual-band SAR data on its own can provide very high value due to its ability to provide imagery independent of clouds and also at night. When combined with optical data (when it is available on cloud free days), this can provide powerful information that feeds the 'big data' analytics engines to create many different types of information products.

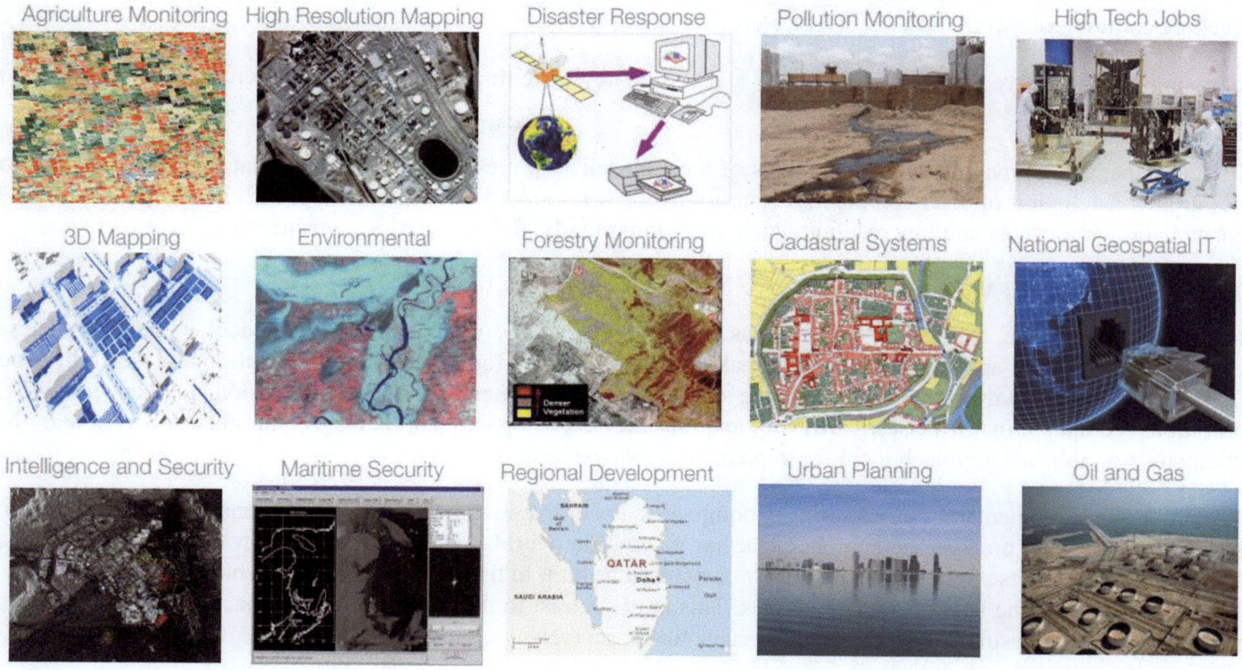

Figure 4. Applications and Benefits of the Constellation.

The Constellation will also serve as a platform for advancing the research and development of a host of new applications and services. The combination of on-board processing capability, flexible and highly-configurable Optical and SAR sensor acquisition modes, and rapid data delivery will be used to support experiments involving novel acquisition geometries and imagery exploitation, and reducing latency in the delivery of actionable information to end users.

Just a small subset of the potential application areas that are currently being explored are:

Maritime Surveillance: The SAR and Optical sensors could be used in a cross-cueing scenario. The leading SAR sensor would scan the ocean, the on-board processor would perform real-time ship detection and the leading Optical sensor would be commanded to acquire very high resolution images of selected ships for positive identification and detection of any pollution discharges. Only the OTH-Gold (Over-The-Horizon) messages, together with their corresponding image chips from both the SAR and Optical sensor, would then need to be downlinked with no additional ground processing required.

Oil Spill Tracking: The X-band SAR provides superior oil-to-sea contrast. The fully polarimetric L-band SAR combined with Optical data provides oil classification in terms of its makeup (plant or mineral) and thickness.

Camouflage Detection: The penetration capabilities of the SAR sensor would be exploited to detect vehicles or other manmade structures hidden beneath natural or artificial foliage. When the Optical sensor fails to detect the same vehicles or structures, this indicates a likely camouflage situation.

Decoy Detection: The material classification capabilities of the SAR sensor would be exploited to differentiate real vehicles from wood, paper or plastic decoys that can easily fool an Optical sensor.

Disturbed Earth Detection: The penetration capabilities of the SAR sensor would be exploited to detect changes to the soil. When the Optical sensor fails to detect the same disturbances, this indicates a likely recent change.

Port Monitoring: The SAR and Optical sensors would be used to count shipping containers and determine stockpile volumes.

Car Counting: The SAR sensor would be used to count parked vehicles located around factories, shopping malls, gathering places and transportation hubs. The medium inclined orbit would be exploited to yield car counts at constantly changing times of day, enabling monitoring of worker shifts, determining peak shopping times and tracking commuter levels.

The applications above are just a snippet of a range of new, exciting and even undiscovered potential from the Constellation.

It's clear that the innovation and applications of the Constellation are considerable, but what does it really take to design a constellation of such complexity? The following section will describe the approach needed to design this constellation at low cost and low risk but which, most importantly, doesn't sacrifice performance.

THE ORIGINAL 'NEW SPACE' WAY

Over the past decade, the space industry has seen a spike in the number of 'New Space' companies. These companies employ an approach that aims to make access to space affordable through innovation and flexibility and in recent times, miniaturisation. So with the challenging requirements of the Constellation, and an emerging, ambitious company in UrtheCast, why not exploit the 'New Space' approach? What makes SSTL and the company's established approach ideal for implementing this mission?

Within the environment of numerous new companies operating in the 'New Space' arena SSTL continues to successfully put into practise the approach that has made it the world leading provider of low cost satellite systems for over 30 years. These practices result in low cost missions built to time and schedule but which, most importantly, focus on delivering the key mission objectives. This is in contrast to the 'New Space' approach where the focus on reducing costs may result in a mission performance that is severely compromised.

The UrtheCast Constellation calls for a mission that is quality-centric, but with an approach that ensures the cost, schedule and risk is minimised. The SSTL approach is a successful combination of management, technical and operational elements developed specifically to allow the company to supply low cost space missions rapidly and without sacrificing quality. SSTL is well known for the considered application of advanced COTS technology to its satellites and indeed this is one of the key elements of its success. Another key element to SSTL's achievements has been the focus of projects on identifying and meeting key operational objectives. Secondary 'nice-to-have' objectives and derived requirements are managed closely to keep the project within timescale and budget. This involves closely working with customers to determine their key criteria for a successful mission – an approach which has been demonstrated through the close working relationship to date between SSTL and UrtheCast in designing the proposed constellation of Earth observation satellites. This requirements management approach ensures that the final mission design results in a useful performance whilst concurrently optimising important factors such as system mass, size, manufacturing timescales and cost. This is opposed to the 'New Space' approach which is following a trend that looks to minimise mass and size, but limits the useful performance obtained as a result.

The ability to manufacture satellite missions in short timescales also allows SSTL to frequently launch missions, proving its technologies and techniques in orbit and providing flight heritage for future missions. This will help to reduce both the development time and the risk involved when dealing with a constellation of this scale with several new and innovative technologies.

SSTL is both vertically and horizontally integrated, executing missions from pre-feasibility studies to in-orbit operations, and manufacturing systems from the component level upwards. In addition, although each project draws on expertise from throughout the company as needed, SSTL's space systems are designed by integrated teams, consisting of a full-time project manager and a dedicated 'core team' of project engineers and assistants, providing a foundation for project activities. This results in well-informed and flexible trade-offs between system, subsystem and equipment level design decisions. Operating in this way, SSTL can reduce levels of equipment-level qualification, formal documentation and quantitative reliability analysis, replacing them with system-level validation, strong internal communications and demonstrated in-orbit heritage, enabling the company to deliver the high-quality product that is required for the Constellation at a low price.

To support the practices discussed earlier, SSTL has a rigorous systems engineering process which has been employed in over 40 missions to date and ensures that the system it is supplying will meet the mission objectives, including the desired availability and lifetime.

To start with, each new mission is managed as an evolution from a previous, existing mission – the so called 'heritage baseline' approach; every SSTL satellite mission since UoSAT-2 (1984) has been derived from a preceding SSTL mission through a controlled process of changes with each mission representing an evolutionary step. Put simply, the heritage baseline meets certain operational requirements in a certain environment, and the goal of the new project is to extend it to meet new operational requirements in a new environment. This is a fundamentally different task from designing a new product from the ground-up to meet the new requirements and results in substantial cost and time savings. For example, although the operating environment for the Constellation is relatively new (i.e. a very low altitude, high drag orbit), SSTL is still able to utilise the 'heritage baseline' approach for both the Optical and SAR spacecraft, implementing modifications where deemed necessary (e.g. a high delta-V propulsion system).

Changes in requirements between the previous mission and the new mission are identified and risks arising from these requirement changes are carefully managed. Analytical or physical validation of existing designs minimises new developments and where new developments are necessary, they adhere to SSTL's proven methods.

SSTL also employs timely and thorough testing to provide the greatest level of product assurance possible within the constraints of each project. SSTL's testing approach focuses testing where it counts most – reducing key risks early in the project, then verifying and validating performance at system level prior to launch.

SSTL tests each item of equipment following manufacture in order to exercise interfaces and verify functionality and key operating parameters prior to system integration. SSTL's assembly, integration and validation (AIV) phase covers an extensive period of functional, verification and validation tests spanning the equipment, subsystem and system levels. As equipment units are brought together to form the integrated system, they are tested individually, in groups, and ultimately as a complete system. This period of integration and testing verifies interface and subsystem functions. The AIV phase also provides an opportunity for a mission-level end-to-end test involving the ground segment hardware and software interacting with the space-segment in a meaningful (yet affordable) dress rehearsal for in-orbit operations. This greatly reduces and optimises the time spent getting the spacecraft to a fully operational state once in orbit – a big advantage especially when considering the size of the constellation. Following AIV, the spacecraft undergoes system-level Environmental Testing (EVT), SSTL's final and most important source of pre-launch quality assurance.

In order to reliably ensure that the mission objectives are met for a specified lifetime, the SSTL approach focuses on providing system robustness, for example, through the use of redundancy. For most SSTL missions, including the Constellation, a high degree of cold parallel redundancy is employed – sometimes utilising equipment of different designs to avoid systematic failures. In addition, SSTL's previous missions provide invaluable knowledge for reliability enhancement with experience from missions in orbit fed back directly to all product teams. SSTL also aims for a safe system and mission design; in which transient events (e.g. radiation-induced upset) or temporary upsets to maintainable systems do not cascade to cause loss of mission or decreased lifetime.

The SSTL approach has been successfully demonstrated in the 43 missions it has launched to date and will continue to be the corner stone in the development of the UrtheCast system ensuring that the SSTL-UrtheCast partnership derives maximum utility from its ambitious and exciting planned constellation of Optical and SAR satellites.

STRATEGIC PARTNERSHIP

The annual turnover of the UK space industry is currently £11 billion, employing over 37,000 people[2]. A target has been set for this to reach £40 billion per annum by 2030. There are several different initiatives and recommendations on how to reach this target, including a big drive to improve competence and innovation in the downstream applications sector, an improvement of the core knowledge base for the future engineers and entrepreneurial business leaders and a significant strengthening of the UK's export activities both upstream and downstream.

The space industry is arguably at its most exciting phase ever, with the 'New Space' age pushing public demand and perception to an all-time high. With the USA currently at the forefront of capitalising on the 'New Space' demand, providing constellations that deliver affordable access to space, there is a view that the UK has remained stagnant, not profiting from this newfound interest through export opportunities.

The announcement earlier this year that SSTL is to team with UrtheCast as the implementation partner for the Constellation mission clearly highlights that this is in fact not the case. This announcement followed over a year of close co-operation between the SSTL and UrtheCast teams to design the high performance low Earth orbiting platforms that will fulfil the ambitious requirements of the UrtheCast Constellation mission.

For over 30 years, the UK industry has shown adaptability and capability to meet consumer demands and to lower costs (i.e. the original 'New Space' approach) for many Earth observation missions. This has led to the UK gaining a formidable reputation in the upstream space sector, leading the advancement and export of satellite technology. The partnership and the resulting Constellation is proof that the UK has been actively pursuing commercial export opportunities in the 'New Space' domain and further enhances the UK's status as a hub for Earth observation excellence. It is hoped that this will demonstrate and improve the perception of the UK as a world leader in space innovation and low cost, high performance satellites, stimulating economic growth in the UK space sector by capitalising on the 'New Space' demand.

CONCLUSIONS

The 16-satellite UrtheCast SAR and Optical Constellation clearly offers an Earth observation capability unrivalled in the industry. The benefits gained from the individual performance of each satellite and the novel concept of operations are wide-spanning and include significant improvements to monitoring, change detection, situational awareness and activity characterisation capabilities as compared to traditional space-based remote sensing systems.

The advantages of SAR sensors are well known, providing reliable image acquisition at any time of day or night, and any weather conditions. It is therefore possible to guarantee, as a minimum, a SAR image in case the Optical image is not adequately illuminated by the Sun. Another advantage of SAR is that it provides texture and roughness information that characterises the scene content.

The advantages of Optical sensors are equally well known, where the spectral information provides easy to interpret and classify imagery. With a rich set of imagery and metadata acquired over a longer dwell time, accurate 3D model reconstruction and motion vector analysis of the scene is possible. Furthermore, this information is useful in generating accurate and higher-value products.

The Constellation will give much greater context regarding the nature of the location and activities being viewed. Rather than just seeing static numbers of people, vehicles or marine traffic within the targeted area, imagery analysts can better detect temporal patterns and assess their significance in the context of the scene which, with the combination of both planes, enables them do so in a time frame that matters to people.

In short, the Constellation exemplifies the old adage of "one plus one equals three" by combining the best of both Optical and SAR sensors yielding more than can be achieved from either sensor alone.

By choosing SSTL as the strategic implementation partner of the mission, UrtheCast can exploit the low cost satellite systems approach pioneered by the company, without the need to sacrifice performance. This partnership highlights and acknowledges the formidable reputation of the UK space industry, stimulating economic growth by encouraging involvement in the exciting 'New Space' age.

References

1. King, M.D. January 2003. "Spatial and Temporal Distribution of Clouds Observed by MODIS Onboard the Terra and Aqua Satellites" *IEEE Transactions, Geoscience and Remote Sensing*, Vol 51, Issue 7.

2. Sadlier, G. July 2015. *The Case for Space 2015*, London Economics, London.

Reinventing Space Conference
BIS-RS-2015 -10

Resource Prospector (RP): A Cost-Effective Lunar Resource Pathfinder

Daniel Andrews
NASA-Ames Research Center
Moffett Field, CA, 94035
USA

13th Reinventing Space Conference
9-12 November 2015
Oxford, UK

S. Hatton (ed.), *Proceedings of the 13th Reinventing Space Conference*,
https://doi.org/10.1007/978-3-319-32817-1_4

Resource Prospector (RP): A Cost-Effective Lunar Resource Pathfinder

D. Andrews[1]
NASA-Ames Research Center, Moffett Field, CA, 94035, USA

I. Abstract

The Resource Prospector (RP) is an in-situ resource utilization (ISRU) technology demonstration mission under study by the NASA Human Exploration and Operations Mission Directorate (HEOMD). This clever mission is currently planned to launch in 2020 and will demonstrate extraction of oxygen, water and other volatiles, as well measure mineralogical content such as silicon and light metals, like aluminum and titanium, from within lunar regolith. Efficient expansion of human presence beyond low-Earth orbit to asteroids and Mars will require the maximum possible use of local materials, so-called in-situ resources. The moon presents a unique destination to conduct robotic investigations that advance ISRU capabilities, as well as provide significant exploration and science value.

This mission is equally important, however, for how it executes as a risk-tolerant, cost-effective mission. RP follows the path-finding approaches of the Lunar Crater Observation and Sensing Satellite (LCROSS) mission[1]. The LCROSS mission confirmed the presence of water-ice on the moon, but also established a new lightweight-approach to project and mission execution which was considerably cheaper and faster than traditional NASA missions.

RP (Figure 1) has been designated as a "Class D" mission, just as LCROSS. This mission classification is the most risk-tolerant class of mission within the NASA risk framework and as such, is given more latitude to accept higher-levels of residual risk. The intention is that by saving monies normally spent attempting to assure a single mission's success, more missions can be funded. A well-designed portfolio can accept occasional mission failure, as its still gets more done for the same investment of resources. This classification enables tailoring the NASA Policy Requirements (NPRs) to "lighter-weight" approaches to mission management and execution.

Figure 1. Resource Prospector team

RP is also assessing both international and commercial partnerships as a means to maximize return on the investment. International partnerships can provide both capabilities synergies and cost-sharing opportunities, while the evolving "new space" commercial options are revealing new approaches to acquiring cost-effective services, including the benefits of "bundling" services. Even the world of launch vehicles is changing, offering much less expensive access to space, especially if NASA is able to be flexible in how it approaches mission assurance. Finally, leveraging investments being made elsewhere within a program portfolio, can enable cost-savings by enabling two applications with one investment.

RP would be the next pathfinder mission to both enable exploration capabilities for future missions, and continue to evolve cost-effective approaches for NASA.

---

[1] RP Project Manager, Programs & Projects Directorate, Moffett Field, CA 94035

II. Resource Prospector Overview & "RP15" Approach

Resource Prospector[2] is a Phase A project, managed within the Advanced Exploration Systems (AES) Division of NASA's Human Exploration and Operations Mission Directorate (HEOMD), currently planned for launch in 2020. The activities which apply to the RP flight plans and hardware are collectively referred-to as "RP" – the mission which will fly. However, in the 12-month span of fiscal year 2015, the RP team has stepped-up to also build a terrestrial Engineering Test Unit (ETU) "surface segment" to be used to mature technologies, perform risk reduction, and practice how we would operate the actual mission. This "RP15" ETU rover/payload surface segment will be derived from conceptual Needs, Goals, and Objectives (NGOs), which were agreed-to with NASA-HQ, to create a working rover/payload ETU in a single year! The great challenge of building this ETU will promote learning and enable risk reduction activities to take place.

Both RP15 and RP seek the same functionality in general; however, RP15 is limited by both programmatic and terrestrial constraints. RP (the flight mission) is designed to prospect the lunar surface, create a map revealing the nature and distribution of the volatiles, and perform an early demonstration of materials processing while on the lunar surface. RP15 enables early testing of some of the most important capabilities required to execute the flight plan of RP. RP mission functionality and RP15 test functionality is illustrated below.

1. Mapping the Surface. The RP15 terrestrial rover will provide mobility enabling roving over surfaces and slopes analogous to what will be expected on the polar regions of the moon. The force of gravity is different for our RP15 terrestrial ETU, but is still relatable to what will be experienced on the moon. Further, the rover system is being designed to enable testing in gravity off-load facilities to see how it performs in a true 1/6g lunar environment. This roving platform will carry both the Neutron Spectrometer System (NSS) and Near Infrared Volatiles Spectrometer System (NIRVSS). During RP15 field testing the NSS is included to replicate packaging constraints, but will not be functional; however, the NIRVSS instrument will be functional, enabling sensing/measuring of the indigenous soil to practice the prospecting part of the mission. Figure 2 illustrates how this scanning would work during actual lunar roving.

Figure 2. RP scanning for subsurface volatiles

2. Entering Permanent Shadows: We envision RP15 will navigate terrestrial "Permanently-Shadowed Regions" (PSRs) by either testing at night in a rock yard, or by testing in a high-bay with darkened conditions. The degree of fidelity (regolith simulant, volatile doping, etc.) is negotiable based on resources available, but this testing could aid in understanding navigation, positioning, and measurement difficulties in rover-only lighting conditions. Figure 3 illustrates the RP rover charging in the sun prior to entering a PSR on the moon.

Figure 3. RP prior to entering a PSR

3. Exposing Regolith: A Drill system is also included on the RP15 ETU, enabling actual drilling operations from the rover. This will enable testing procedures and methods for drilling and capturing tailings from the drill bit, but acknowledge this testing will be in a 1g environment. We expect to learn about constraints on the rover system while drilling on slopes, which side-loads the drill bit, potentially causing binding. The rover suspension system will be required to compensate for loads which move during these challenging drilling conditions, enabling extrapolation of what might be found in the 1/6g environment of the moon. Further, the NIRVSS instrument will measure the

volatiles excavated from the soil to the surface. We prepared drilling sample tubes, buried in the lunar rock yard, wherein the rover can drive over the sample tube, drill, excavate and measure the volatiles.

4. Capturing and Heating Regolith: The drill which provides the excavation capability illustrated above has the capability to acquire samples from as deep as 1m and transfer the material into a sample handling system for processing. The samples are then deposited into the Oxygen and Volatile Extraction Node (OVEN) Subsystem, evolving the volatiles by heating the regolith in a sealed chamber and extracting oxygen and hydrogen from the regolith sample. RP15 will be carrying an ETU version of OVEN, capable of performing similar functions to the flight version of the same subsystem. This includes the ability to heat the regolith samples to 150-450 degC to liberate the volatiles for measurement. Figure 4 illustrates the RP rover performing a drill operation and bringing material from 1m below the surface up into the OVEN system.

Figure 4. RP drilling for subsurface volatiles

5. Identifying the Volatiles: The RP Lunar Advanced Volatile Analysis (LAVA) Subsystem is also included in RP15's functionality. This system will analyze the effluent gas/vapor from the OVEN Subsystem using gas chromatograph and/or mass spectrometer sensor technologies. Volatiles evolution from material in the doped sample tubes will enable verification of the design approaches planned for RP.

The RP15 build of the RP flight system will be greatly reduce system risk, driving-down much-more expensive redesign costs later in the development flow.

III. Reinventing the Approach

As illustrated above, the RP mission is fairly complex and yet the team has been challenged to approach mission development differently, similar to what was accomplished on the LCROSS mission. To do so requires reinventing the approach to how missions are traditionally done, and requires the customers to understand some of the ramifications of alternative approaches.

Spaceflight is historically expensive. The best starting point for making a mission more cost effective and streamlined is to challenge the notion of what is required to accomplish the mission. Space missions suffer from three circular maladies: They traditionally do not deal in volume production so they are expensive; because they are expensive, they "cannot fail"; since they cannot fail, their success must be assured… which is expensive! The result is mission designers spend a lot of resources attempting to assure a mission succeeds because they are too expensive to tolerate failure. However, there are ways to help strike a more efficient balance, an area in which LCROSS was a pathfinder.

Performance: The first step in containing cost and risk is to eliminate the pursuit of maximum performance. Maximum performance requires non-traditional, sometimes heroic activities which do not come cheap. Further, maximum performance can carry exotic parts requirements which do not usually come cheap. If the mission doesn't require being a technical marvel, be mindful of your mission design sophistication. LCROSS was not a Faster, Better, Cheaper mission; it was a Faster, "Good-enough", (and therefore) Cheaper mission. LCROSS strived to have high-heritage, low-complexity, and just *good-enough* mission. The spacecraft was simple by design and as a result, was a low-risk approach.

Risk-Tolerance: Further enabling cost savings is the risk classification of the mission. NASA carries four mission risk classification levels: A, B, C & D, wherein Class A is the least risk-tolerant (expensive flagship or human spaceflight missions), and Class D is the most risk tolerant (smaller robotic missions). LCROSS and RP are both designated Class D missions. Within NASA this risk classification enables taking-on greater technical challenges, or performing to tighter programmatic constraints like cost and schedule. This risk-tolerant classification enables "single-string" designs, which means that there is no required redundancy in the design, (acceptable for the system to fail if a single event within the system occurs). This approach allows the mission designer to save mass, cost, and schedule by streamlining both the design process and the complexity of the system, no longer requiring redundant systems.

There is an important tie connection to the Performance topic above. A mission simply accepting additional risk is not necessarily desirable unless it is really trapped in a tight schedule or cost box. More ideal would be to directly attack mission and design complexity. Low-complexity designs are far more likely to work the first time and continue to work than sophisticated designs. Since the design is likely also "single string", having fewer ways the design can fail helps compensate for not having expensive backups systems. Put another way, as a system becomes more complex, it is more susceptible to human design and test errors, and less likely to achieve success. The key is to keep the mission scope as small and simple as can be tolerated.

Design to Cost: Another way to look at the performance topic is to consider designing to cost. Instead of approaching the requirements with, "what is technically possible", look to what *capabilities exist* to minimally achieve the mission goals. Instead of pushing the limits of technology or performance, do as much as you can within the capabilities of what exists. This is how LCROSS operated in the conceptual phase of project development. The team understood that custom development is fraught with risk and can be costly by taking a lot of time in the design and testing phases. LCROSS was a *design-to-cost* [3] (DTC) project, working to cost and schedule constraints, the principal drivers for the project. By dealing as much as possible with existing designs, we had a set of capabilities with which to work, and that helped to contain cost and schedule.

RP will be taking similar approaches as LCROSS and applying them where appropriate. However, RP is an example of a mission where the stated goals are fairly challenging, and more sophistication will have to remain in the system. Our challenge will be in deciding where we can be leverage what exists and what simply doesn't exist and requires custom development. In any case, we will always be eyeing simplicity everywhere we can find it.

Risk Retirement: Where custom designs are required, retire risk as soon as possible. Risks that are left to fester later and later into the system design get more and more expensive to root-out later. RP is very much taking this approach with "RP15", discussed earlier. RP15 is an ETU to learn about the design intentions in Phase-A when it is much easier and less costly to change approaches in the interest of cost savings. The RP team has effectively gone through an entire design and test cycle with RP15, yielding many, many useful results, including simplifications, and design intentions which didn't work the way we thought. RP15 enabled us to retire both schedule and technical risk early, and should enable us to streamline moving through the typical mission development phases, reducing schedule and reducing cost.

Risk retirement also means evaluating the degree to which testing is required – if at all. If you are working in a cost-constrained environment, you cannot afford to do more than what is essential to meet your overall goals. With a keen understanding of the nature of your risks, estimate the likelihood and consequence of the risk occurring and attempt to have rough parity with your other mission risks. True, there may be some risks which require special attention, but in general, the overall posture of the project risks should be pretty self-consistent. There is no value in having one system have a 98% confidence of working, when another system, maybe your communications system, has only a 70% confidence… If your communications system fails on the mission, it doesn't likely matter that the other systems are working beautifully, as you are no longer able to communicate with the spacecraft! This risk parity approach could enable saving some time and money by actually choosing to do less throughout the design.

Here's an excellent example from the LCROSS mission involving environmental testing. The LCROSS spacecraft, like any spacecraft, needed to go through environmental testing to provide some confidence that when the mission is launched and activated, it will power-up and successfuilly execute the mission. The problem is that environmental testing can be very expensive, as it sometimes involves very large, expensive facilities, large power sources and some exotic gasses and temperature which can make the testing quite expensive. Thermal-Vacuum (TVAC) chamber testing is probably the most expensive because this type of testing requires a large facility (large enough to house your spacecraft), which can drop to temperatures and pressures the mission is likely to see in space, as well as during launch. There might even be additional tests combined with the TVAC test which bring additional costs. For example, if you wished to simulate lighting conditions during TVAC testing you need a high-powered, solar-spectra light panel shining on one side of the spacecraft to testing power systems as well. The possibilities are endless, as are the costs. *So how much testing is enough?* Traditional missions can run as many as 7 or 8 full TVAC cycles, or more, to attempt to make sure the systems will behave as planned, taking weeks or months to conduct this testing (depending on the size of the facility), racking-up quite a bill. LCROSS studied white papers looking for guidance on TVAC thermal cycles and discovered that while many TVAC cycles can get you 95% confidence on your system, most all of the workmanship problems on the spacecraft are revealed *in the first full cycle of testing*[4]! Because the LCROSS design was intentionally making high-use of proven parts, tested at subsystem levels, or already proven with the residual risk accepted, the largest risk we carried going into TVAC testing was building or assembling errors. Inspections can help alleviate that risk, but if you really want to know if it's ready, TVAC testing is the way to go. Given the white paper conclusion about workmanship issue detection, LCROSS proceeded to plan for a single cycle of TVAC testing, later to evolve into 1.5 cycles. Here's what we did: we closed the chamber at room temperature and pressure and then pumped-down to vacuum, and proceeded to heat the spacecraft. The idea was that any contaminants or volatiles which might still be on the spacecraft would be "baked-off" during this hot cycle. We soaked at that elevated temperature getting the spacecraft to a steady-state condition, and then proceeded to cool the chamber-down to levels we'll see in space. We lowered the temperature *at a pace which matched the launch assent profile* so that we could actually have the spacecraft dry-run what it will see during launch – a clever, additional verification. We then "cold-soaked" at that temperature since this will be the environment in which the spacecraft will spend most of its mission time. The spacecraft was then returned to ambient conditions, allowed to stabilize and it was removed from the TVAC chamber. All this testing took place in less than a week. It's estimated that this short test could have saved several-hundred-thousand $USD.

Resource Prospector is a different kind of mission than LCROSS in that it is having to work with more custom designs, given the complicated nature of the mission. The RP team will have to look carefully at trading between subsystem level testing and whole-spacecraft level testing and determine the nature of the residual risk going into TVAC testing. This evaluation will define how much time and how many cycles are needed to relieve RP risk looking toward flight.

<u>Stakeholder Reporting:</u> All of the cost-saving ideas presented thus far have been related to technical topics, but cost drivers are not always technical in nature. If the stakeholders carry reporting-burdens which are heavy with detail and frequent in delivery it might be surprising to some just how much effort is being spent pulling all of that together. Further, you are taking the team off of their real purpose which is to design the system – or worse, you hire many more people to handle reporting activities in an attempt to minimize bothering the designers – further driving cost! It's a vicious cycle which is sometimes difficult to quantify, but when you witness a truly lightweight reporting environment, you can feel and quantify the difference.

LCROSS and RP customers felt that typical NASA reporting and oversight caused more problems than they solved for monitoring Project teams and their execution. Lightweight reporting for LCROSS and RP includes a single, monthly report with simple "Quad charts" to capture the data. When you compare this level of reporting to missions I've witnessed which hold monthly 3-day MMRs (Monthly Management Reviews), wherein the whole team essentially stopped productive work to build charts and participate in the monthly reviews. This reporting becomes a significant "product" of the team. Every stakeholder community will be different and every PM will have different levels of depth required to manage a project, but our LCROSS approach flows from the idea that every piece of work should be helping move the project forward; otherwise you are wasting resources. Every activity spent on non-valuable work is displacing some other activity, and it is important that reporting not take on a life of its own. Frequently, reporting becomes something which must be *fed* and the reporting is not allowed to focus on what is needed *to enable the project.*

IV. Reinventing the Source

There are traditional sources for space missions and then there are alternative source which might enable a flight project to be more efficient. The use of the term, "source" is loosely used here since it can apply to hardware, instruments or even people. This section will illustrate some of the opportunities employed in the past and under consideration for RP.

<u>Unexpected Sources:</u> LCROSS was a pathfinder for cost-constrained missions, certainly making use of its Class D

designation to stay within the cost box, but also recognizing that we wanted the flight project to be successful. Part of our answer was to look to Commercial Off The Shelf (COTS) instruments and some flight-proven instruments, such as LCROSS' visible camera. We looked to well-established instruments from the commercial and industrial world to see if we could use them on the LCROSS mission. The instruments would ideally be ruggedized to improve their chances of survival in the LCROSS launch and space environments, confirming that fact by testing them in relevant environments (vacuum, temp extremes, vibe, etc.). This relationship with COTS vendors was interesting and synergistic, as the vendors were very interested in seeing their instruments get tested by NASA, and were quite accommodating in providing support when we found issues. It was a classic win-win. In the end, most instruments did very well in flight environments. There were some small issues which were easily addressed such as one instrument test failed because a small bolt in an electronics box came loose during vibe testing. No adhesive had been applied to the bolt threads to help secure it for a dynamic loads environment. Once the adhesive was applied, the device passed testing just fine. In another case, an internal cable came loose in the instrument, because it was not staked down. We reduced the length of unsupported cable by staking it, thereby decreasing the cable strain experienced during the launch environment.

This approach was applied across the entire suite of LCROSS instruments including a thermal camera (MID-IR1), which has been used in motorsports applications; Near-IR spectrometers (NSP1 & NSP2) used in beer-making and carpet fiber analysis for assessing recyclability; UV visible spectrometers (UVS) from standard bench-top laboratory equipment; a visible camera routinely used in shuttle launch imagery; and Near-IR cameras (NIR-cam) used in fiber optic communications applications.

We applied this approach elsewhere on LCROSS, making use of surplus flight hardware to save money. For example, we discovered an existing, surplus TDRSS satellite propulsion tank, which with some minor modifications, became the main propellant tank on LCROSS! We even used a TRIANA satellite Inertial Measurement Unit (IMU) in the LCROSS attitude control system design.

In the end, this suite of instruments and flight hardware was cleverly applied on the LCROSS mission, which in the end, worked flawlessly, while saving a lot of money and time. RP is now taking a similar approach, where feasible, to decreasing cost and technical risk by flying a modified version of an LCROSS spectrometer, which has now also been utilized on the LADEE mission. The body of knowledge on this instrument is quite high and so the risk of deploying on RP is small. We're also taking battery technologies, cells and designs from the Robonaut robot and scaling the design to work for RP. These batteries have already been approved for deployment on the International Space Station (ISS), in and around humans, which saves a considerable amount of labor attempting to qualify a new battery design. Additionally, there are numerous rover steering system designs inspired by the Chariot human rover. Further, the basis of the RP flight software (FSW) comes from the same software built and successfully flown on the LADEE mission. Each of the leverages come with cost savings because the development, and even some of the testing, already exists.

<u>Commercial Competitiveness:</u> Related to the use of COTS hardware, this approach carries the benefit that the marketplace drives costs down and reliability up. NASA is famous for designing sophisticated, capable instruments for its missions, but pays dearly for them in performing early development, investing in all the non-recurring costs to bring a concept to reality… but what if there is an instrument out there which accomplishes 80% of what was needed on the mission? Sometimes that last 20% is essential for the mission, but many times it is not and a satisfactory and wildly cheaper instrument deployment could be had. Commercial entities are most effective at drilling-in efficiencies once a technology is demonstrated, which ultimately translates into cost savings for all subsequent users – even if not originally designed for your application.

Another frequent topic affecting both cost and risk is the use of Electrical, Electronic, and Electromechanical (EEE) parts in spacecraft designs. Fifty years ago when electronic parts were starting to become mainstream, the quality of those part's manufacture was found to be highly-variable and at times "garbage". These parts certainly were not something to be relied-upon to run important systems, such as military spaceflight hardware. This reality gave birth to "Military grade" parts manufacture. The requisite design hardening, test-to-failure parts-screening, and lot testing to make sure these parts could be relied-upon, soon followed. All these quality and reliability improvements, however, drove costs skyward making military and spaceflight hardware necessarily expensive.

An interesting thing then happened over the next couple decades… the commercial world increasingly used electronics in their products, increasing consumer product sophistication, but also increasing the demand for good quality electronic componentry. One bad choice of a EEE parts vendor could mean the end of a product line, or even a whole consumer company, if the products failed. This created a natural pressure on the EEE supply chain to increase the quality control and overall reliability of even "commercial" parts – not because the free world depended on the part working, but because the company's reputation did! Can you imagine if it was routine for a cell phone manufacturer today to have 10% of their phones not work out of the box, or break if they were slightly mishandled? That company would vanish from the commercial marketplace in no time at all… and that possibility has driven-in unbelievable quality to commercial grade parts. However, many spaceflight missions still automatically require military-specifications (MilSpec) parts as a default. Now don't misunderstand this point: I am not saying all MilSpec parts can be replaced by commercial grade parts. I'm simply illustrating that there are lower-cost, higher-availability options which should be considered. Space missions which will endure long times in space with long exposures to radiation are likely going to still require radiation-hardened "Rad Hard" parts; however, if you have a limited-life missions, or can protect your electronics through other means, commercial grade parts are something in which to take a look. LCROSS used commercial-grade components in most of its instruments, because these were commercial instruments, intended for use in a laboratory or in automotive applications – not space. However, LCROSS subsystem environmental testing was able to show these instruments were tough-enough, and if encountered a radiation fault, would simply reboot and then resume measurements. If the whole instrument were compromised, the overall instrument suite was robust-enough to tolerate the loss. This is exactly how we made this commercial approach work to LCROSS cost advantage.

Buying in Bulk: The value of buying in bulk is probably self-evident, but there is considerable savings when acquiring goods in bulk since the supplier is able to consolidate activities. In my experience, NASA hasn't made effective use of this approach. This has recently become apparent in my detailed surveying of NewSpace commercial entities and their ability to profitably-support lunar activities; including launchers, LEO-delivery, Translunar LLO-delivery and lunar landing. While at the time of this writing lunar landing has not yet been accomplished by a private entity, many companies are aspiring to do just that, and as you might expect, are offering capabilities that are significantly discounted if you buy a large percent of the payload capacity, with maximized savings if you buy the entire manifest!

This pricing reality introduces some interesting twists if the buyer is willing to be a broker of the bulk manifest buy. I encountered scenarios from some companies where I could nearly pay-off the cost for my part of the payload, by selling-off the excess payload at small-payload rates. The customers who bought those smaller payload opportunities would be satisfied with the opportunity, and I would recover enough funding to nearly pay for my entire payload – for taking-on the role of broker.

Buying in Bundles: A related topic to buying-in-bulk, is buying-in-bundles. This variant isn't about getting a discount for a bulk quantity purchase; it's about savings which come from bundling of various services. For RP, I performed market research on the readiness of the commercial marketplace to provide lunar landing services. This is different than the traditional model, where NASA buys a launch vehicle and pays a company to affix the spacecraft atop the launch vehicle. NASA works with the vendor to launch and release the spacecraft and the NASA mission proceeds to its destination under NASA control. This bundled services model is much closer to a trucking service, commercially acquired. In this case, NASA acquires a lunar delivery service wherein NASA provides the specifications of the desired landing location and parameters and a commercial entity bids on that *service.* NASA simply shows-up with its payload, and the commercial entity takes it, says thank you, and then promises to deliver it to location (X,Y) on the surface of the moon.

With this approach the commercial vendor has to perform all the coordination of services so that it is an end-to-end capability to NASA. This also means that the commercial company can works deals, bulk-buys, and consolidate savings and pass them onto NASA in ways NASA could have never realized. One stark example is the cost of launching spacecraft. NASA launches establish provisions, riders, and requirements on launches of its spacecraft which drive commercial pricing quite-high – as much as 40% higher in some cases. However, if NASA were to enable the commercial bundler to work deals with launch providers without the NASA additional requirements, there are considerable savings which can be passed-along to NASA as an end-buyer.

I've started to see hints of even deeper commercial bundling opportunities coming. NASA wouldn't simply buy lunar landed mass services, but could buy *lunar rover hosting* services, where NASA simply places its instrument on a commercial rover and the commercial entity does everything from launching, to landing, to deployment, to roving! NASA simply takes data from their instrument and the commercial partner provides "the pipe" to send that data back to Earth for NASA use. The idea of buying "data services" wherein NASA provides nothing but data requirements and *the whole mission is bundled* is likely right around the corner.

V. Rediscovering Partnering

Partnering is a term which is currently enjoying political favor. It sounds good to be partnering with others to achieve shared/blended objectives which can serve many different political needs; however, my interests are more with the direct benefits to the spaceflight project as a means to reinvent how to think of shared benefit.

The Potluck: One obvious benefit of partnering is the "potluck". In a potluck dinner, all the invitees who come for dinner bring a dish to share themselves, and once everyone arrives a wide-variety of food is available for everyone's benefit, yet the only financial investment by each attendee is the cost/labor associated with the dish they brought. This analogy holds for partnering on a mission.

Launch Partnering: It's well understood that launch vehicles are expensive, especially when the mission is having to carry the entire financial burden of the launch. However, this is another place where partnering can help. LCROSS and LRO shared an Atlas V launch vehicle, as it enabled the Exploration Systems Mission Directorate (ESMD) to get two missions from the cost of one launch vehicle. Similarly, two or more customers can split their payloads on a single launch and benefit with economies of scale of a larger launch vehicle. In the extreme case, very small spacecraft such as cubesats can either fit in the margins (small available space) of larger launches, or broker out an entire launch with dozens to hundreds of cubesats manifested on a single launch, maximizing the economics.

Pooled Capabilities Partnering: Pooling similar requirements into a single requirement set can bring savings to both missions. Two lunar missions, for example, both need a launch vehicle and a lander to begin their missions on the surface of the moon. The launch vehicles and landers will need propulsion systems, communications systems, power systems, structural systems, etc., and all those systems will need to be tested before deployment. Each mission will need a launch window, a launch site, frequency allocations, mission designers, etc., in order to successfully get to the surface of the moon. All of those facets (and many more) need to be paid-for. Now if those missions can be comanifested on the same launch and come to agreement on the nature and requirements of their trans-lunar journey, many of those required capabilities will be bought once, but enjoyed by two separate missions. True, two missions carry more mass and will "cost" more propellant and structural design, but all-in-all, there is very real savings enjoyed by the two missions on all the common systems, in combining their requirements into a single requirements set for the launch and lunar delivery services provider.

Mission Element Partnering: RP's plan is to partner on its lunar lander needs. NASA would provide the rover, ISRU payload, and launch vehicle, while an international or commercial partner would provide the lander. The RP lander partnership concept was originally driven by purely political needs, but it carries with it very real cost benefits from the potluck metaphor. The intention would be to find a partner wherein building a lunar lander is something they would like to do to satisfy their own exploration goals. This lander would carry the RP surface segment rover, which would then prospect the surface of the moon, satisfying NASA's requirements. NASA brings very real money investment to that mission and the lander partner would as well, and both parties share other's data/results, making the investment doubly worthwhile. Additionally, one partner may need instrument A to take measurement A, but the other partner might be able to make use of Instrument A to take measurements B and C (i.e. a completely different use of Instrument A). It is not only cost-efficient, but also saves mass, volume and a number of technical measures. This is a current topic under discussion by RP and its partners regarding spectrometer measurements.

VI. Conclusion

RP will be the first ISRU demonstration on another planetary body, taking first steps to be able to "live off the land". RP is in Phase A Formulation, having successfully passed its NASA Mission Concept Review (MCR) in the Fall of 2013. It is actively involved in international partnership discussions, to maximize return on this novel mission, within budgetary constraints. NASA will be providing the ISRU payload instrumentation, including drill, the roving mobility platform and launch vehicle. The lander will be provided either through an international partnership, or through a commercial relationship for an earlier, smaller demonstration prior to a full RP mission. The mission is scheduled for a 2020/21 launch, and NASA is working to a cost-effective \$250M budget (not including the launch vehicle).

Figure 5. RP15: Designed and built in a single year

Most recently, the Resource Prospector team completed the design and build of "RP15"[5], a mission-in-a-year build of an entire terrestrial rover/payload system. RP15 is a Phase A "deep dive" ETU development, moving from concept to working hardware roving in a lunar-analogue rock yard within a single year. Through subsequent testing in 2016, RP15 will help reduce risk and improve resource prospector designs and approaches.

The great work of this RP team can be followed on the web at: www.nasa.gov/resource-prospector and on Twitter @NASAexplores.

Acknowledgments

The authors would like to thank the hard-working and immensely capable Resource Prospector team, across the many participating NASA Centers. Simultaneously working a Phase A mission towards flight, while also taking a terrestrial ETU rover/payload system from concept, design, build, test, and operations in a single year, is an impressive feat. This work greatly reduces mission-level risk and matures concepts well beyond Phase A maturity. Additionally, we would like to express gratitude to our customers within the Advanced Exploration Systems Division of NASA-HEOMD as well as the Game Changing Division of NASA-STMD, who asked this team to step-up to this important challenge, while encouraging lightweight approaches to mission management.

References

1 Andrews, D. R., "LCROSS – Lunar Impactor: Pioneering Risk–Tolerant Exploration in a Search For Water on the Moon," 7th International Planetary Probe Workshop (IPPW-7), Barcelona, Paper IAC-11-A5.1.4, 2011.

2 Andrews, D., Colaprete, A., Quinn, J., Chavers, D., Picard, M., "Introducing the Resource Prospector (RP) Mission", American Institute of Aeronautics and Astronautics (AIAA) SPACE 2014, San Diego, CA, USA.

3 Atkins, K., "How to Plan and Manage Reserves Effectively," IEEE Aerospace Conference Proceedings, 2004.

4 Quintero, A., Welch, J., "Perceptiveness of Thermal Vacuum Testing" 18th Aerospace Testing Seminar, March 16-18, 1999.

5 Andrews, D., Colaprete, A., Quinn, J., Bluethmann, B., Trimble, J., "Resource Prospector (RP) – Early Prototyping and Development", American Institute of Aeronautics and Astronautics (AIAA) SPACE 2015, Pasadena, CA, USA, Aug 2015.

Firefly – New Generation of Low Cost, Small Launch Vehicles Designed to Serve the Rapidly Growing Small Satellite Market

Andy Bradford
Firefly Space Systems

Tom Markusic
Firefly Space Systems

Shey Sabripour
Firefly Space Systems

13th Reinventing Space Conference
9-12 November 2015
Oxford, UK

S. Hatton (ed.), *Proceedings of the 13th Reinventing Space Conference*,
https://doi.org/10.1007/978-3-319-32817-1_5

BIS-RS-2015-64

FIREFLY – NEW GENERATION OF LOW COST, SMALL LAUNCH VEHICLES DESIGNED TO SERVE THE RAPIDLY GROWING SMALL SATELLITE MARKET

Tom Markusic, PhD, Shey Sabripour, PJ King, Andy Bradford
Firefly Space Systems, 1320 Arrow Point Drive Cedar Park, TX 78613, USA.
Andy.bradford@fireflyspace.com

In an ever evolving space industry where small satellites are recognised as representing one of the largest growth areas in the market, the need for launch vehicles and associated services aimed primarily at small satellites and their specific requirements (technical, economic and programmatic) is clear.

The 'Firefly Alpha' vehicle has been designed to cater specifically for small satellites, having the capability to launch up to 400kg in to Low earth orbits, with options to launch single payloads, multiple small payloads (e.g. cubesats) or combinations of the two. The Firefly Alpha vehicle is intended to be the first of a series of small launch vehicles, with a roadmap in place to introduce larger vehicles (that are still primarily targeted at small satellites) over the next 5 years, with re-usable elements also to be phased in.

I. INTRODUCTION

Firefly Space Systems is developing the Alpha launch vehicle to provide low-cost, high-frequency launch capability for the rapidly growing and critically underserved small satellite industry. A 'simplest-soonest' approach has been adopted in the development of the Alpha vehicle, recognising the urgent need for a simple, reliable and robust low cost vehicle to serve the needs of the small satellite market.

Each Firefly vehicle is engineered with cross-industry design insights and leverages high Technology Readiness Level (TRL) design elements to reduce risk and guarantee reliability. The technologies employed in our Alpha flagship vehicle provide a clear pathway for future incremental improvements in vehicle capability.

This unique development approach aims for monthly rocket launches within 3 years and a reusable launch vehicle within 5 years

II. THE MARKET OPPORTUNITY

Before embarking on the costly development of any new launch vehicle it is clearly required that the market for that vehicle (current and future) is well understood. Forecasts and predictions regarding the expected growth in the small satellite market over the next five years can vary dramatically. Firefly have therefore carried out their own market analysis in order to fully understand the potential of the market and assure that the vehicle being designed addresses that market optimally.

From 2009 until 2015, the market growth for small satellites has been strong, exhibiting a growth of over 28% per year. Research reports from various public sources, forecasts and surveys suggest that this trend is likely to continue at this end of the satellite market. A summary of the forecasted recent and near-term growth is shown in Figure 1

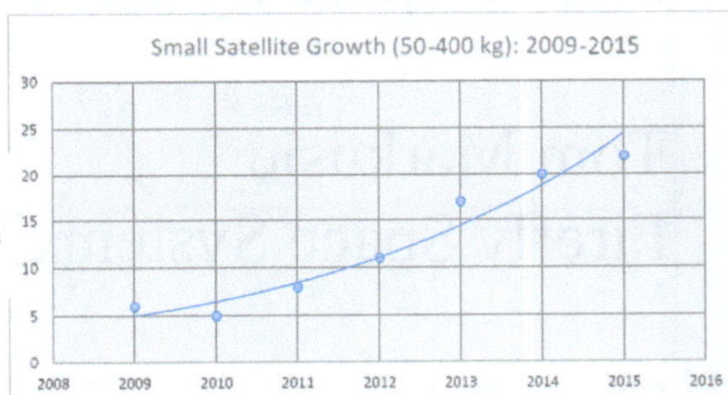

Figure 1: Small Satellite Market Growth

It can therefore be safely and robustly concluded that the market currently exists for dedicated small satellite launch vehicles (if introduced at the right price), and that demand is likely to grow significantly with a need for up to 40 launches possible in the 2018 timeframe, growing further through the 2020 timeframe. Even if only half of this market is accessible by a dedicated, commercial small launcher such as Firefly Alpha, there is still clearly a need for this type of vehicle.

III. MARKET REQUIREMENTS

The requirements on launch services associated with small satellites are often not the same as those associated with larger, more complex and higher value spacecraft.
It is not true to say that small satellites are always simpler, cheaper and quicker to build than larger spacecraft and missions, however in the main small satellites do tend to be built within low cost and short duration programmes and missions when compared to larger missions and spacecraft.

Given this the typical launch requirements for small satellites tend to be (in order of priority, with vehicle reliability assumed as a given/obvious requirement regardless of mission class or size):

- Low cost/ programme compatible price point
- Availability & schedule reliability
- Simplicity of interfaces
- Availability
- Simplicity of ancillary and associated launch activities and services

All of the above requirements have been taken in to account in the definition and development of the Firefly Alpha vehicle. The key requirement is clearly low cost (of the entire launch service and process), however the availability and schedule reliability requirement is perhaps the most interesting and less obvious driving requirement. In a market that has been historically very well served by auxiliary or 'piggy back' launch opportunities, there is an increasing need for schedule reliability, due to the growing commercial nature of the small satellites that are constituting the majority of the growth in the area. It is typically extremely unlikely for an auxiliary launched satellite to be able to have any control over the launch schedule, and this has generally been accepted as a consequence of paying such a low price for those types of launch opportunities. However, the ever increasing number of commercial small satellite based missions, which require some kind of assurance on launch dates, are not able to easily absorb the delays typically associated with auxiliary/clustered launch opportunities on larger launchers. This recent change in the nature of business and applications served by small satellites is a key factor in the emergence of the need for cost effective, dedicated small satellite launch vehicles.

IV. THE FIREFLY ALPHA VEHICLE

Overview

Firefly Alpha is a two stage expendable launch vehicle utilising simple and low cost technologies and production techniques in order to achieve an extremely low recurrent price. The first variant, Alpha 1.0, utilizes efficient technologies such as composite tanks, a plug cluster aerospike, and traditional bell nozzle engines with hydrocarbon fuel. The Alpha vehicle has been developed with the following key factors in mind:

- **Payload Capacity and Orbit** (to optimally address small-sat market)
- **Schedule** (time to market)
- **Launch Cost**
- **Launch Schedule** (dependable launch frequency)
- **Reliability**

Design Drivers & Major Trades

The design of the Alpha vehicle has been driven by three primary factors:

1. **Propellant**
2. **Engine cycle**
3. **Staging**

Propellant: LOx/hydrocarbon propellant combinations are the most mature, simplest to handle, and are highly available. Hydrocarbon fuels under consideration by Firefly are Rocket Propellant-1 (RP-1) and methane. Methane is the preferred propellant for the upper stage, while RP-1 is the preferred fuel for the first stage. The upper stage will utilize RP-1 on the initial '1.0' version of Alpha.

Engine cycle: Both pump-driven and pressure-fed engine cycles were considered. A pressure-fed system best matches the key top level approach principles and was therefore selected. The downside of pressure-fed systems is that propellant and pressurant tank masses become large, leading to a low vehicle payload mass ratio. To address this Alpha will field a pressure-fed system that can be readily augmented with a GG-cycle turbopump. In addition, the pressure-fed Alpha vehicle counters the tank mass growth by utilizing all-composite, high-pressure tankage

BIS-RS-2015-64

Staging: Two stages, at a minimum, are required to achieve orbit with LOx/hydrocarbon propellants. While adding additional stages would increase the payload mass fraction, high level considerations and FOM's (Figures of Merit) predicate simplicity over performance, and thus the choice of two stages for Alpha.

An analysis of launch failure history between 1980 and 1999 by Aerospace Corporation[1] showed that 91% of known failures are attributed to three causes: engine failures, stage separation failures and, to a much lesser degree, avionics failures. Stage separation mechanism failures were the largest contributor to failed launches. This evidence was a key input to the decision process regarding the number of stages to be embarked in the Alpha vehicle design.

A rigorous trade-off process has resulted in the Alpha vehicle utilising an RP-1 first stage, a methane upper stage, common dome tank partitions and helium pressurization.

Vehicle Design

The Firefly Alpha Vehicles' main characteristics are illustrated in Figure 2. The vehicle is approximately 80 feet (24m) tall when fully assembled.

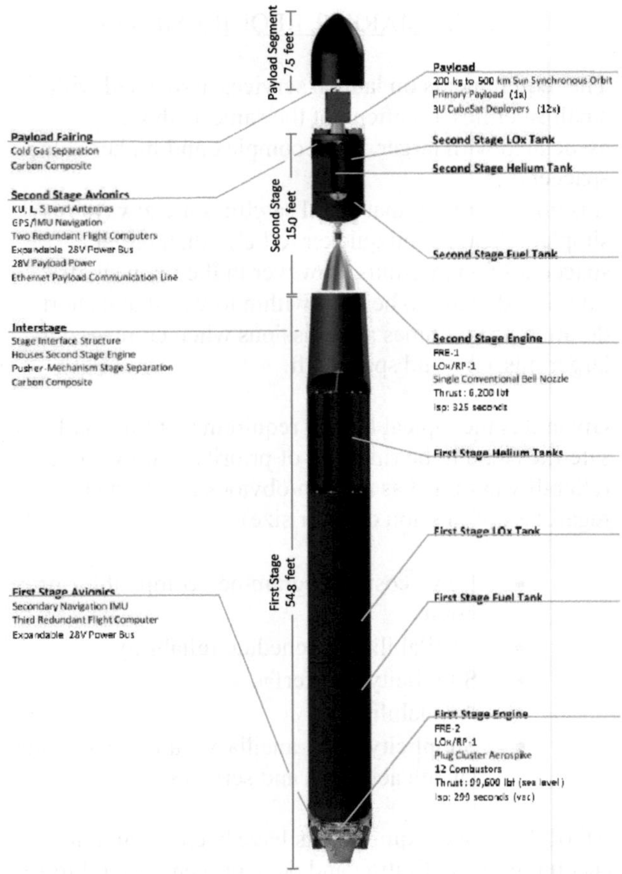

Figure 2: Firefly Alpha Overview

Propulsion

The "kernel" of the Alpha propulsion system is a relatively small, regeneratively-cooled, conventional de Laval geometry combustor utilizing the historically-proven pintle injector design. The upper stage of Alpha will be powered by a single combustor, whereas the first stage will be powered by a cluster of twelve combustors, arranged in a plug-cluster aerospike configuration. The layout of the second stage engine is shown in Figure 3.

Figure 3: Alpha Single Engine Combustor

Using the same combustor on both stages dramatically simplifies propulsion development effort and also reduce production costs through economies of scale and high throughput manufacturing. Additionally, the use of an engine cluster on the first stage greatly increases vehicle reliability, as the vehicle will have an engine-out tolerance for much of the boost phase.

Structures

Firefly's all-composite rocket structures enable its mission capabilities and high-throughput launch and manufacturing efficiency. Carbon fibre composites are an ideal pathway for launch platforms to drive down weight and achieve higher payload mass fractions. Historically, Aluminium has been the lightweight choice for airframe construction.

Today, structural weight and stiffness requirements exceed the capabilities of conventional aluminium and its derivative alloys. Over the past decade, composite materials have developed to offer competitive mechanical properties at a fraction of the weight. Firefly utilizes extensive in-house experience to facilitate the design and development of carbon composite airframes and structures. In addition, Firefly leverages external partnerships with specialist carbon composite organisations who together have vast expertise in Composite Overwrapped Pressure Vessel (COPV) development, cryogenic material testing, and controlled and scalable composite construction

Avionics

Firefly Alpha's Avionics design philosophy is consistent with the approaches employed through Propulsion and Structures. Firefly Avionics will architect a simple, low-cost, reliable system, drawing on COTS parts where available. Where unavailable or cost-prohibitive, Firefly will develop in-house solutions.

Firefly Alpha avionics are designed to exceed industry reliability standards while optimizing the time to market. In compliance with FAA regulations, Firefly's avionics architecture physically separates flight termination functionality while maintaining redundant subsystems. These versatile subsystems run on real-time, priority-based software that enables subsystems to perform housekeeping functions without sacrificing hard-coded redundancy. Furthermore, an autonomous guidance and termination system is implemented to reduce the number of subassemblies by eliminating the need for an on-board C-band tracking system, enhanced GPS receiver, dedicated antenna, and other RF components. Firefly avionics are enabled through seamless integration of the navigation and termination software with established range-specific flight corridors, mission flight paths, and updated mission state vectors.

Figure 4: Firefly's Standard AG&T Avionics Hardware

Firefly's integrated avionics architecture maintains a spatiotemporal separation (FAA requirement) and redundancy and combines functions into a mission-configurable hardware/software set. A typical avionics module (shown with a standard business card to indicated scale) is shown in Figure 4. Each module, or "cartridge", is approximately 85x55x20mm (i.e. credit card sized).

V. FIREFLY ALPHA PERFORMANCE

The performance of the Firefly Alpha vehicle has been analysed, and the payload delivery capability of the vehicle to various orbit altitudes and inclinations is summarised in Table 1.

	Payload Delivery Performance to Altitude				
Orbit Type & Inclination	*300km*	*400 km*	*500km*	*600km*	*700 km*
Equatorial (0^0)	390 kg	380 kg	370 kg	360 kg	350 kg
60^0	310 kg	300 kg	290 kg	280 kg	270 kg
Polar (90^0)	240 kg	230 kg	220 kg	210 kg	200 kg
Sun Synchronous (SSO)	220 kg	210 kg	200 kg	190 kg	180 kg

Table 1: Firefly Alpha Payload Delivery Capability

Flight Profile

A typical mission flight profile for the Firefly Alpha vehicle can be seen in Figure 5. As can be seen from the figure, the second stage is re-startable, and therefore is capable of performing the multiple types and durations of burns required to achieve multiple mission profiles.

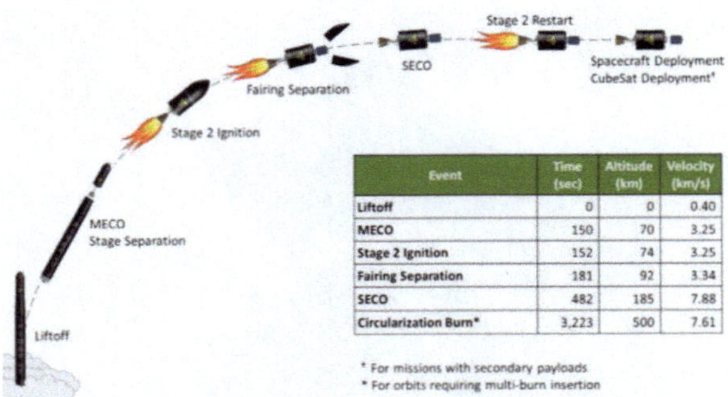

Figure 5: Firefly Alpha Flight Profile

Payload Accommodation and Interfaces

The standard Firefly Alpha payload bay and fairing provides a payload volume suited ideally to the types of satellites expected to be launched. The payload fairing volume, with dimensions, is shown in Figure 6. Multiple payload scenarios are envisaged and catered for through the inclusion of a number of standard satellite interface systems, including the most common small satellite separation systems used today (details can be found in the Payload User Guide[2]).

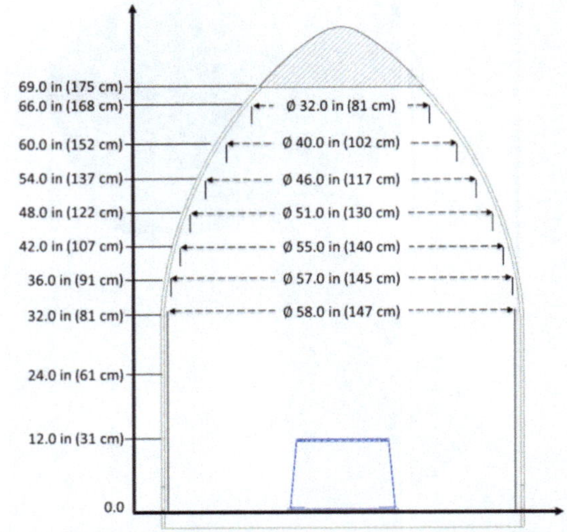

Figure 6: Standard Firefly Alpha Payload Envelope

Expected and offered payload configurations cover the following main scenarios:

- Single satellite (primary passenger) only
- Prime Passenger plus auxiliary payloads (cubesats)
- Cluster configurations featuring (e.g.) multiple cubesats.

An example of the second scenario is depicted in Figure 7

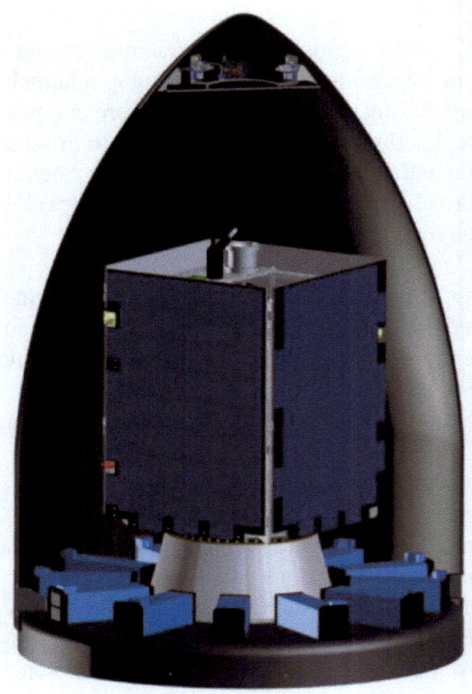

Figure 7: Alpha payload fairing with primary small satellite and cubesats accommodated

BIS-RS-2015-64

Electrical Interfaces are also kept simple and to the minimum in the baseline offering, in line with typical small satellite customer requirements. Power lines for battery trickle charging are included but it is assumed that the majority of customer satellites will be launch in an off state and therefore more comprehensive electrical interfaces are not likely to be needed by the majority of customers.

Hypergolic propellants such as hydrazine are also not expected to be featured on the kinds of satellites most likely to be launched by an Alpha vehicle, and therefore hazardous fuelling operations post satellite to adapter/fairing integration are also not expected to be included typically.

VI. FIREFLY MISSIONS & CAMPAIGNS

In line with the overall low cost approach being followed for the Alpha development programme, Alpha missions and campaigns are designed to be streamlined and efficient in terms of duration and overall cost. It is has been recognised that the costs associated with missions such as mission planning, meetings, fit checks and interface discussions, as well the launch campaign itself, can be significant, especially in the context of a low cost small satellite mission. Therefore a baseline mission planning and implementation approach has been defined which minimises these costs and activities whilst maintaining excellent customer flexibility and service. For early launches a target of no more than 3 weeks total duration for a launch campaign has been defined, with a desire to reduce this further as mission experience matures with time. It is also recognised that many small satellite customers require or at least prefer a 'no frills' approach to launch planning and campaign activities to minimise the overall cost of the end to end launch service. With this in mind a clear baseline 'no frills' launch service offering has been defined with a number of anticipated optional extras that can easily be added at the customer's request.

VII. FIREFLY ALPHA SCHEDULE AND DEVELOPMENT PLAN

A robust development and qualification plan for the Alpha vehicle has been defined, informed heavily by the experience gained by the core team members' significant experience in previous launch vehicle development programmes. Key technology development and demonstration has been the focus of the early development phase with later phases focussing on the demonstration of the main functional blocks and assemblies. The entire first stage, for instance, will be tested and demonstrated through sub-orbital flight test campaigns in 2017.

A significant milestone was completed in September 2015 with the successful hot firing of the FRE-1 engine[3] which is at the core of the first and second stage engines. The first test series successfully demonstrated start-up, shutdown, and steady state combustion. The test also served to prove the complete functionality of Firefly's new test site. A photograph of the hot fire test is shown in Figure 8.

Figure 8: First Hot Fire of the FRE-1 Engine core in September 2015 at Firefly's new engine test facility near Austin, Texas

With the early tests complete, the next phase of engine tests will emphasize performance tuning and longer duration "mission duty cycle" runs. The first hot-fire tests of the FRE-2 aerospike engine are expected to take place in early 2016. Further development activities will continue through 2016 and 2017, culminating in first full flights at the end of 2017, with commercial flights scheduled to begin in Quarter 1 2018.

VIII. FUTURE FIREFLY UPGRADES AND VEHICLES

The Firefly Alpha vehicle is intended to be the first of a series of launch vehicles developed to address the growing small satellite market. The performance characteristics and feature of the 'Alpha 1.0' vehicle have been described in this paper; further Alpha derivatives (1.1, 1.2 etc.) are planned with performance enhancements being introduced in an incremental and low risk manner, in line with the 'simplest-soonest' approach being followed. One such planned upgrade is the adoption of methane for the second stage in place of the RP-1 fuel to be used in Alpha 1.0.

In parallel, future, larger Firefly vehicles are in development, featuring larger payload capacity and also reusable elements.

IX. CONCLUSIONS

Firefly Alpha is designed from the ground up to meet single digit $M costs per flight to Low Earth Orbit (LEO). This vision requires not only a novel approach to launch vehicle design and production, but also a fundamentally simpler, faster, payload integration technology, paired with advanced avionics, autonomous launch operations, and mission planning and execution components. Firefly's vehicle architecture is driven by customer-driven high level technology considerations. Key propulsion, structures, and avionics systems are based on established technologies. However, Firefly will be the first to incorporate some technologies into flight hardware. This will enable Firefly to field gamechanging launch vehicles while keeping development risks as low as possible.

Firefly Space Systems is privileged to be at the forefront of the NewSpace movement. Firefly will deliver well-engineered, well-built launch vehicles that are unprecedented in their focus on meeting the demands of the emerging small-satellite market and dramatically enhance Firefly customers' experience and business competitiveness.

REFERENCES

[1] 'Space Launch Vehicle Reliability, Aerospace Corporation, October 2008. (see http://www.aero.org/publications/crosslink/winter2001/03.html)
[2] Firefly Payload User Guide, Version 1, July 2015
[3] First Rocket Engine Test a Success for Firefly Space Systems (Press Release), September 10th 2015

Reinventing Space Conference
BIS-RS-2015-23

HeL1oNano: *The first CubeSat to L1?*

Cyrille Tourneur, Matthew Stuttard, Dr. Markos Trichas
Airbus Defence & Space

Dr. Jonathan Eastwood
Imperial College London

Dr. John Bellardo
Cal Poly

13th Reinventing Space Conference
9-12 November 2015
Oxford, UK

S. Hatton (ed.), *Proceedings of the 13th Reinventing Space Conference*,
https://doi.org/10.1007/978-3-319-32817-1_6

BIS-RS-2015-23
Reinventing space 2015
HeL1oNano
The first CubeSat to L1?
Cyril Tourneur, Matthew Stuttard, Markos Trichas – Airbus Defence & Space
Jonathan Eastwood – Imperial College London
John Bellardo – Cal Poly
Imperial College London
CAL POLY
SAN LUIS OBISPO
AIRBUS
DEFENCE & SPACE

HeL1o Nano: *The mission concept (1)*

- *Mission summary*
  - A technology demonstration and Heliophysics science mission
  - Based on 6U cubesat of 14 kg placed on a Lissajous orbit about Sun Earth Lagrange point 1 (SEL1)
    - *Twin spacecraft option to improve reliability*
  - One year on-station operations after a 6 months low cost transfer to SEL1
  - Will provide differential environment measurement baselines with existing space assets already located at SEL1 *(DSCOVR, ACE, Wind,..)*
  - Developed in the frame of a USA & European cooperation
  - Two launch options for end 2018:
    - *with ORION/MPCV on board SLS EM1*
    - *or with JWST on board Ariane 5*

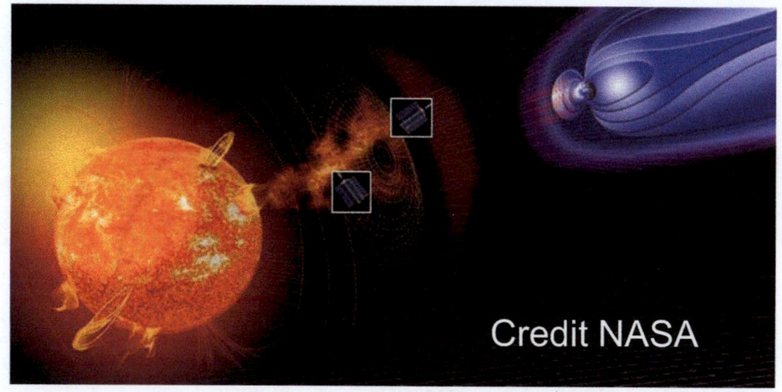

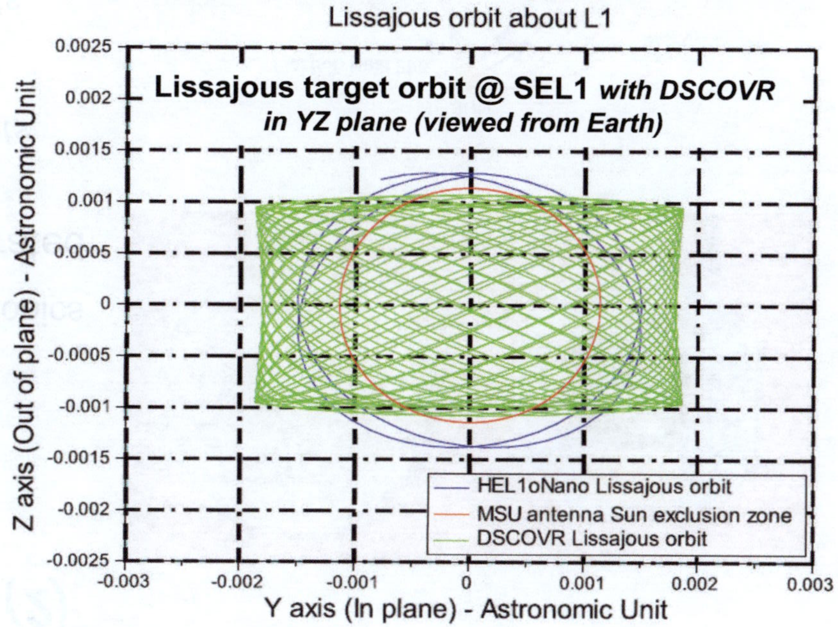

HeL1o Nano: *The mission concept (2)*

- *Mission objectives*
  - Technology demonstration and nanosatellite TRL augmentation for beyond LEO missions
    - Especially propulsion, deep space transponder and avionics
  - Miniaturization of space weather instruments operated beyond the Earth magnetosphere
    - *e.g.: Magnetometers and plasma monitoring instruments*
  - Heliophysics knowledge augmentation
    - *Through time coordinated 3D differential measurements*
    - *In coordination with other space assets already @ L1*
    - *During transfer to SEL1 and once on station*
  - Gather lessons learned and better understanding of solar related space weather events
    - *For assessment of Coronal Mass Ejections and Stream Interaction Regions*
    - *For future space weather monitoring systems*
    - *For future Human Space Exploration missions*

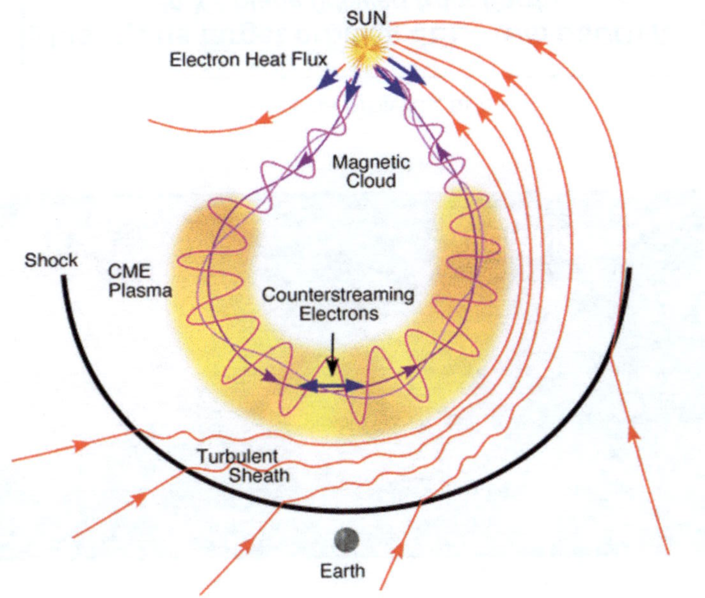

Credit D.Eddy & T.Zurbuchen

Imperial College London CAL POLY SAN LUIS OBISPO AIRBUS DEFENCE & SPACE

Science goals

- *Main thematics*
  - Coronal Mass Ejections (CME)
  - Heliospheric Magnetic Field (HMF)
  - Stream Interaction Regions (SIR)
- *Through multipoint differential 3D magnetic field and energetic charged particle measurements*
  - Vs other spacecraft @ SEL1 (DSCOVR, Wind ACE,.)
  - From start point (EML or SEL2), during transfer and on station
- *Essential benefits*
  - Better knowledge of conditions at the sub-solar magnetosphere for better understanding of sub-storm triggering mechanisms
  - Improved characterization of solar wind phase fronts to determine their curvature and evolution so as to improve precision of solar wind propagation calculations

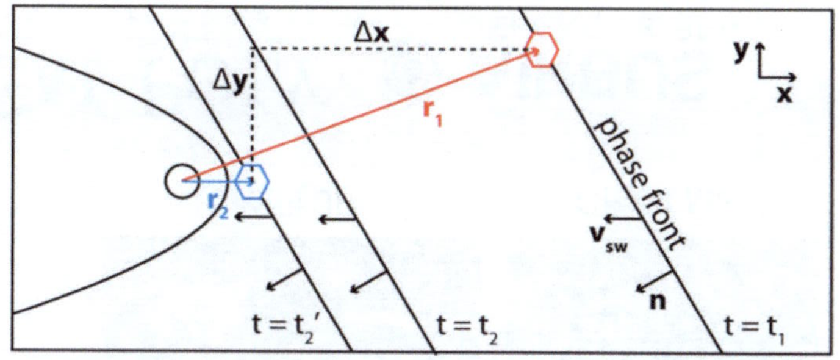

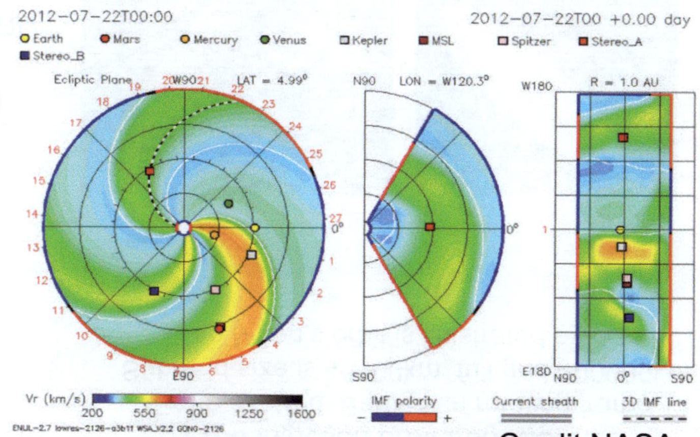

Credit NASA

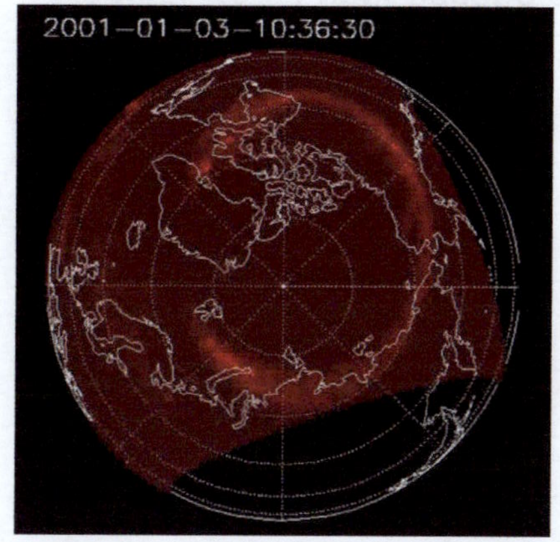

Imperial College London CAL POLY SAN LUIS OBISPO AIRBUS DEFENCE & SPACE

Concept of Operations

- *Low cost transfer to SEL1*
  - With SLS EM1/ Orion/MPCV:
    - Injection on a Lunar DRO then low cost transfer to Earth Moon Lagrange points (EML 1 or 2) then “free” transfer to SEL 1
  - With Ariane 5 /JWST:
    - Injection on SEL2 manifold then heteroclinic “free” transfer from SEL2 to SEL1
  - $\Delta V < 100$ m/s and final transfer ~ 6 months
- *Communications*
  - Offline CONOPS concept
    - *Science data are permanently recorded on board*
    - *Download every day*
  - Choice of chemical propulsion enables science during transfer
  - Ground network:
    - *Deep Space Network (DSN) antenna during transfer and critical operations: 30 mn session per cubesat per day*
    - *Morehead University (or other) antenna for routine operations: 90mn session per cubesat per day*
    - *ESA DSN (Estrak) as an option*

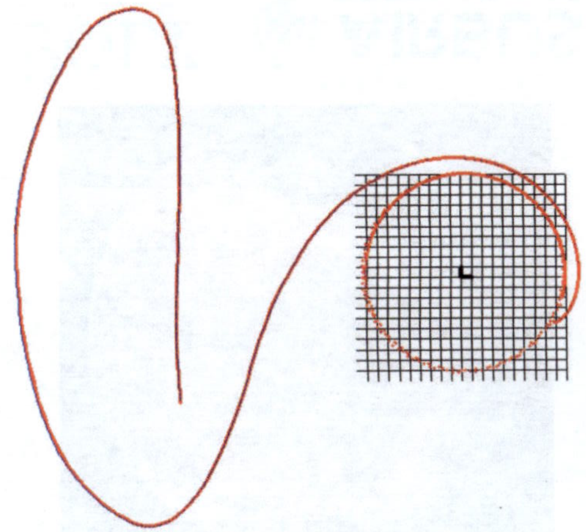

Transfer from Lunar escape orbit to Free injection orbit about SE L1.
Earth centred, Earth-Sun rotating frame.
Sub-grid size is 40000km, full grid 400000km.
The Moon’s orbit is illustrated by dots

Credit JPL

Credit MSU

Imperial College London | CAL POLY SAN LUIS OBISPO | AIRBUS DEFENCE & SPACE

CubeSat outline (1)

- *Payload*
  - Primary payload: MR magnetometer from Imperial College
    - Mounted on a short boom
    - Three axis measurement
    - Dynamic range: Typ. +/- 125 nT per axis – can be adjusted
    - Accuracy: 0.2 nT (NB requires refinement)
    - Time cadence: 4 seconds
    - One year operation heritage on CINEMA (LEO)
  - Auxiliary payload: Particles detector
    - Not yet selected amongst candidate groups in Europe & USA
- *Platform main features*
  - IRIS V2 rad hard deep space transponder + LGA
  - Chemical + cold gas propulsion
  - 2 x 18 U deployable solar arrays
  - Rad tolerant avionics
    - AOCS controller & OBC
    - Reaction wheels and star tracker
  - Radiation protection
    - Dedicated shielding for dose
    - Customized avionic for SEU

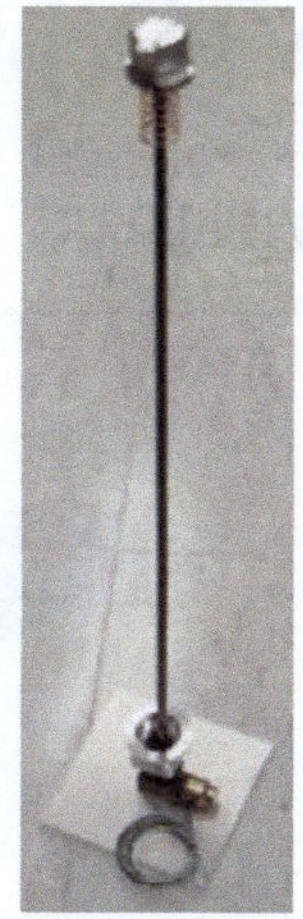
Boom credit
TiNi aerospace

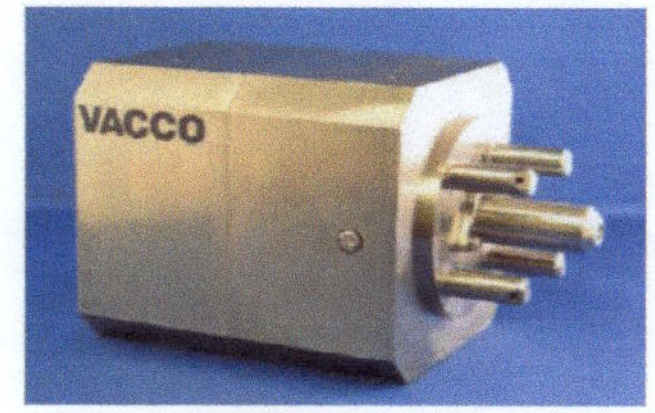

Credit VACCO

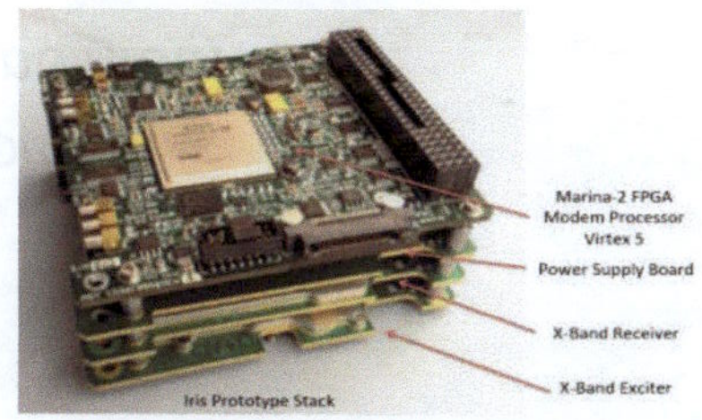

IRIS transponder - Credit JPL

Credit MMA

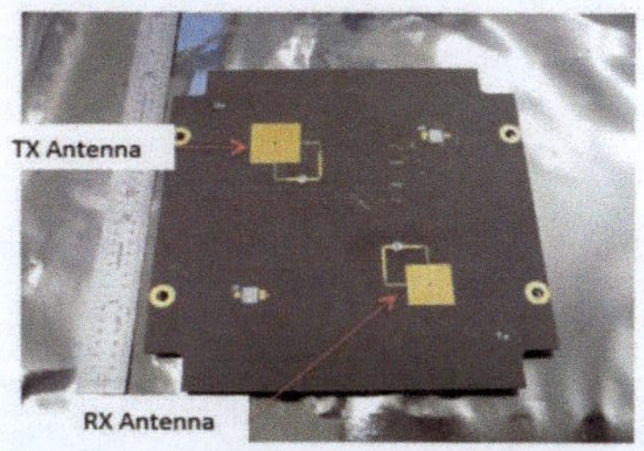

IRIS antenna - Credit JPL

Imperial College London   CAL POLY SAN LUIS OBISPO   AIRBUS DEFENCE & SPACE

CubeSat outline (2)

- *Key budgets*

#	Budget
Form factor	6U: 10 x 20 x 30 cm
Mass	14 kg launch mass + 5kg for canister
ΔV	100 m/s nominal – up to 150 m/s
Attitude	• Spun about Earth pointing axis • 0.1° APE
Data rate @ SEL1	• 33 kpbs with DSN @ 5° elevation *(LGA Earth pointed)* • 9.6 kbps with MSU @ 5° elevation *(LGA Earth pointed)*
Power generation	Up to 72 W

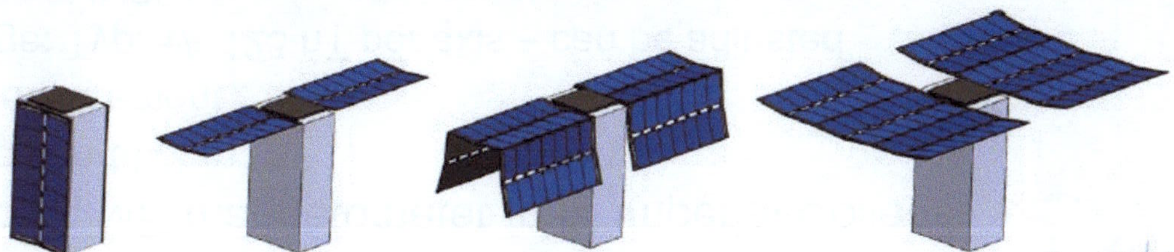

Credit MMA

Imperial College London — CAL POLY SAN LUIS OBISPO — AIRBUS DEFENCE & SPACE

Teaming – as of October 2015
• EU & US Core team
– CalPoly
– Cubesat design, development & operations
– Imperial College
– Magnetometer
– Science requirements
– Science operations
– Airbus
– Coordination & sponsor
– System engineering
• Key partners
– NASA, ESA & UKSA
– Launch & sponsor
– JPL
– IRIS V2 deep space transponder
– Deep Space Network
– Mission analysis & navigation operations
– Morehead State University
– Deep space antenna (21m dish) for on station operations
– Key cubesat hardware providers for a demanding deep space mission
– Vacco, Blue Canyon Technologies, Tyvak, AstroDev, Clyde space, Isis,
CAL POLY
SAN LUIS OBISPO
Imperial College London
AIRBUS GROUP
UK SPACE AGENCY
JPL
Jet Propulsion Laboratory
California Institute of Technology
tyvak
Nano-Satellite Systems Inc.
A Terran Orbital Company
MOREHEAD STATE UNIVERSITY
AIRBUS DEFENCE & SPACE

Thank you !

jonathan.eastwood@ imperial.ac.uk

bellardo@calpoly.edu

matthew.stuttard@airbus.com - cyril.tourneur@airbus.com - markos.trichas@airbus.com

Imperial College London — Cal Poly San Luis Obispo — Airbus Defence & Space

Innovative Small Launcher

A.J.P. van Kleef, B.A. Oving
Netherlands Aerospace Centre NLR

C.J. Verberne, B. Haemmerli
Nammo Raufoss AS

M. Kuhn, I. Müller, I. Petkov
German Aerospace Center DLR

S. Hatton (ed.), *Proceedings of the 13th Reinventing Space Conference*,
https://doi.org/10.1007/978-3-319-32817-1_7

Innovative Small Launcher

A.J.P. van Kleef, B.A. Oving
Netherlands Aerospace Centre NLR
Voorsterweg 31, Marknesse; +31 885114834, +31 885114233
Arnaud.van.Kleef@nlr.nl, Bertil.Oving@nlr.nl

C.J. Verberne, B. Haemmerli
Nammo Raufoss AS
Enggata 37, Raufoss; +47 61152970, +47 90737496
cj.verberne@nammo.com, Bastien.Haemmerli@nammo.com

M. Kuhn, I. Müller, I. Petkov
German Aerospace Center DLR
Pfaffenwaldring 38-40, Stuttgart; +49 71168628408, +49 7116862448, +49 71168628131
Markus.Kuhn@dlr.de, Ilja.Mueller@dlr.de, Ivaylo.Petkov@dlr.de

ABSTRACT

The market for small satellites is expected to increase substantially in the coming years, but there is little capacity to launch them affordably. No operational dedicated launcher for small satellites exists today. Small satellites, launched as secondary payloads, are entirely dependent on the constraints set by the primary payload, such as launch date and target orbit. Launch costs of less than €50,000 per kg of payload are required in order to directly compete with piggy-back ride shares. With a dedicated launcher a higher cost per kg can be accepted for payloads which need to be delivered timely and accurately to a desired orbit.

A consortium of 13 companies and institutes are joining forces in a Horizon 2020 work programme to design a dedicated small launcher to be built in and launched from Europe. The project is called "SMall Innovative Launcher for Europe" (SMILE) and is currently in its preparation phase for the Grant Agreement with the European Commission. Kick-off is planned for 1/1/2016. The SMILE project aims at a combined research approach into a new innovative European launcher for an emerging market of small satellites up to 50 kg using a multidisciplinary design and optimisation approach strengthened by the demonstration of critical technologies for cost-effective solutions and complemented with the design of a European-based launch capability from Andøya (Norway). For the intended market, cost reduction is essential. One option to reduce cost is to apply reusability of one or more of the stages. Cost can also be reduced by applying commercial industry-grade components. Another means of cost reduction is through volume production. Finally, the production process can be optimized for cost, e.g. automated manufacturing for composite parts and 3D-printing for metallic parts. Critical launcher technologies in various expertise areas will be developed in SMILE, but this paper focusses on the rocket engine developments and their impact on cost reduction and design since the engines are the most critical and expensive parts of a launcher. For the rocket engines, both hybrid engines and reusable liquid engines are assessed.

Hybrid engines combine some of the advantages (simplicity, both in functioning and in hardware) of solid engines with those of liquid engines (inherent safety, throttling). The chosen combination of propellants (H2O2/HTPB) gives good performances on a wide range of mixture ratio, thus allowing a great versatility of the mission. Besides, it offers the advantage of being already available in industrial quantities, while being completely green (only CO2 and H2O produced). The engines and their propellants are also safe to handle (nontoxic constituents) and safe to operate (the two propellant ingredients stored separately). Those characteristics, coupled with a simple fluid system, will substantially reduce hybrid propulsion life cycle cost. In order to keep the price of the propulsion system as low as possible, reusability of components is a key feature leading to cost reductions through volume production and increased reliability through automated production. In that sense, a Unitary Motor is thought of as a building block that can be clustered to deliver the required thrust for a micro-launcher.

Liquid propulsion is a reliable technology which is favourable due to its flexibility as the engines can be throttled at a wide range and easily re-ignited. The combination of LOX/kerosene as green propellants seems to be very promising. Compared to other propellants, both oxidiser and fuel are low-cost, worldwide available and storable. The envisaged engine design is thereby based on ceramic materials. Due to their oxidation resistance, high specific strength and low thermal expansion behaviour at high temperatures, such ceramic materials are specifically suited for liquid propulsion components. Applying fibre-reinforced ceramics, the material's characteristics can be further improved yielding into damage tolerant and reliable structures, being insensitive against thermo-shocks as well as thermal cycling. Compared to classical ITAR-controlled metal alloys (as the current main material for thrust chamber assemblies), the envisaged ceramic materials are lightweight and not subject to ITAR-controls. In combination with 3-D printed components and the potential use of CFRP (carbon-fibre reinforced plastics) housing structures, the engine's structural weight can thereby be significantly reduced. It is expected that a combination of LOX/kerosene operation in a clustered design with multiple sub-scaled engines based on ceramic materials and a transpiration cooling technique enables a considerably improved engine lifetime. This could indeed pave the way for prospective reusable liquid rocket propulsion.

The combination of applied research on both the two propulsion technologies will allow the use of the right technology at the right place to offer a launcher delivering the required performance at the lowest price possible. Ultimately, the choice of the propulsion system for all the stages of the rocket will be a trade-off between performance, launch objectives and cost.

KEYWORDS: Small Dedicated Cost-Effective Launcher, Innovative Hybrid and Liquid Rocket Technologies

BACKGROUND

The new generation ARIANE 6 and VEGA C launchers will guarantee Europe's independent access to space for the high-end market of satellites in terms of mass and size with a competitive edge in the world market of launchers. These launchers however are significantly less attractive for smaller satellites. The initiative therefore addresses reliable, affordable, quick, and frequent access to space for the emerging market of small satellites up to 50 kg, fulfilling the needs from the European space Research and Technology Development (RTD) community as well as commercial initiatives to put satellites into specific LEO orbits within a preferred time window. Herewith a market niche is addressed, which is projected to grow significantly in the coming decades and presently lack the availability of a dedicated European launcher.

The market for small satellites is expected to increase substantially in the coming years, as shown in market analyses of among others SpaceWorks Enterprises Inc (SEI, Nano/ Microsatellite Market Assessment 2015, August 2014) and shown in Figure 1. The excellent prospects for the small satellite market are confirmed by EuroConsult (Prospects for the Small Satellite Market, Feb 2015) with an estimate of more than 500 small satellites (nanosats, microsats, and minisats) to be launched in the next five years. Currently, the U.S. is the most active country in small satellite deployment with almost half of the 620 satellites launched in the past 10 years with Europe as the second-largest region. Historical analysis suggests the current supply of launch vehicles will not sufficiently serve future nano/microsatellite market demand.

Nanosats and microsats nowadays have to share a ride on a large rocket for a primary customer, which often causes conflicts with respect to the timeline and the orbit properties. Now that smaller satellites become technologically more advanced and mature, a call for 'affordable' dedicated launches is expedient for small satellite operators.

This situation has led to several initiatives of small launchers for various payloads in the range of 1 to 150 kg: India (Reusable Launch Vehicle, ISRO), New Zealand (Electron, Rocket Lab Ltd.) and USA (SuperStrypi, Aerojet Rocketdyne; LauncherOne, Firefly, Virgin Galactic; Lynx, XCOR; ALASA, DARPA). But also within Europe, efforts are ongoing: France (Eole, CNES), Norway (North Star, Nammo/Andøya Space Centre), Spain (Arion, PLD Space), Switzerland (SOAR, S3) and UK (Skylon, Reaction Engines Ltd.).

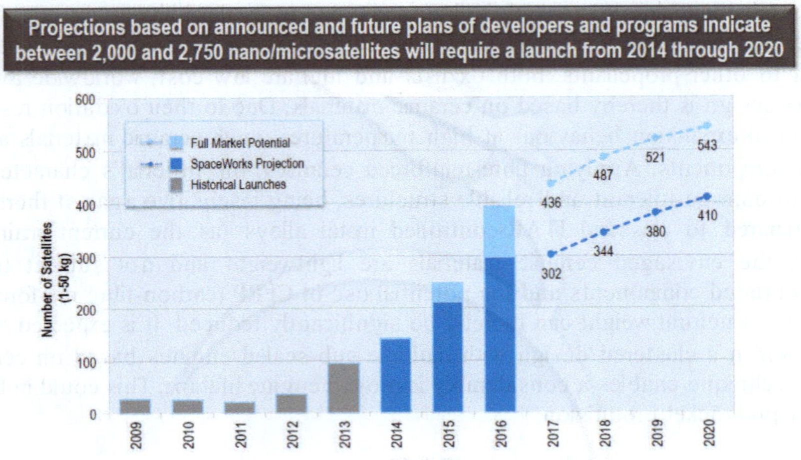

Figure 1. Future launch market for small satellites 1 - 50 kg (courtesy: SEI).

Although the above mentioned launch initiatives focus on the small satellite customer market, none of these focus on delivering the market's "sweet spot" to orbit and focus on specific payload launch ranges (e.g. 1-10 kg or 100kg+). Based on the market analyses the range up to 50 kg payload capacity can be considered the "sweet spot" for a small satellites launcher. Such a launcher will provide a proper launch capability for a single 50 kg satellite (i.e. commercial, scientific, and governmental) as well as for a flexible configuration of multiple smaller satellites (i.e. education, in-orbit demonstration) up to a total mass of about 50 kg. The above mentioned initiatives are in different states of development and are providing no launch services at this moment.

SMILE PROJECT

No operational dedicated launcher for small satellites exists today. Small satellites, launched as secondary payload, are entirely dependent of the constraints set by the primary payload, such as launch date and target orbit. Launch costs of less than €50,000 per kg of payload are required in order to compete directly with these piggy-back ride shares which are the current economically viable access to space for small satellites. With a dedicated launcher a higher cost per kg can be accepted for payloads which need to be delivered timely and accurately to a desired orbit. A shorter project schedule from concept to launch and better science are the arguments most commonly mentioned to support this. Hence, a consortium of 13 partners from 8 European countries are joining forces in a Horizon 2020 work programme to design a dedicated small launcher to be built in and launched from Europe. Together, the consortium coordinated by the Netherlands Aerospace Centre NLR covers all aspects of marketing, developing, and operating a cost-effective launcher with a well-balanced mix of companies, SMEs, and institutes.

The project is called "SMall Innovative Launcher for Europe", SMILE, and is currently in its preparation phase for the Grant Agreement with the European Commission with a planned Kick-off date of 1/1/2016. The project duration is set to three years. The SMILE project aims at a combined research approach into a new innovative European launcher for an emerging market of small satellites up to 50 kg using a multidisciplinary design and optimisation approach strengthened by the demonstration of critical technologies for cost-effective solutions and complemented with the design of a European-based launch capability from Andøya (Norway).

Aiming for commercial launch prices of less than 50,000 €/kg up 50 kg payload capacity, the total maximum cost for a launch shall be well below 2.5 M€. This target cost drives the design, construction, and operation of the launcher. After 2020, it is anticipated that the market for launching small satellites is in the order of several hundred per year and growing. A total capacity of up to 50 launches per year is foreseen. Using a flexible configuration of the launcher-payload interface structure, several combinations of small satellites up to 50 kg can be served.

The launcher will use advances in technology to achieve cost reduction, including design for series production, reusability, and the use of COTS components. Critical technologies enabling affordable and independent access to space will be developed in this project. To be able to meet the target price, the design will be based on existing advanced technologies as a starting point, and drive the development of required new technologies forward as part of the program. The overall objectives of the SMILE project therefore are:

- To design a concept for an innovative, cost-effective European launcher for small satellites
- To design a Europe-based launch capability for small launchers, based on the evolution of the existent sounding rocket launch site at Andøya Space Center
- To increase the Technology Readiness Level (TRL) of critical technologies for low-cost European launchers
- To develop prototypes of components, demonstrating this critical technology
- To create a roadmap defining the development plan for the small satellites launcher system from a technical, operational and economical perspective

Figure 2 shows a high level system view approach for three parallel paths throughout the project. The path towards a conceptual design of the launcher is split into an architectural design phase and a detailed design phase. Likewise, the critical technology path is split into two phases: a preparation phase and a demonstration phase for developing prototypes. The ground segment depends on the launcher design, but will also supply requirements to the launcher, and its phases follow the launcher development.

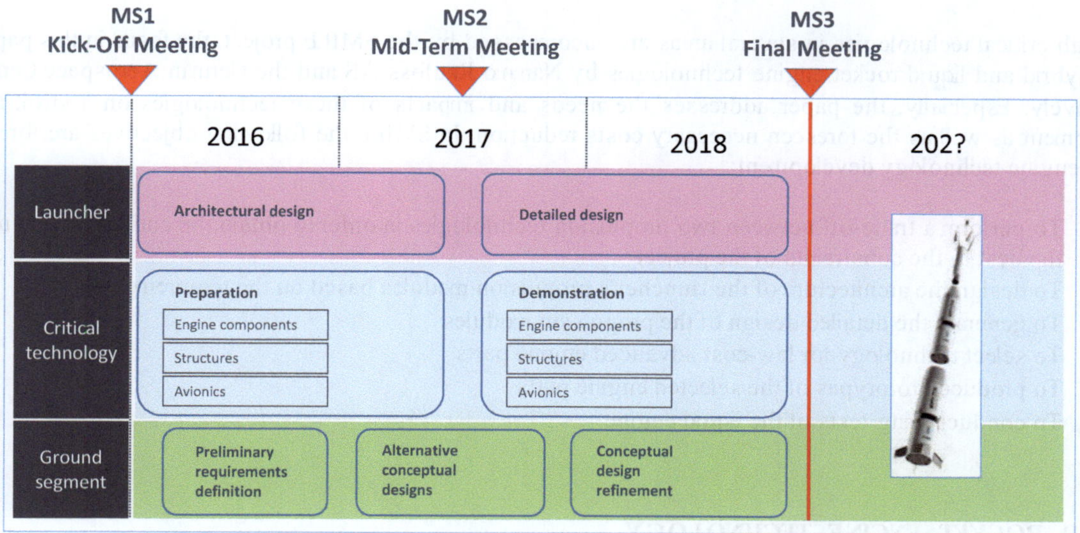

Figure 2. High level system view (source: Andøya Space Center).

In order to fulfil the project's objectives the consortium has identified a number of technologies that are capable of upgrading the actual state-of-the-art of this type of vehicles. These include:

- Hybrid engine technology
- Liquid engine technology with transpiration cooling
- Advanced low-mass and low-cost materials
- Series production of low-cost composite structures
- Printing technology for low-cost metal components
- Advanced, reliable COTS technology for miniaturised, low-power avionics
- European-based launch facility

At the end of the project the target Technology Readiness Levels (TRL) for the critical technologies shall be according to Table 1.

Table 1: Target Technology Readiness Levels (TRL) for the critical technologies

Item	TRL
Launcher concept	2
Hybrid rocket engine	7
Liquid rocket engine	5/6
Advanced materials	3
Automated manufacturing of composites	5
Printing technology	8
Advanced avionics	4
European launch facility concept	2

In order to enhance the continuity of the project's objectives, a roadmap will be set-up by assessing scenarios and critical future steps at technical, financial, and organisational levels. A business development shall include a technology roadmap towards a TRL 9 launcher. Furthermore, it presents a strategy to achieve commercially feasible launch services, including cost – benefit analysis.

Although critical technologies in several areas are encompassed by the SMILE project, the focus in this paper is on novel hybrid and liquid rocket engine technologies by Nammo Raufoss AS and the German Aerospace Centre DLR respectively. Especially, the paper addresses the needs and impacts of these technologies on a small launcher development as well as the foreseen necessary costs reduction. In SMILE the following objectives are foreseen for critical engine technology development:

- To perform a trade-off between two propulsion technologies in order to obtain the configuration answering the best to the constraints of the project
- To design the architecture of the launcher's propulsion modules based on the requirements
- To generate the detailed design of the propulsion modules
- To select technology for low-cost advanced engine parts
- To produce prototypes of the selected engine parts
- To conduct firing tests of the liquid engine

HYBRID ROCKET ENGINE TECHNOLOGY

Current State Of The Technology

Up to now, only two kinds of engines have been used for operational launchers: liquid engines (such as the European Vulcain II, the Russian RD-180 or the American Merlin 1A) and solid engines. The latter are mainly used as boosters for the big launchers (Ariane 5's SRBs) or for the first stages of medium launcher (Vega's P80, Pegasus system) or sounding rockets.

Liquid engines offer high versatility, through thrust regulation and restart capabilities, and high performance (high specific impulse), but are somewhat limited in thrust and their high complexity (with a turbo-pump feeding the combustion chamber with propellants) makes them quite costly, both in terms of mass budget and development cost. On the other side, while solid engines offer simplicity and high performances in terms of thrust, they have the drawbacks of being inherently hazardous (the oxidizer and fuel are intimately mixed in the grain), uncontrollable (impossible to stop once ignited), and tailored to one specific task.

Hybrid propulsion development started at the same time as for the other two. The goal was to combine the advantages of both types of engine (inherent safety, versatility, being able to throttle, and simplicity) at low cost. Unfortunately, knowledge at that time didn't allow hybrid engines to compete in terms of performance, especially because of a low regression rate of the fuel (leading to only small thrust capabilities, or complex fuel grain geometry).

In the last decade however, hybrid propulsion has matured, mainly through research and technology programs. Full scale flight weight rocket motors are now totally conceivable at low price, and with capabilities and performance allowing a competition with liquid or solid engine.

Nammo Raufoss AS (Nammo), a Norwegian based defence company, has since 2003 invested in the hybrid rocket propulsion technology. Based on hydrogen peroxide (H_2O_2), a completely green oxidizer, and HTPB fuel, Nammo has moved the technology forward through the following projects:

- The upscaling of the hybrid technology to a 30kN-class engine under the ESA funded Future Launcher Preparatory Program (FLPP)
- The establishment of a new 500kN Green Propulsion Test Stand
- The development of a throttleable hybrid engine for a Lunar Lander under the European Community funded 7th Framework Program, SPARTAN
- The development of a so-called "Hot Gas Reaction System" (HGRS), a new (mono-propellant) Reaction Control System for Ariane 5ME, Ariane 6 and Vega to replace the hydrazine alternative

The combination H_2O_2/HTPB offer the advantage of being already available in industrial quantities, while being completely green (only CO_2 and H_2O produced), safe to handle (nontoxic products) and safe to operate (two propellant completely segregated). Those characteristics, coupled with a simple fluid system, will substantially reduce hybrid propulsion life cycle cost compared to other propulsion systems. Moreover, with the use of a catalyst bed to decompose the H_2O_2, the engine can be stopped and restarted at will, without the need of an external igniter (which is the case with liquid engine). This could prove crucial for small launchers that want to launch multiple payloads on different orbits.

With Nammo's hybrid architecture, it is possible to develop an engine with performances high enough to suit the needs of small satellites launchers, at a much lower price tag.

The Unitary Motor: The Building Block Of The Hybrid Rocket Propulsion System

The current state of the hybrid technology at Nammo is represented by the Unitary Motor (UM), a novel concept of hybrid rocket engine developed by Nammo under an ESA-FLPP contract. It uses high concentration hydrogen peroxide (87.5% H_2O_2) as oxidizer and hydroxyl-terminated polybutadiene (HTPB) rubber as fuel. Its working principle is shown on Figure 3. The incoming liquid oxidizer, with a mass flow of about 11 kg/s, is first decomposed over a catalyst into hot steam and gaseous oxygen to a temperature of 670°C. It then goes through the injector and enters the combustion chamber in hot gaseous form, where ignition of the hybrid combustion occurs without any dedicated ignition device due to the high oxidizer heat flux, sufficient to vaporize the solid fuel. The vortex flow-field in the chamber generated by the injector helps in maintaining a high heat flux to the fuel surface and in achieving appropriate mixing of the reactants for a high combustion efficiency. The hot product gases are then expelled through a nozzle, generating close to 30 kN of thrust.

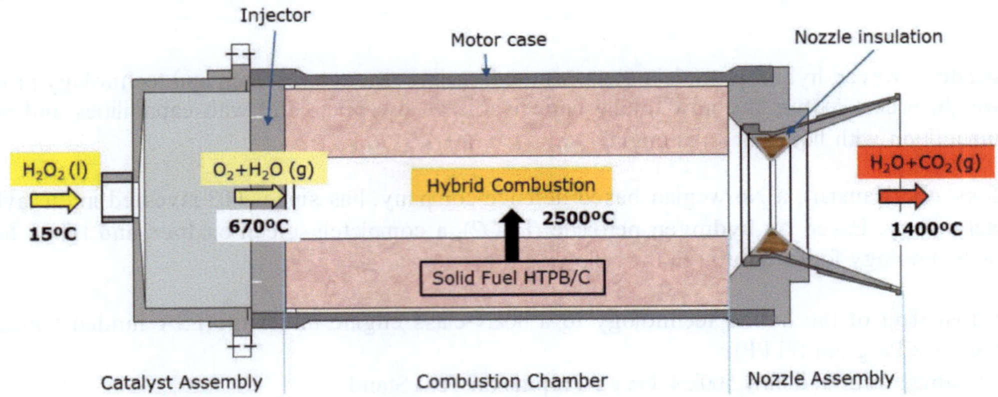

Figure 3. Working principle of the Unitary Motor.

Compared with solid rocket motors, the Unitary Motor designed by Nammo has a rich set of attractive features, even when compared with other versions of hybrid rocket engines with which its shares the inherent properties of hybrid propulsion. These features are:

- Self-ignition increasing engine start reliability and enabling an unlimited restart capability
- Wide range throttling with limited performance losses
- Green life cycle and exhaust properties
- Solid inert fuel and high-density green storable oxidizer
- High engine combustion efficiency, performance and stability
- Simplicity of a single circular port and single feedline configuration
- Low development and operational costs

Some of these features are common with liquid rocket engines, but compared with liquid rocket engines, the architecture of the UM is much simpler and the same features are obtained for a fraction of the cost.

The design of the UM has been split in two phases. First, a Heavy-Wall configuration (HWUM) has been designed, manufactured and tested in the fall of 2014. The goal was to assess the up-scaling of the hybrid technology (i.e. inner ballistic, regression rate of the fuel) without the constraint of a flight-weight engine. The HWUM demonstrated great behaviour in terms of both performance and stability from the first test firing (see Figure 4 and Figure 5), and continued to do so throughout the rest of the campaign. This allowed Nammo to complete the HWUM development test campaign in only 6 hybrid firing tests and one iteration on the motor configuration. The HWUM ground tests were concluded with the delivery of a very satisfactory motor design yielding the performance desired (see Table 2) for the next stage in the program.

Figure 4. HWUM during 3rd firing on November 18th, 2014.

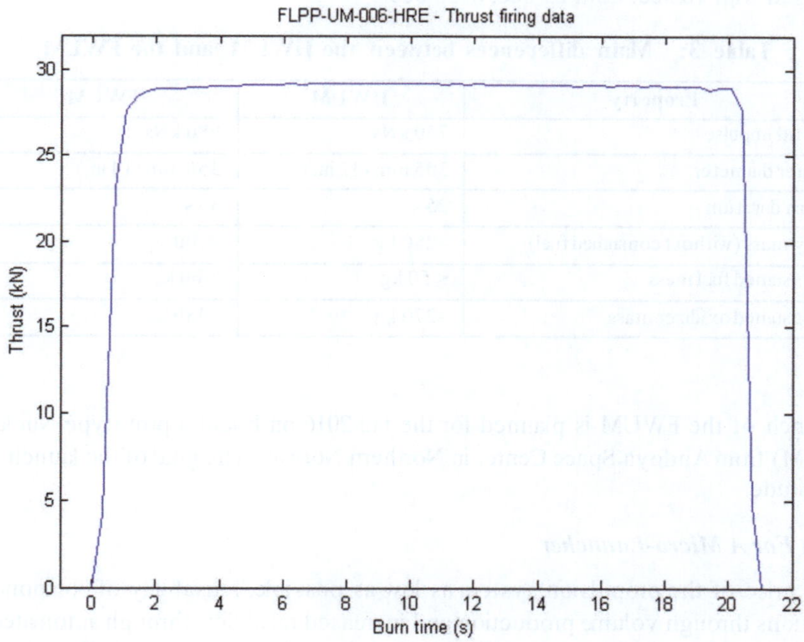

Figure 5. Thrust measured during the 4th HWUM firing, on November 28th, 2014. The measured data has been filtered down to 3 Hz sampling for visualization in this paper.

Table 2: Comparison of the 5th HWUM test experimental results (December 09th, 2014) with the motor design target. In this table all mean values are averaged over the entire motor burn duration and all values have been rounded independently

Firing	FLPP-UM-007-HRE	Design model target
Burn Duration	25 s	25 s
Mean oxidizer mass flow	10.8 kg/s	10.8 kg/s
Mean fuel mass flow	1.9 kg/s	1.6 kg/s
Mean oxidizer to fuel ratio	5.75	6.75
Mean chamber pressure	36 bar	35 bar
Mean specific impulse (ground level)	234 s	230 s
Mean engine efficiency	95 %	94 %
Total impulse (ground level)	750 kNs	700 kNs

Based on the results from the Heavy Wall Unitary Motor firings, a Flight Weight Unitary Motor (FWUM) has been designed. This design is currently being manufactured and the test campaign should start in November 2015. The design of the FWUM mainly replaces over-dimensioned parts with optimized parts, but it will also increase the capabilities of the Unitary Motor. Based on discussions with the user community, the capabilities of the UM are adjusted to a larger total impulse capability of 1000 kNs approximately. Based on the demonstrated performance of the HWUM, this can be achieved within an outer diameter of 14 inches, which is the standard sounding rocket payload diameter in use at Andøya Space Center and Europe in general. The updated design data is given in Table 3. Although also feasible, no attempt has been made to achieve a higher thrust level for the FWUM, but rather a longer burn time. It is increased with 10 sec. from 25 sec. to 35 sec.

Table 3: Main differences between the HWUM and the FWUM

Property	HWUM	FWUM
Total impulse	750 kNs	980 kNs
Outer diameter	305 mm (12 in.)	356 mm (14 in.)
Burn duration	25 s	35 s
Dry mass (without consumed fuel)	>280 kg	<100 kg
Consumed fuel mass	< 50 kg	> 60 kg
Consumed oxidizer mass	~270 kg	~380 kg

A demonstration launch of the FWUM is planned for the fall 2016 on board a prototype Nucleus sounding rocket (based on a single UM) from Andøya Space Center in Northern Norway. The goal of the launch is to reach the space frontier at 100 km altitude.

Hybrid Rocket Stage For A Micro-Launcher

In order to keep the price of the propulsion system as low as possible, reusability of components is a key feature leading to cost reductions through volume production and increased reliability through automated production. In that sense, the Unitary Motor is thought of as a building block that can be clustered to deliver the required thrust for a micro-launcher. The North Star rocket family, a Norwegian initiative of sounding rockets and micro-launchers, is based on that principle, with the utilization of two high thrust motors, the UM and its future upgrade the UM2, for the first stages and a third high performance engine with a more moderate thrust requirement and longer burn-time needed to obtain orbit insertion on the upper stage. Figure 6 presents the concepts of the different rockets of the North Star Family and Figure 7 the preliminary design performance of the different propulsion stages.

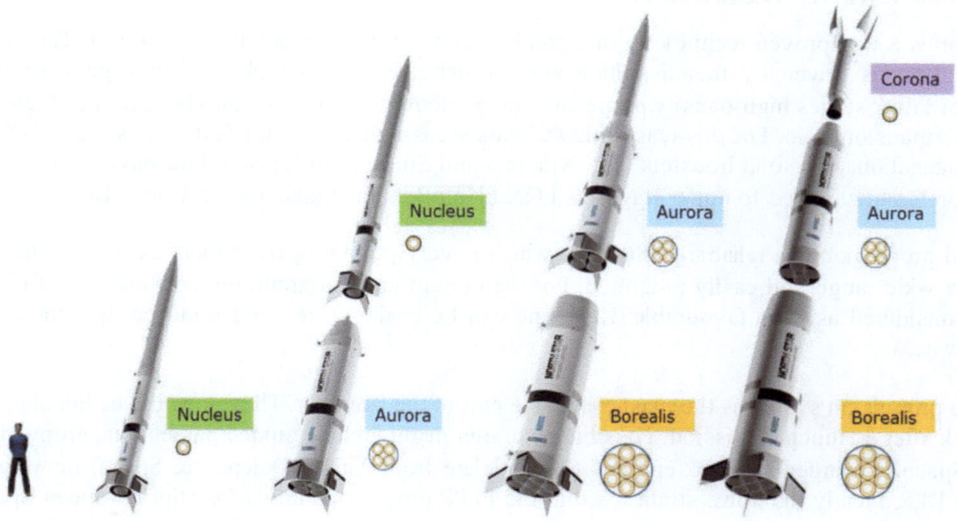

Figure 6. The North Star Rocket Family (source: Andøya Space Center).

Rocket stage	Motorization	Indicative impulse
Nucleus	1x Unitary Motor 1	Thrust: 28 kN Burn time: 35 s Total impulse: 1 MNs
Aurora	4x Unitary Motor 1	Thrust: 114 kN Burn time: 35 s Total impulse: 4 MNs
Borealis*	7x Unitary Motor 2	Thrust: 450 kN Burn time: 64 s Total impulse: 30 MNs
Corona*	1x High Performance Hybrid Motor	Thrust: 5 kN Burn time: 70 s Total impulse: 0.35 MNs

*The characteristics of the stages Borealis and Corona will depend on the flight performance of the Nucleus and later on, the Aurora stages.

Figure 7. The North Star Rocket Family (source: Andøya Space Center).

In SMILE, the same principle will be used with the added value of combining hybrid stages of clustered Unitary Motors with liquid stages. Based on the results and performances obtained during the FWUM test campaign and the demonstration launch, the sizing of the different propulsion stages of the micro-launcher will be achieved by clustering the Unitary Motor. The fluid feeding system (bringing the liquid oxidizer to the motors) will have to be design and sized accordingly and the performances (i.e. thrust, specific impulse, weight and size envelope) will be provided to the other members of the consortium for the global design of the launcher. It is strongly believed that both the inherent lower price of the hybrid technology and the clustering of elements enabling a more cost-effective production will be a large contribution in bringing the global cost of the launcher within the required range of 50.000€/kg.

LIQUID ROCKET ENGINE TECHNOLOGY

Liquid propulsion is a well proven technology that can be operated with different types of propellants. Hereby, the choice of propellants is driven by their resulting specific impulse, thrust-levels, and tankage-to-propellant mass ratios. Hence, for lower stages high-density propellants are preferred which yields into both reduced tankage volume and geometrical expansion ratio. For this reason, LOX/kerosene is rather used for first stages than LOX/LH2; in the latter case a combination with solid boosters (e.g. Ariane 5 and Space Shuttle) would be aimed for the launch or the propellants are preferably applied to upper stages as LOX/LH2 offers the highest specific impulse.

In general, liquid propulsion is a reliable technology which is very promising due to its flexibility as the engines can be throttled at a wide range and easily re-ignited. For the current configuration, the combination of LOX/kerosene propellants is considered as very favourable. Kerosene can be easily stored and refuelled, is a cheap fuel, and is available worldwide.

In any case, the propulsion system is the most expensive part of the launcher. Thus, it would be beneficial to retrieve the engines back after a launched mission. Possible solutions might include guided parachutes, propulsion-assisted boosters (like SpaceX), winged fly-back engines (like Adeline from Airbus Defence & Space) or winged fly-back boosters where DLR already did some studies within the FLPP programme funded by the European Space Agency (ESA). Once the engines are retrieved, they have to be inspected in order to have them refuelled and put into operation again.

Ceramic Based Design

In contrast to solid, hybrid or classical liquid engine approaches, liquid engines based on a ceramic design are very promising candidates with respect to such reusability aspects as they offer:

- Improved lifetime
- Thermo-shock resistance
- Thermal-cycling ability
- Reliability and damage tolerance
- Reduction in structural weight
- Oxidation resistance
- High specific strength at elevated temperatures
- Low thermal expansion

Hence, this specific kind of propulsion system using ceramics is well suited and applicable as it can be thermally cycled without degradation which is not the case for metallic approaches.

At DLR, there is a long experience on liquid rocket propulsion. The Institute of Structures and Design in Stuttgart is thereby focusing on ceramic-based designs which are based on the transpiration cooling technique. This is very favourable as all ceramic materials, such as non-oxide and oxide ceramic matrix composites (CMCs), can be manufactured in-house[4, 5, 6]. The transpiration cooling principle enables to highly increase the chamber wall lifetime while permitting a slight decrease of specific impulse. Compared to classical metallic solutions, it is possible to substantially reduce the engine's structural weight, depending on applied ceramic materials[7] and proposed design. In general, transpiration cooling consists of two mechanisms, as depicted in Figure 8: A small portion of the coolant is penetrating the combustor walls and thereby convectively extracting heat from the hot wall; in addition, a coolant layer forms at the inner combustor wall which protects the wall from hot combustion flow.

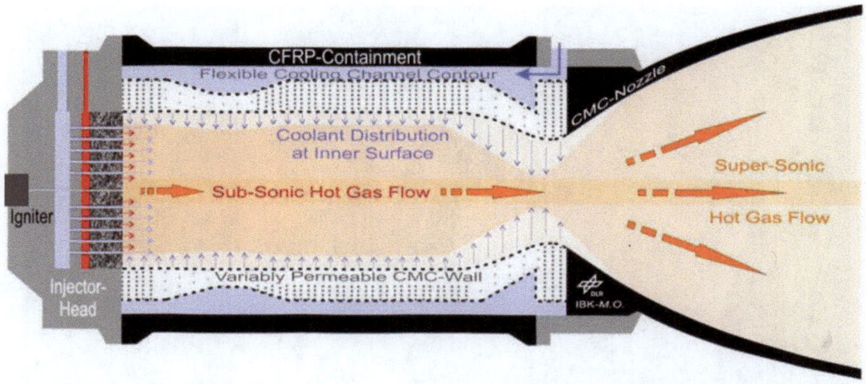

Figure 8. Schematic of a transpiration cooled ceramic thrust chamber.

First initial experiments on transpiration cooled segments for liquid rocket propulsion have been performed at the end of the 1990s. All testing was performed at various high-performance rocket engine test benches of DLR Lampoldshausen, up to 90 bars combustion chamber pressure. They solely focused on hydrogen-oxygen propellants, including cryogenic conditions as well. The development resulted in sophisticated design approaches which were investigated in different projects.

Between 2008 and 2012, four separate test campaigns were performed within the DLR projects *KSK* (Keramische Schubkammer, ceramic thrust chamber) and *KERBEROS* (Keramische Bauweisen für Experimentelle Raketenantriebe von Oberstufen, Ceramic Design of Experimental Rocket Engines for Upper Stages), as given in Table 4. The different configurations included the variation of wall and nozzle materials, injectors (API: advanced porous injector from DLR Lampoldshausen; TRIK: coaxial injector by DLR Stuttgart), contraction ratio, coolant blowing ratio, characteristic chamber length, etc. Further details can be obtained from[8, 9, 10].

Table 4: DLR ceramic thrust chamber test campaigns 2008-2012

	KSK-KT	KSK-ST5	MT5-A	WS1
Year	2008	2010	2012	2012
Test bench	P8	P8	P6.1	P6.1
Propellant combination	LOX/LH2	LOX/LH2	LOX/GH2	LOX/GH2
Injection temperature (fuel)	≈ 55 K	≈ 55 K	≈ 135 K	≈ 150 K
Injection temperature (oxidizer)	≈ 155 K	≈ 155 K	≈ 125 K	≈ 140 K
Coolant	H_2	H_2	H_2	H_2
Wall material	C/C	Al_2O_3 and C/C	Al_2O_3 and C/C	Various
Nozzle material	Copper	C/C	C/C	C/C
Injector	API	API	TRIK	TRIK
Chamber diameter (d_c)	50 mm	50 mm	50 mm	50 mm
Throat diameter (d_t)	31.6 mm	31.6 mm	20 mm	20 mm
Characteristic chamber length (l^*)	0.86 m	0.68 m	1.75 m	1.83 m

Figure 9 shows test operation of the ceramic thrust chamber during the test campaign MT5-A. Especially in combination with the transpiration cooling technique and the use of CFRP housing structures, the engine's structural weight can be significantly reduced. On the other side, sophisticated CMC materials enable replacing ITAR-controlled metal alloys (as the current main material for combustion chambers) in the future.

Figure 9. Liquid rocket engine test MT5-A at P6.1 test bench in Lampoldshausen (LOX/GH2).

Furthermore, the general feasibility in GOX/kerosene combustion environment was successfully demonstrated in the EC project ATLLAS (coordinated by ESA and funded within FP6, 2006-2009). All tests were performed at the high-pressure rocket combustion chamber test bench at Technische Universität München (TUM), see Figure 10. Various CMC materials were tested, whereas oxide CMCs seem to be very suited for this kind of application as the material is able to withstand hot gas oxygen attacks. Figure 11 shows two of the integrated CMC liner materials: C/C (non-oxide) and WHIPOX (oxide). With respect to cooling performance, hydrocarbon-based coolants such as Jet A-1 kerosene turned out to be very efficient.

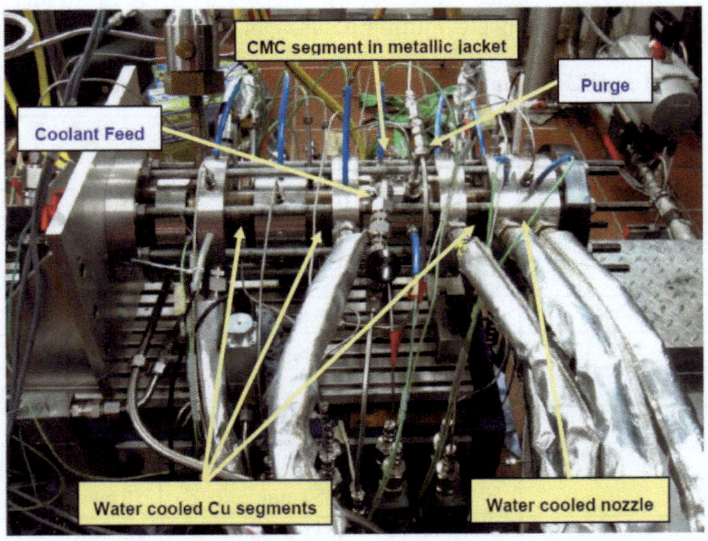

Figure 10. Liquid rocket engine test at TUM test bench (GOX/Jet A-1).

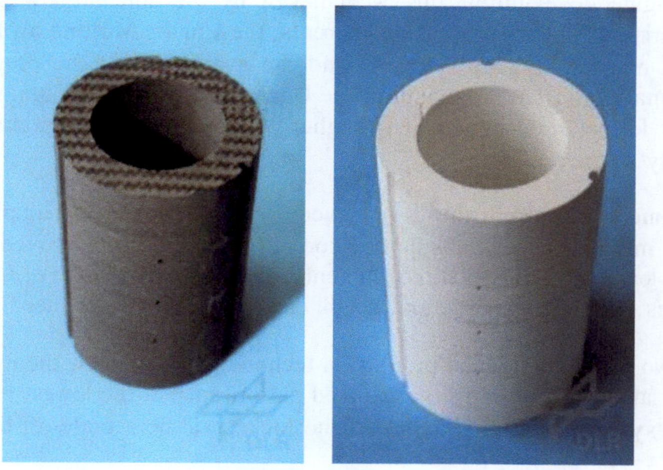

Figure 11. Ceramic inner liners for TUM test (left: C/C, right: WHIPOX).

Engine Reusability

Already in the late 1960s, Pratt & Whitney developed the transpiration cooled XLR-129 rocket engine with a chamber pressure of approximately 100 bars. The engine was extensively tested and based on the results; a transpiration cooled design was developed for the Space Shuttle Main Engine (SSME). Transpiration cooling was selected in order to fulfil the NASA criteria of 100 time engine reusability[11]. This engine development of Pratt & Whitney is the only known experimental study dealing with transpiration cooled engine life cycle, durability and re-usability to date. Based on the published results of Pratt & Whitney and theoretical considerations, the lifetime of transpiration cooled chambers is expected to be at least 10 times higher than that of regeneratively cooled chambers.

It has to be mentioned that at this time, transpiration cooling research was mainly conducted considering metallic materials. In case of local hot spots, such metallic structures tend to melt and cause a catastrophic failure. This is in clear contrast to ceramic-based materials which do not exhibit such behaviour. Additionally, ceramic-based designs enable improved lifetimes due to their positive thermal-cycling ability and thermo-shock resistance.

Envisaged SMILE Approach

It is expected that existing design approaches could be transferred to LOX/kerosene operation. In doing so, a ceramic-based thrust chamber assembly will be designed. Whereas the injector head might be made via SLM (selective laser melting)-techniques, the combustor component will be designed of ceramic liners actively cooled by transpiration. Here, both fuel and oxidiser are considered as potential coolants. In addition, a ceramic nozzle section is foreseen.

A clustered design is considered which would result in multiple turbopump-fed sub-scaled engines, depending on the mission scenario. DLR's engine enables reliable low-cost components to fit into the envisaged target price of 50,000€ per kg of payload with a future potential of reusability.

CONCLUSIONS

There is a need for a dedicated and affordable small satellite launcher. A major challenge for the launcher design is to become cost efficient within all technology development areas in order to offer future customer launch prices of less than €50,000 per kg of payload. The SMILE project will take up this challenge by aiming at a combined research approach into a new innovative small launcher for an emerging market of small satellites up to 50 kg using a cost-effective design approach. Cost reduction is achieved by applying reusability of one or more stages, applying commercial industry-grade components and through volume production including cost-optimized manufacturing process. In this paper the cost effectiveness for the rocket engine development is addressed.

For the hybrid rocket engine development this is achieved by the inherent low life-cycle cost of the hybrid technology and the clustering of unitary propulsion elements, the Unitary Motor. Low life-cycle cost is achieved by a simple architecture, the non-toxicity, the inertness and the availability of the propellants and the overall low development and operational costs. The clustering of the Unitary Motor will also bring the cost down, thanks to a higher volume production for each component. This higher volume could also legitimate an automated production leading to a better reliability of the product.

For the liquid rocket engine development this is achieved by an operation of multiple LOX/kerosene sub-scaled engines based on ceramic materials and a transpiration cooling technique for improved engine lifetime and reuse. In combination with reliable low cost 3-D printed components and the potential use of CFRP (carbon-fibre reinforced plastics) housing structures, the engine's structural weight can be significantly reduced.

The combination of the two hybrid and liquid propulsion technologies will allow the use of the right technology at the right place to offer a launcher delivering the required performance at the lowest price possible. Ultimately, the choice of the propulsion system for all the stages of the rocket will be a trade-off between performance, launch objectives and cost.

Acknowledgments

Gratitude goes out to the SMILE consortium supporting the project and this paper (Netherlands Aerospace Centre NLR, Nammo Raufoss AS, German Aerospace Center DLR, Terma, Andøya Space Center ASC, INCAS, Airborne Technology Centre ATC, Heron Engineering, Innovative Solutions In Space ISIS, 3D Systems LayerWise, PLD Space, Tecnalia and BoesAdvies).

References

1. Onno Verberne, Martina G. Faenza, Adrien J. Boiron and Bastien Haemmerli, "Development of the North Star Sounding Rocket: Getting ready for the first demonstration launch", *to be presented at the AIAA 51st Joint Propulsion Conference*, Orlando, Florida, July 2015.
2. Adrien J. Boiron, Martina G. Faenza, Bastien Haemmerli and Onno Verberne, "Hybrid Rocket Motor Upscaling and Development Test Campaign at Nammo Raufoss *", to be presented at the AIAA 51st Joint Propulsion Conference*, Orlando, Florida, July 2015.
3. Onno Verberne, Adrien J. Boiron and Terje Vesterås: "FLPP Large Scale Hybrid Rocket Demonstrator", *Space Propulsion 2014*, Cologne, Germany, May 2014
4. B. Heidenreich, S. Hofmann, R. Jemmali, M. Frieß, D. Koch: "C/C-SiC Materials Based on Melt Infiltration - Manufacturing Methods and Experiences from Serial Production". *HT-CMC 8*, Xi'an, China, 2013.
5. E. Klatt, S. Hackemann, M. Kuhn, M. Friess, H. Voggenreiter: "Creep Behavior of Oxide CMC Based on Polymer (OXIPOL)", *HT-CMC 7*, Bayreuth, Germany, 2010.
6. M. Schmücker, A. Grafmüller, H. Schneider: "Mesostructure of WHIPOX all oxide CMCs", *Composites Part A 34*, 2013, doi:10.1016/S1359-835X(03)00100-3.
7. M. Kuhn, M. Ortelt, H. Hald, C. Kirchberger, G. Schlieben, H.-P. Kau: "CMC Materials for Combustion Chamber Applications", *3rd European Conference for Aerospace Sciences (EUCASS)*, 6-9 July 2009, Versailles, France.
8. A. Herbertz, M. Ortelt, I. Müller, H. Hald: "Transpiration-Cooled Ceramic Thrust Chamber Applicability for High-Thrust Rocket Engines", *in 48th Joint Propulsion Conference*, Atlanta, USA, 2012, AIAA-2012-3990.
9. H.Hald, A. Herbertz, M. Ortelt and M. Kuhn: "Technological Aspects of Transpiration Cooled Composite Structures for Thrust Chamber Applications", *16th AIAA/DLR/DGLR International Space Planes and Hypersonic Systems and Technologies Conference*, Bremen, Germany, 19–22 Oct 2009.
10. A. Herbertz, M. Selzer: "Analysis of Coolant Mass Flow Requirements for Transpiration Cooled Ceramic Thrust Chambers", *Transactions of the Japan Society for Aeronautical and Space Sciences*, Space Technology Japan, 12 (ists29), Pa_31-Pa_39, 2014, ISSN 1347-3840.
11. D. Mulready, D.: "Advanced Engine Development at Pratt & Whitney", *Society of Automotive Engineers*, Inc., Warrendale, USA, 2001.

The Google Lunar XPRIZE – Past, Present and Future

Andrew Barton, Nathan Wong, Derek Webber
XPRIZE Foundation
800 Corporate Pointe, Suite 350, Culver City, CA 90230, USA;
+1 310 741 4971, Andrew.barton@xprize.org

ABSTRACT

The Google Lunar XPRIZE is a US$30 Million prize purse competition to foster a new space economy through low-cost, efficient access to the Moon. Aside from the promise of re-opening the lunar surface to science and exploration, the Google Lunar XPRIZE is also a unique example of how the incentive prize model can be used to spur innovation in the space industry. This paper examines some of the most important ways that that innovation is being stimulated and summarises the progress and impacts to date.

In early 2015 XPRIZE awarded US$5.25 million in Milestone Prizes across five teams that demonstrated advanced progress towards flight readiness in the key challenge areas of landing, mobility and imaging. The paper examines XPRIZE's objectives, plans and implementation of the Milestone Prizes as well as their impact on the overall competition.

An important component of the XPRIZE approach is robust and professional media engagement around the prize. This paper summarizes major PR activities, media coverage and key audiences targeted particularly in the course of the Terrestrial Milestone Prizes but also extending to the overall prize.

As we approach the end of 2015 the prize is now entering its final phase in which only teams with confirmed launch contracts can be considered viable competitors so this paper also summarizes the mission plans and business models of the teams with launch contracts or which look most likely to obtain launch contracts.

Looking to the future, assuming a successful completion of one or more Google Lunar XPRIZE missions, there will be certain profound impacts on lunar science and exploration, as well as the broader space industry. The paper summarises XPRIZE's current expectations and plans around these impacts.

Finally this paper also summarizes some early work by XPRIZE on an overarching roadmap for the future of the space industry, including a summary of future space prize concepts that have been investigated to date within XPRIZE and with our stakeholder community.

I. Introduction

The $30M Google Lunar XPRIZE is an unprecedented competition to challenge and inspire engineers and entrepreneurs from around the world to develop low-cost methods of robotic space exploration. To win the Google Lunar XPRIZE, a privately funded team must successfully place a robot on the moon's surface that explores at least 500 meters and transmits high-definition video and images back to Earth [1].

Currently there are 16 eligible teams from around the world competing to win the US$20,000,000 Grand Prize. In addition to the Grand Prize there is a Second Place Prize of $5 Million and bonus prizes for achieving additional milestones such as surviving the lunar night or imaging heritage items. In October 2015, Team SpaceIL from Israel became the first team with a verified launch contract for a launch on a Falcon 9 secured through Spaceflight Industries.

S. Hatton (ed.), *Proceedings of the 13th Reinventing Space Conference*,
https://doi.org/10.1007/978-3-319-32817-1_8

II. Significant Prize Updates

General Registered Team Statistics

According to data collected in July 2015, the teams have spent US$60.9 Million ($15 Million increase from 2013) cumulatively on Google Lunar XPRIZE related expenditures and have contributed 945 person years towards their missions. As a result of the Milestone Prizes a significant amount of investment was made towards Google Lunar XPRIZE missions. Figure 1 shows the increasing rate of funds spent per team overlaid with information on the amount of person years spent per team.

Figure 1: Graph of expenditure and person hours per (averaged) team by year.

Awarding of Terrestrial Milestone Prizes

$5.25 Million in Terrestrial Milestone Prizes were awarded in January 2015 in recognition of key technological advancements towards their Google Lunar XPRIZE mission. [2] The Judging Panel awarded prizes based on monitoring visits, which included design reviews, software validation, and hardware testing in three categories: Landing, Mobility and Imaging. Table 1 summarizes the winning of Terrestrial Milestone Prizes. The five teams awarded Milestone Prizes were Astrobotic (Landing, Mobility, and Imaging), Moon Express (Landing and Imaging), Part-Time Scientists (Mobility and Imaging), Team Indus (Landing), and Hakuto (Mobility). A summary of the prizes awarded can be seen in Table 1.

The teams made undeniable technical progress including the construction and testing of new hardware representative to the systems that they plan to launch to the Moon. In the Imaging and Mobility categories, all teams awarded prizes were able to demonstrate a substantial proportion of the in a hardware configurations, form factors and materials that were highly representative of the equipment that would eventually operate on the lunar surface. For the three teams selected in the Landing System category the progress necessarily included more substantial analysis and simulation but in each case, at least one major sub-system of the lander was built and tested under key operations conditions. More detailed information on the objectives, approach, implementation, and impacts can be found in section III of this paper.

Category	Accepted to Accomplishment Round	Milestone Award (US$)
Landing	Astrobotic Indus Moon Express	1,000,000
Mobility	Astrobotic Hakuto Part-Time Scientists	500,000
Imaging	Astrobotic Moon Express Part-Time Scientists	250,000

Table 1: Summary of Terrestrial Milestone Prize Announcement

Competition Extension to 2017

After seeing substantial progress in the Terrestrial Milestone Prizes and still no teams with a verified launch contract and extension of the grand prize deadline was made. In order to secure an extension beyond 2015 at least one team had to have provided XPRIZE and Google with notification of a launch contract by December 31. For the team with that contract, they would have until December 31, 2017 to complete their mission. If at least one team provided notification of a launch contract by December 31, 2015 all remaining teams will have until December 31, 2016 to also provide a notification of launch contract in order to remain in the competition until the end of 2017. [3] This conditional extension better reflected the realities and timing of securing a launch contract compared to the previous hard stop deadline. On October 7, 2015 Team SpaceIL became the first team to have a verified launch contract triggering this extension for their launch in 2017 in addition to the extension for all teams into 2016.

SpaceIL Verified Launch Contract Announcement

The SpaceIL launch contract became the first verified by the XPRIZE Foundation and officially extended the prize deadline to Dec 31, 2017. SpaceIL will launch on a SpaceX Falcon 9 launch vehicle as a co-lead payload on a mission fully manifested by Spaceflight Industries. With this launch SpaceIL looks to send the first Israeli mission to the Moon in addition to the world's first private lunar mission. SpaceIL is a non profit organization with benefactors including the Dr. Miriam and Shedlon G. Adelson Family Foundation and Morris Kahn's Kahn Foundation.

The SpaceIL craft will launch with over 20 other spacecraft that will be deployed in a sun synchronous orbit. After those other payloads are deployed the Falcon 9 upper stage will place the SpaceIL craft in a highly elliptical orbit with apogee around 40,000 km. The SpaceIL vehicle will take over from there providing the rest of the propulsion to take it to the Moon, land and move 500 meters. [4]

Unlike most other teams, SpaceIL does not have a commercial vision beyond the prize however it does intend to continue working towards it non-profit goals related to education. SpaceIL has informed us however that some of their technical partners do intend to leverages their work on the prize for future commercial business.

Part-Time Scientists Partner with Audi

German automaker Audi is supporting Google Lunar XPRIZE Team Part-Time Scientists with "expertise in lightweight construction and e-mobility, with Quattro permanent all-wheel drive and with piloted driving" In addition to this technical support Audi will also be helping with testing and quality assurance of the Part-Time Scientists "Audi Lunar Quattro" rover. [5]

Hakuto Signs Contract to Ride with Astrobotic

Hakuto, the only Japanese competitor in the Google Lunar XPRIZE announced in February that they had signed a contract with US based team Astrobotic to fly its two rovers (Moonraker and Tetris) to the lunar surface. [6] Both teams are interested in landing in the Lacus Mortis region due to the potential for lunar lava tube skylights. As part of the Terrestrial Milestone Prizes Astrobotic performed precision landing tests with a Masten vehicle to validate their software is able to land them safely close to an opening on the lunar surface and Hakuto performed advanced mobility field-testing with the rovers tethered together. Both of these tests will be crucial for completion of their respective lunar missions. In addition to the Hakuto Rovers Astrobotic has capabilities of delivering additional payloads to the lunar surface. Currently they have contracts including Google Lunar XPRIZE Teams, a Japanese sports drink, Elysium burial services, and more. Astrobotic is looking for additional payloads to fill up its initial flight manifest.

Team Angelicvm of Chile also announced a similar ridesharing arrangement with Astrobotic during the 2015 Team Summit.

Moon Express Shifts Development to Florida with Lease of Space Launch Complex 36 from Space Florida

In January 2015 US based Moon Express signed an agreement to use Space Launch Complex 36 at the Kennedy Space Center. Previously this site was used by the Air Force for Atlas V launches up until 2005. After the decommissioning of the launch site Space Florida, the state's aerospace economic development agency took over and began looking for tenants. Moon Express will be able to use the space for development and a continuation of flight tests performed in the Milestone Prizes at the Shuttle Landing Facility. [7]

Astrobotic invited to White House Demo Days

In August 2015 the White House hosted the first ever Demo Day showcasing innovators around the country. Astrobotic was invited to this event to share their story and demo their technology for the President of the United States of America. [8] President Obama had the following to say about Astrobotic, "And then there are the folks at Astrobotic Technology in Pittsburgh. They are shooting for the Moon. Literally. With plans to land a rover on the lunar surface within the next couple of years. Which is pretty exciting. I wouldn't, you know, mind seeing how that turns out!

In America, that's who we've always been. We explore next frontiers, we're pioneers with a vision for tomorrow, whether it's Lewis & Clark, Sally Ride, we're the nation of Franklin and Edison and Carver and Salk and Gates, and the folks here today are heirs to that legacy. They're the driving force in a 21st century economy."

This event provided a great opportunity for a Google Lunar XPRIZE team to show their work and progress at the highest level of government.

Omega Envoy and Team Puli perform field testing

Hardware demonstrations were not limited to the Terrestrial Milestone prizes with Omega Envoy performing testing of their Sagen Rover at the Kennedy Space Center Swamp Works. The testing focused on the effects of lunar dust on the mobility and imaging system. [9]

Team Puli took their rover to Austria to test it at the Austrian Rock Glacier Mars Simulation. During this simulation the Puli Rover searched and mapped predefined areas to provide input for the human portion of the mission. [10] This test showed the robustness of the rover design on rough terrain and provided a test for mission operation procedures.

Educational and Public Outreach Initiatives

In the past year the Google Lunar XPRIZE has continued its education and public outreach initiatives. The primary educational outreach tool is the Moonbots program. The 2015 MOONBOTS Challenge was an international on-line competition that challenged youth from 8 to 17 years old to form a team (2-4 members) to design, create and program their own robots. The competition was divided into two phases. In Phase One, teams submitted a short video or written story about what inspires them about the Moon, from old tales to the potential of lunar exploration. Thirty teams were selected by a panel experts to move onto the next stage: Phase Two. In Phase Two, teams received a robotics kit of their choice: LEGO ® MINDSTORMS® EV3, VEX IQ Superkit or MECCANO Meccanoid G15 KS to create their own robot to rove on a simulated lunar landscape based on the moon tale the team created in Phase One of the competition. Teams were also asked to create and upload a video showing how they have demonstrated their MOONBOTS challenge to children and adults in their community as part of their STEM outreach effort. Four Grand PRIZE winners won a trip to Japan to meet with teams from all over the world competing in the Google Lunar XPRIZE. [11]

In Phase 1 there were 235 teams from 29 different countries with over 1200 people participating with an average age of 12. The 30 teams who made it to phase 2 came from 11 different countries and had 41% female involvement.

In addition to the Moonbots program the Google Lunar XPRIZE full dome planetarium show is still showing in planetariums around the world and a flat screen version was released on YouTube for anyone to watch.

III. The Milestone Prizes Program

Objectives

The main objective of the Milestone Prizes was to assist teams towards their Google Lunar XPRIZE mission goals by giving teams additional incentives to complete difficult technical milestones, and thereby make tangible progress towards flight readiness. The substantial cash prizes in the Milestone Prizes strengthened the business plans of the winning teams, bringing forward some of the prize money that otherwise would have required them to overcome the extraordinarily high risks and costs of procuring a launch vehicle.

A secondary purpose of the Milestone Prizes was to provide additional recognition for leading teams to strengthen awareness of the team to prospective investors and increase the awareness and impact of the Google Lunar XPRIZE as a whole.

Approach

The Milestone Prizes guidelines were defined to reward teams for verifiable technical steps. Specifically, the prizes were structured to incentivize teams to identify key technical risks and to retire them via testing activities. Three categories of prizes were made available Landing, Mobility and Imaging for teams that demonstrated advanced progress in their mission preparations. The Landing category was for advanced progress towards a spacecraft that could land safely on the Moon. The Mobility category was for moving across the lunar surface and the Imaging was for the system that acquires high definition video and photographs on the lunar.

To compete for Milestone Prizes teams first had to submit a set of technical documents called a Milestone Definition Data Package (MDDP), which was assessed by the Judging Panel (an independent panel of 9 industry experts) according to the scoring criteria that took into account the team's overall mission concept, the team's assessment of technical and the development & verification plan. At the end of the assessment process the Judging Panel decided which entries were be accepted to proceed into the Accomplishment Round. A total of 11 entries were accepted (across 5 teams), but of those only 9 were awarded prizes (see table below).

Implementation

During the Accomplishment Round the Judging Panel monitored each accepted team as it completed the activities specified in its MDDP(s). The primary mode of monitoring was via participation in key review meetings with the teams. Judges also attended certain key tests. In total, across the 5 teams, the Judging Panel travelled to witness 48 review meetings and tests in addition to receiving regular teleconference updates from teams.

The Judging Panel decided the winners of the Milestone Prizes after they had verified that the teams had satisfactorily completed of all their MDDP activities. The panel had the option to decide on winners before the end of the Accomplishment Round if all the MDDP activities had been satisfactorily completed (which was the case for one team).

After extensive deliberations, the judging panel chose to award a total of $5.25 million dollars across the 5 teams and 3 technical categories, as shown in Table 1.

An award recognition ceremony was held on January 26th 2015 at the California Academy of Sciences in San Francisco to an audience of 200 people from Google, XPRIZE, teams, judges and press.

The prize money awarded to teams in the Milestone Prize(s) was "brought forward" from the eventual GLXP Grand or Second Place Prize winnings of the respective teams. If a team won a Terrestrial Milestone Prize and then goes on to win either the GLXP Grand or Second Place Prize, the equivalent monetary value of the Terrestrial Milestone Prize will be deducted from that team's Grand or Second Place Prize

Participation in the Milestone Prizes was optional, and teams that did not compete for, or were not awarded, Milestone Prizes can still fulfill the requirements to win the GLXP Grand or Second Place Prizes.

Impacts

The two US teams that were awarded Milestone Prizes were able to leverage the recognition of technical progress and the cash injection to further their business models with customers and/or investors. In addition to those benefits, each of the non-US teams became overnight success stories for commercial space in their respective countries – India, Germany and Japan, and for each of those teams the Milestone Prizes represent the first time that an incentive prize was awarded for achievements in space technology.

News coverage of the Milestone Prizes generated more than 550 stories equating to more than 5.5 billion media impressions with 55% of coverage from outside the US. The news of the SpaceIL launch announcement and competition extension garnered an even higher level of media impressions, currently scored at 5.8 billion. These figures, although highly impressive, should be taken in the context of modern social media trends whereby single stories and messages can rapidly propagate across numerous platforms and the consumption of such media can be very fleeting in nature. The strong interest in the media can also be explained by the high interest amongst the general populace in space-related news, a factor that XPRIZE will continue to leverage in the future.

Assuming a successful completion of one or more Google Lunar XPRIZE missions, there will be certain profound impacts on lunar science and exploration, as well as the broader space industry.

IV. Future Space Prizes

XPRIZE has commenced a program to road map the future evolution of space technology, and the general societal trends as our species cements itself as a space-faring civilization. Broadly speaking, the road mapping process will consist of the following key steps:

1. Identification of potential futures states of the space industry
2. Selection of desired states
3. "Back-casting" from those desired states to the present day
4. Identification of key technical challenges
5. Down-selection of prize-able technical challenges

6. Outlining of multiple prize concepts

The 2015 Google Lunar XPRIZE Team Summit in October held in Tokyo presented a valuable opportunity to gather the minds of many space innovators and other industry participants, and XPRIZE took advantage of this to run a space prizes design workshop. Further events involving other segments of the XPRIZE community and the space community at large are foreseen for 2016.

XPRIZE currently aims to produce a first version of its Space Prizes road map some time in 2016 and will periodically refine it thereafter.

V. Conclusions

This year for the Google Lunar XPRIZE has included many changes including the first verified launch contract of a Google Lunar XPRIZE team and the extion of the competition to Dec 31, 2017. Many technical advances have occurred thanks to the completion of the Terrestrial Milestone Prizes. As a result of these prizes teams have been able to secure much needed sponsorships and partnerships to help them complete their lunar mission. Although frontrunners have been identified through the Terrestrial Milestone Prizes the remaining teams continue to work through their mission and hardware design and are continuing to make progress. In addition to technical progress some teams are focusing on business development as the most important aspect of their team.

XPRIZE is consolidating its efforts to define and execute a longer term roadmap incorporating numerous new space prizes in the future that will support humanities use of space as well as its permanent settlement.

To follow the competition updates, please visit http://lunar.xprize.org

VI. References

[1] XPRIZE Foundation. 2015. "Google Lunar XPRIZE." Retrieved 9/17/2015 from: http://lunar.xprize.org/about/overview

[2] XPRIZE Foundation. 2015. "Google Lunar XPRIZE." Retrieved 9/17/2015 from: http://lunar.xprize.org/about/milestone-prizes

[3] XPRIZE Foundation. 2015. "Google Lunar XPRIZE." Retrieved 9/17/2015 from: http://lunar.xprize.org/press-release/deadline-30-million-google-lunar-xprize-extended-end-of-2016

[4] XPRIZE Foundation. 2015. "Google Lunar XPRIZE." Retrieved 10/30/2015 from: http://lunar.xprize.org/press-release/israeli-google-lunar-xprize-team-first-sign-launch-agreement-private-mission

[5] Engadget. 2015. "Audi is helping to make a moon rover for Google's Lunar Xprize." Retrieved 9/17/2015 from: http://www.engadget.com/2015/06/26/audi-lunar-rover-xprize/

[6] XPRIZE Foundation. 2015. "Google Lunar XPRIZE." Retrieved 9/17/2015 from: http://lunar.xprize.org/press-release/two-google-lunar-xprize-teams-announce-rideshare-partnership-mission-moon-2016

[7] Space.com. 2015. "Moon-Mining Company to Use Historic Florida Launch Complex." Retrieved 9/17/2015 from: http://www.space.com/28345-moon-express-cape-canaveral-launch-complex.html

[8] Pittsburgh Post-Gazette. 2015. "Pittsburgh-based Astrobtic invited to first-ever White House Demo Day." Retrieved 9/17/2015 from: http://www.post-gazette.com/business/tech-news/2015/08/02/Pittsburgh-based-Astrobotic-invited-to-first-ever-White-House-Demo-Day/stories/201508020203

[9] XPRIZE Foundation. 2015. "Google Lunar XPRIZE." Retrieved 9/17/2015 from: http://lunar.xprize.org/teams/omega-envoy/blog/team-omega-envoy-completes-successful-mobility-and-visual-tests-nasa

[10] Puli Space. 2015. "Puli Rocks." Retrieved 9/17/2015 from: http://www.pulispace.com/en/media/news/240-puli-rocks

[11] XPRIZE Foundation. 2015. "Google Lunar XPRIZE." Retrieved 9/17/2015 from: http://lunar.xprize.org/education/moonbots

BIS-RS-2015-50

SEAHAWK: A Nanosatellite Mission for Sustained Ocean Observation

Hazel Jeffrey, Hessel Gorter, Alasdair Gow, Craig Clark
Clyde Space Ltd.
5B Skypark 5, 45 Finnieston Street, Glasgow G3 8JU; +44 141 946 4440
enquiries@clyde-space.com

Alan Holmes, Craig Herrin, Linda Sasaki
Cloudland Instruments
130 Castilian Drive, Suite 102, Goleta, Ca. 93117; 805-308-6977
alan@cloudlandinstruments.com

John Morrison
The University of North Carolina Wilmington
UNCW Center for Marine Science, 5600 Marvin K. Moss Lane
Wilmington, NC 28409; 910-962-2333
morrisonj@uncw.edu

ABSTRACT

In a recent report, the US National Academy of Science has highlighted the need for sustained, advanced ocean colour research and operations. The report shows that ocean colour satellites provide a unique vantage point for observing the changing biology of our ocean's surface. Space observations have transformed biological oceanography and are critical to advance our knowledge of how such changes affect important elemental cycles, such as the carbon and nitrogen cycles, and how the ocean's biological processes influence the climate system. Many coastal applications—such as monitoring for Harmful Algal Blooms (HABs), ecosystem-based fisheries management, and research on benthic habitats including coral reefs and coastal wetlands—require greater spatial resolution than is currently available to resolve the complex optical signals that coastal waters produce. To combat this a team of scientists and engineers in the UK and United States have come together to develop a high resolution ocean colour sensor capable of integration with a custom designed 3U nanosatellite, termed Seahawk.

The aim of this paper is to describe the technical and science objectives of the mission, as well as outlining the technical challenges and initial design of the Ocean Colour payload and the supporting spacecraft platform. Clearly this is a very ambitious mission with particular challenges not only in the development of an advanced payload of this type – capable of fitting within the constraints of a 3U CubeSat – but also in the platform in terms of power, attitude control, and data processing. The SeaHawk team expects that these demanding requirements will result in a CubeSat that challenges the very limits of what is possible on this size of spacecraft.

KEYWORDS: CubeSat, Ocean Colour Monitoring, Multispectral, SOCON, SeaHawk, HawkEye, Moore's Law, SeaWiFS

S. Hatton (ed.), *Proceedings of the 13th Reinventing Space Conference*,
https://doi.org/10.1007/978-3-319-32817-1_9

1. INTRODUCTION AND BACKGROUND

History

Ocean colour data provides information on the presence of sediments and phytoplankton, and the type and concentration of phytoplankton in particular can be a key indicator of the environmental conditions in our oceans. As a result, the key role that ocean colour data and ocean colour monitoring can play in our understanding of the environment is becoming ever clearer to the wider scientific community and, with over 70% of the Earth's surface covered by ocean, satellites are a crucial element in collecting useful ocean colour data. Currently there are only a few instruments performing ocean colour monitoring around the world, operating aboard large satellites such as NASA's Moderate-resolution Imaging Spectro-radiometer (MODIS), with one of each on board the Terra (EOS-AM-1) satellite, and previously operational on Aqua (EOS PM). It has been unfortunate that a number of vital ocean colour sensors have been lost in recent years, particularly ESA's MEdium Resolution Imaging Spectrometer (MERIS) on board Envisat, and NASA's Sea-viewing Wide Field-of-View Sensor (SeaWiFS). SeaWiFS was launched on board the OrbView 2 satellite designed by Orbital Sciences (both pictured in Figure 1) in August 1997. SeaWiFS observed in 8 wavelengths across the visible and near-infrared (IR) regions with a spatial resolution of 1.1 Km. SeaWiFS could provide daily re-visits over specific areas, important for monitoring changes in oceanic health. Unfortunately in December 2010, after 13 years of successful operation, communications with the satellite came to an end.

Figure 1- (Left) SeaStar illustration, (Right) SeaWiFS Instrument. Image Credit: http://oceancolor.gsfc.nasa.gov/SeaWiFS

Considering the lack of operating instruments, combined with the growth in demand for ocean colour data only capable of being collected through satellite observation, there is a clear need for further ocean colour space missions.

Sustained Ocean Colour Observations with Nanosatellites (SOCON)

Ocean colour instruments like SeaWiFS have proven to be extremely successful in supplying ocean colour measurements and monitoring global ocean colour, but their large pixels make measurements of lakes, rivers, estuaries, and the coastal zone difficult. Observation of these regions is important as much of the world's population lives in coastal regions. For the last few years a team of scientists and engineers, many of whom were part of the team responsible for delivering SeaWiFS, have been working to define and fund a mission that can satisfy the need for higher-resolution coastal and estuarine waters. A nanosatellite mission was of particular interest due to the low cost and rapid development timescales they offer. From these discussions the idea of using nanosatellites with integrated ocean colour sensors to emulate the SeaWiFS performance at higher resolution arose - and led to the development Sustained Ocean Colour Observations with Nanosatellites (SOCON) project, led by Professor John Morrison from the Center for Marine Science at the University of North Carolina Wilmington (UNCW). The goal of the project is to enhance the ability of the earth sciences to observe ocean colour in high temporal and spatial resolution modes, through the use of a low-cost, next generation, miniature ocean colour sensor flown aboard a CubeSat. To achieve this, two identical 3U CubeSats, named SeaHawk, will fly integrated miniature ocean colour sensors, referred to as HawkEye.

The SeaHawk platform will be provided by Clyde Space Ltd. With 10 years' experience, Clyde Space is one of the leading and most experienced designers of nanosatellites globally. With a focus on volume manufacturing of nanosatellites, approximately 40 spacecraft will be constructed at the Clyde Space facilities throughout 2015 and 2016, including the 3U PICASSO CubeSat for the European Space Agency (ESA). An Earth observation mission with a hyperspectral imaging payload, PICASSO is a baseline reference informing the SeaHawk platforms' similar design.

The design of the HawkEye ocean colour sensor is carried out by Cloudland Instruments, a highly experienced team and experts in their field, based in Santa Barbara, California. Specifically, the sensor optical design is led by Cloudland CEO Alan Holmes, who was previously a Systems Engineer for the SeaWiFS design. HawkEye is a multispectral sensor with spectral characteristics comparable to the 8 bands provided by SeaWiFS. The sensor is capable of collection of near-synoptic ocean colour data in open-ocean to coastal-margin to littoral to terrestrial environment, with a spatial resolution between 95-150m and swath of 250-400km in a 400-550km orbit.

The project officially kicked-off in April 2015. Currently the design is headed towards an expected Preliminary Design Review (PDR) in November 2015 with the Critical Design Review expected to be held around April 2016, depending on the outcome of the PDR. It is the aim of all members of the project team to have both CubeSats fully integrated, tested and ready for flight by 2017.

Moore's Law in Space

The funding for the SOCON project was provided through a grant from the Gordon and Betty Moore Foundation, which has a particular interest in SeaHawk from the perspective of technology demonstration, a "proof of concept" of a "game-changer" in the methodology of collecting and disseminating Ocean Colour Data via use of Nanosatellites.

Gordon Moore, who was also one of the founders of the Intel Corporation, is well-known as the author of Moore's Law, which refers to the trend whereby computer processing capability doubles every two years. The comparison between the previous SeaWiFS mission and SeaHawk & HawkEye is a cogent example of how the technological advancements driven by Moore's Law can be leveraged to deliver valuable scientific data from low-cost nanosatellites, on rapid development timescales.

The Orb View 2 spacecraft took more than a decade to develop, while the SeaWiFS instrument delivered a ground resolution of 1.1km. By contrast the HawkEye sensors – designed to operate within similar observation bands as SeaWiFS – will be produced in 2-years, and deliver a ground resolution of around 120m, whilst maintaining a Signal/Noise Ratio approximately 50% that of SeaWiFS. The CubeSat will be 530 times smaller (0.0034 vs 1.81m^3) with 115 times less mass (3.4 vs 390.0 kg), and all for a budget approximately 5 - 10% that of SeaWiFS.

2. MISSION AND SCIENCE OVERVIEW

The objective of this program is to demonstrate ocean colour observation in high temporal and spatial resolution modes, through the use of a miniature ocean colour sensor flown aboard a CubeSat. This system will have significantly higher spatial resolution than standard satellite systems, and be capable of providing observation of sub-mesoscale variability, giving insights into ocean mixing dynamics that are poorly understood. High spatial resolution imagery will also improve our ability to monitor fjords, estuaries, coral reefs and other near-shore environments where anthropogenic stresses are often most acute, and where there are considerable security and commercial interests. Due to the low volume, mass, and particularly cost, it would become practical to fly constellations of these spacecraft, opening up opportunities to significantly improve temporal sampling.

In 2011, an ad-hoc study committee of U.S. National Research Council [1] convened to review the need to sustain global ocean colour radiance measurements for research and operational applications. They concluded "to support the goals and priorities outlined in the National Ocean Policy [2] and Ocean Research Priorities Plan [3] continued monitoring of the ocean's ecosystems on a global scale is essential. The continuity, global coverage, and high temporal and spatial resolution of ocean colour products make remote sensing a critical tool for monitoring and characterising ocean biology and marine ecosystems."

Ocean colour is one of the most useful remote-sensing missions to society, for both science and operations. Phytoplankton is overwhelmingly the largest contribution to primary production in the oceans and plays a vital role in the global carbon cycle. Present in all marine surface waters, their domain covers 71% of the Earth's surface, though their abundance can vary dramatically with location. The ocean colour community requires a broad and secure commitment for an integrated constellation of ocean-colour sensors to provide continuing ocean-colour data of the highest quality, thus ensuring that this capability will exist uninterrupted into the future: "Ocean colour provides our only window into the ocean ecosystem on synoptic scales. It is the sole method we have available to take a global view of the marine biosphere. Climate change is accelerated by the enhanced greenhouse effect, an increase in atmospheric carbon dioxide caused by the activities of man. We need to understand the processes that control atmospheric concentrations of carbon dioxide and other greenhouse gases. The Earth is a planetary system in which the land, oceans and atmosphere interact closely with each other requiring observing systems that show the linkages. The Earth's carbon cycle includes two-way flows between all three components - the role of the ocean is especially important. We are accustomed to thinking that to be Green is to be environmentally responsible and protective of the Earth. But the Earth is a Blue planet, three-quarters covered by water: we need to be aware of what is happening to both parts of the Earth's ecosystem, water as well as land. It is the aquatic biosphere that is monitored uniquely by ocean colour. The aquatic biosphere is under threat from global warming and ocean acidification. We want to know how it is responding to these and other perturbations" [4]

The more we observe the ocean on a synoptic scale, the more we realise that the processes we observe on global scales is replicated down to the smallest observable scale. Spatial heterogeneity, or ''patchiness,'' of marine phytoplankton populations is one of the oldest and most robust observations of open ocean biological oceanography. Structure is found in phytoplankton distributions at scales ranging from meters to the basin scale. At the mesoscale (1–300 km), structure in phytoplankton fields is often associated with physical features such as fronts and eddies. Meanwhile, our major long-term satellite ocean colour observations have been limited, and will continue to be limited in the near future, by available technology to scales of approximately 1 km^2.

3. TECHNICAL CHALLENGES

The Seahawk CubeSat is pushing the boundaries in what was previously thought capable with such tiny spacecraft. The most obvious challenge throughout the design is the extremely tight physical limitations for both the payload design and the CubeSat platform.

Power Generation

The current Concept of Operations (CONOPS) states that we expect the satellite to only observe during sunlight between specific limits, and that the CubeSat shall be capable of performing an observation once per orbit. This is a realistic – but challenging – requirement both in terms of power and data handling. Satisfying this drives a requirement to downlink within the same orbit, in order to avoid overloading the On-Board storage capability. Depending on the ground segment, this could imply a downlink opportunity on every orbit, i.e. both the payload and transmitter would be powered for periods of every orbit, placing an extremely large power demand on the satellite.

Data Handling

To obtain such high resolution and spectacular imagery, the expected data generation is immense. Over the course of a 100-second observation period, it is expected Hawkeye will be capturing hundreds of thousands of pixels every 0.01 seconds. This relates to a total raw data generation of 1.1 Gigabits for the full observation sweep. This is a significant volume of data; to help mitigate this data aggregation will be used to reduce the generated data to a more manageable amount. It is expected the CubeSat will typically be capable of downlinking the aggregated data files once per orbit, if the ground station is available. There are options to reduce the data handling requirement, such as expanding the ground segment, therefore increasing downlink opportunities, or reducing the observation period below 100s at times. In these cases the team would look to downlink the raw higher resolution data, if the power budget is favourable and if a lower observation period satisfies the science requirements.

Attitude Control

The need for fine pointing stability, and particularly pointing knowledge is extremely important. Obtaining high pointing stability will mean a smoother, more accurate image would be captured; however, due to the high resolution of the image, with insufficient position knowledge it will be difficult to determine the exact location of the observed area visually, particularly in the absence of any visible landmarks. Without this knowledge the scientific value of the data could be limited.

It is also important to consider the calibration requirements for the instrument. The main capture mechanism for HawkEye is a set of CCD linear arrays, therefore in order to maintain good image quality over time the effect of dark current must either be accounted for, or measures taken to reduce the effect. One such measure would be performing on-orbit calibration. The team is investigating the feasibility of performing a lunar calibration manoeuvre at least once per month, in order to determine the dark current affects and account for this over time. This manoeuvre would require the CubeSat to fully rotate past the moon accurately enough capture the full moon image within the CCD array field of view.

4. SYSTEM DESIGN

HawkEye

Hawkeye is a small, 1U-sized instrument intended to measure ocean colour from low earth orbit. The intent of the instrument is to obtain performance comparable to SeaWiFS, but with higher resolution. The design will capture images 1800 pixels wide by 4000 pixels tall with a ground resolution of 120 meters, for eight bands similar to SeaWiFS, during a single overpass ("sweep"). It will accomplish this by using 4 linear arrays, each with three 4080x1chroma channels, 10 micron pixels. Each array will sense two bands, with light focused onto each array by two individual F/5.0 triplet lens. Every lens has a band-defining interference filter. The three chroma channels for each array will be summed to improve the signal to noise ratio (SNR) of the result. The summing of the outputs increases the signal by 3X, and the noise by 1.7X, producing an SNR gain over a single array read of

1.7X. The arrays will also be read out 4 times during the time the instrument moves one pixel of ground distance, causing oversampling which in turn increases the signal to noise by another 2X. The oversampling enhances the dynamic range, so the instrument will sense light levels from the blue ocean to the sunlit cloud tops without saturation. 16 bit A/D converters will be used for each channel, and the aggregated result will be 12 bit, but with little noise. Double-correlated sampling will be used to reduce the reset noise of each array; a readout noise level of 25 electrons rms per read has been achieved in early testing.

The instrument will also use an area array CCD, with 648x486 pixels, under a separate short focal length (8.5 mm) lens, boresighted with the linear arrays. The purpose of this lens is to take a snapshot approximately every 400 lines of data, for use in later determining the direction of the motion of the scene across the linear arrays, allowing later resampling of the scene data to an orthogonal grid. We refer to this subsystem as the finderscope. At this time the finderscope has been included within the HawkEye design since it has not yet been determined if the CubeSat attitude can be known, after the fact, to within the 200 arcsecond accuracy required. The finderscope could work in conjunction with the CubeSat ADCS to achieve the precise attitude knowledge required.

It is anticipated that the CubeSat nadir-pointing direction will drift constantly at a slow rate. The finderscope array will determine the pointing direction and yaw angle every 10 seconds. If this proves not to be fast enough the array can be read out more often in orbit, with little impact on the total data volume. The linear arrays generate data at around 10.8 megabits per second (Mbps), and each frame of the finderscope is 2.5Mbit. Therefore one finderscope read every 10 seconds equates to 2% of the data volume. With the primary instrument arrays generating around 11 megabits per second of data for downlinking, over 100 seconds this accumulates to 1.1 gigabits of raw data – a large amount of data to downlink with current CubeSat capabilities. A combination of data compression and 2x2 binning of the data on orbit can be used to allow a good image to be collected in a single overpass while limiting the data generation. High data-rate downlink solutions are currently being investigated for the SeaHawk platform.

The HawkEye instrument will also contain a shutter. No instrument temperature regulation is planned, therefore the CCD dark current will vary considerably over time. The linear arrays will have a low level offset pattern that must be corrected, and the shutter will enable a dark frame to be captured at the beginning of a sweep. The dark data will then be subtracted from each array's data for the rest of the sweep. The shutter will also protect the CCD from inadvertent exposure to focused sunlight between scans. Although it is not known if the sun could cause damage to the linear arrays the expectation is that it could. By contrast the finderscope will not be affected by sunlight, since it uses an extremely small lens with an aperture <1mm and therefore, if required, it could be used to help establish CubeSat attitude control prior to collecting linear data. The dark current for this CCD is low enough that no shutter is needed. The linear array shutter will be a thin blackened metal sheet sliding on two rails. Solenoids fixed to the structure attach to the shutter sheet through a lever arm to uncover the arrays during a sweep. The translational movement will be only about 2 mm. In its non-energised state, a spring will hold the shutter in position to cover and protect the arrays. The drive mechanism is very simple, using redundant solenoids and springs to provide the actuation. Cloudland Instruments will work with Clyde Space to verify that the activation of the solenoid will not perturb attitude control significantly over the 100 seconds per orbit where it is energised. Figure 2 shows side on views of the instrument showing the optical path and various components discussed.

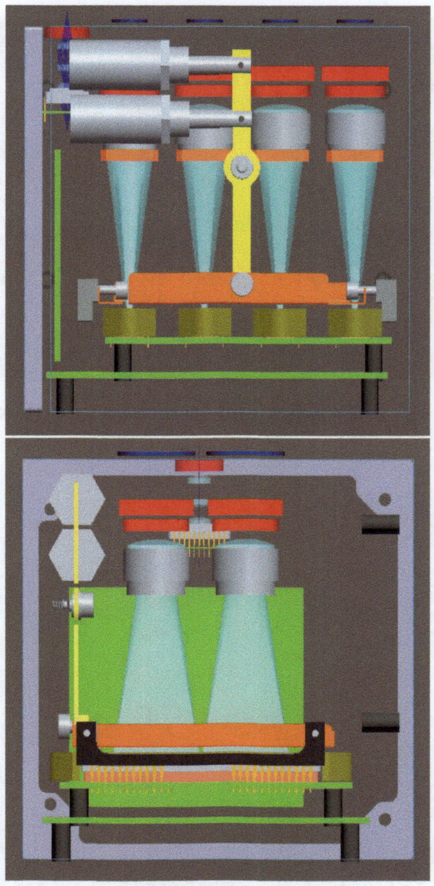

Figure 2 - Two side-on views showing the optical paths

In operation, the CCD arrays are first powered and take dark data, which is nothing more than 50 lines of data with the shutter closed. The shutter is then opened, and the linear arrays readout continuously. Every 625 lines of band data the area array is readout again (without interrupting the band data readout). At the conclusion of the data collection the shutter is closed. The CCD arrays and associated circuitry are powered down to save power while HawkEye's processor formats the data for downlink. The aggregation and accumulation of the 3 lines of data, oversampled 4 times, takes place in real time before storage of the data in non-volatile memory. The processor is well suited to this operation due to its FPGA component.

The optics, filters, and arrays will fit within the constraints of a 10cm^3 unit, allowing for components of the CubeSat structure within this. Figure 3 shows the optics and CCDs mounted in the CubeSat envelope. Figure 4 shows the optics portion in a closer view; however, it does not show the baffles that will be necessary to confine the light from each lens to its associated array, and to suppress the out-of-field hemisphere of illumination striking the package.

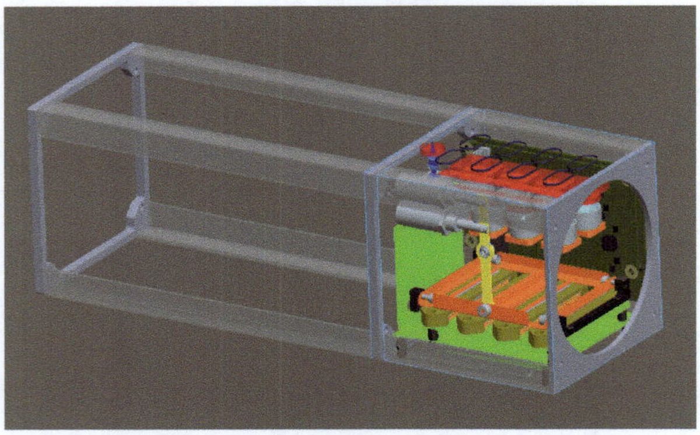

Figure 3 – The Optics and Detectors in a 1U section

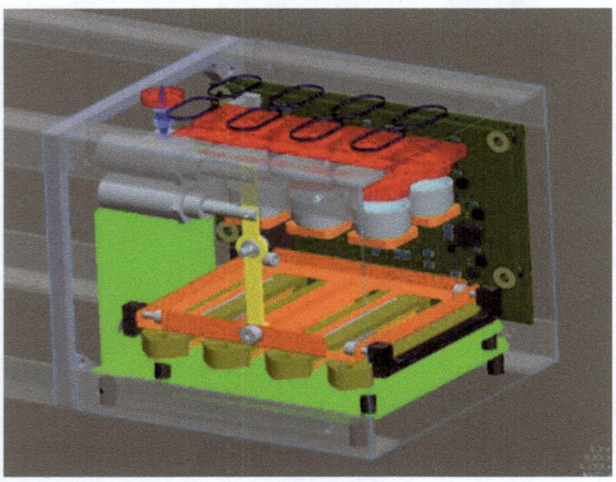

Figure 4 – Two bands are sensed for each array

We envision the baffle will be formed using a series of thin plates between the lens assembly and the arrays, with sharp edges defining the aperture within the interior. Stray light reduction is very important in this application, and will require blackening of lens edges and careful sizing of the optics to prevent the sunlit earth from lighting up internal structures near the optical path that can be seen by the linear arrays.

It will be challenging aligning 4 arrays to an accuracy approaching +/- 10 arcseconds (one half pixel). The CCDs cannot be rotated about their axis to make them coplanar after soldering, but the lenses will be translated in X and Y to achieve centring of each array's midpoint to a common reference. Ground equipment will be used to establish the final alignment of all arrays with each other and the finderscope array. We expect the array misalignments at the ends to be less than +/- 10 pixels.

The performance of this system to L-typical ocean colour radiance levels is shown in Table 1. This is for the full 1800 pixel, 120 meter ground resolution. Table 2 shows the 2x2 binned performance for 900 pixel wide data. The performance exceeds the SeaWiFS specification for Bands 1 through 6.

Table 1 – Chroma Array (10 Micron pixels, No binning)

Standard (Ltypical) Radiance			Radiance					**CubeSat**	SeaWiFS	Ratio,
Band	Wavelength	Bandwidth	Watts/m^2/	Optical	QE	Photons	Signal in	**SNR per**	SNR	CubeSat/
	in nm	in nm	ster/micron	Trans		per watt	Electrons	**pixel**	Spec	SeaWifs SNR
1	412	20	78.6	0.850	0.59	2.07E+18	20390	**487**	499	0.98
2	443	20	70.2	0.740	0.64	2.23E+18	18652	**465**	674	0.69
3	490	20	53.1	0.580	0.72	2.47E+18	13822	**398**	667	0.60
4	510	20	45.8	0.550	0.75	2.57E+18	12210	**373**	616	0.61
5	555	20	33.9	0.520	0.74	2.79E+18	9177	**321**	581	0.55
6	670	20	16	0.580	0.68	3.37E+18	5331	**239**	447	0.54
7	745	20	8.3	0.850	0.46	3.75E+18	3084	**175**	455	0.39
8	865	40	4.5	0.850	0.28	4.35E+18	2335	**149**	467	0.32

Table 2- Chroma Array (10 Micron pixels, 2x2 binning)

Standard (Ltypical) Radiance			Radiance					**CubeSat**	SeaWiFS	Ratio,
Band	Wavelength	Bandwidth	Watts/m^2/	Optical	QE	Photons	Signal in	**SNR per**	SNR	CubeSat/
	in nm	in nm	ster/micron	Trans		per watt	Electrons	**pixel**	Spec	SeaWifs SNR
1	412	20	78.6	0.850	0.59	2.07E+18	20390	**974**	499	1.95
2	443	20	70.2	0.740	0.64	2.23E+18	18652	**931**	674	1.38
3	490	20	53.1	0.580	0.72	2.47E+18	13822	**797**	667	1.19
4	510	20	45.8	0.550	0.75	2.57E+18	12210	**747**	616	1.21
5	555	20	33.9	0.520	0.74	2.79E+18	9177	**642**	581	1.11
6	670	20	16	0.580	0.68	3.37E+18	5331	**479**	447	1.07
7	745	20	8.3	0.850	0.46	3.75E+18	3084	**351**	455	0.77
8	865	40	4.5	0.850	0.28	4.35E+18	2335	**297**	467	0.64

The optical design is complete and the lens performance is excellent. A side view of the lens is shown in Figure 5.

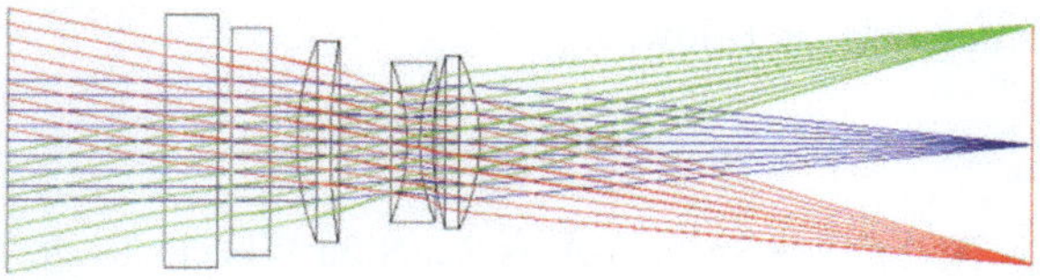

Figure 5 – Optical Layout

The lens focal length will be 45 mm with an aperture of 9 mm. The lenses will be custom, as will be the filters. The challenge for the filters is not the bandshape particularly, but the need to have excellent blocking of out-of-band radiation. This design should be excellent for stray light due to the simplicity of the optical design. The lenses will be antireflection coated. The arrays will be mounted without a window. These linear arrays do suffer from blooming at high signal levels, but this was investigated and it was found that as long as the light levels are below 1.6X saturation the blooming will not spread more than a pixel or two beyond the saturated pixels, while at levels of 2.0X saturation the blooming can corrupt pixels 180 pixels downstream of a saturated region's edge, along the array.

The three sensitive chroma arrays for each band are spaced 9 pixels apart, so to avoid blurring of the final image formed by summing separated arrays the CubeSat must align the array perpendicular to the velocity vector to an accuracy of about +/- 3 degrees (at full resolution). For 2x2 binned operation this relaxes to +/- 6 degrees.

Polarisation sensitivity has been investigated and the linear arrays were found to have a variation of +/- 3% in sensitivity with polarisation at normal incidence. This was surprising but may have something to do with the CCD structures being comparable to the wavelength of light. To reduce this sensitivity further a single quartz prism depolariser will be added to each band in front of the filter. This will reduce this sensitivity to below +/- 0.5%.

A once-per-month lunar calibration will be possible by rolling the CubeSat view past the moon. This will help track changes in the absolute calibration. Linear arrays also have pixel-to-pixel variation of about 0.5% in sensitivity that can be calibrated using flat fields or sphere sources during ground testing. Changes during orbit should be minimal, and any image striping that develops correctable based on data with uniform scenes.

Assembly and test of the two units contracted will be performed in 2016, with delivery late in the year.

Power System

The SeaHawk power system is composed of a number of Clyde Space COTS components which together provide all power provision, protection, distribution and conditioning means for the CubeSat sub-systems and the HawkEye payload module. The main components to make up the power system are the EPS board, battery secondary power source and a combination of high-performance solar panels. The EPS chosen for SeaHawk is the Clyde Space Third Generation FlexU EPS, an Off-The-Shelf subsystem specifically designed to support high-power CubeSats with deployable solar panels. A FlexU EPS is displayed below in Figure 6

Figure 6 - Clyde Space FlexU EPS Board.

The FlexU EPS provides three regulated primary power lines – 3.3V @ 4.5A 5V @ 4.5A, 12V @1.5A – in addition to an unregulated battery line at 4.5A, and provides 10 Latching Current Limit power distribution switch channels. Battery Charge Regulators (BCRs) with Maximum Power Point Tracking (MPPT) consistently optimise power generated from the solar panels. The FlexU EPS also provides protection for the spacecraft against power bus over-current, and battery over- or under-voltage events, as well as featuring a watch-dog timer to reboot the system for error recovery in case of unforeseen events. The Clyde Space EPS range has significant heritage, with an appreciable fraction of all CubeSat missions being powered by a Clyde Space EPS.

The secondary power source selected for SeaHawk is the Clyde Space Third Generation 30Whr standalone battery. Initial analysis shows this battery configuration should provide enough power to operate the relevant systems in eclipse, and exhibits good depth of discharge characteristics. Based on Lithium Polymer technology, the battery cells are arranged 2S3P, with charging EoC voltage of 8.2V.

At this stage of the project the solar array configuration is still to be finalised, pending further analysis. However, a flower arrangement similar to that displayed in Figure 7 with double sided panels is one of the promising configurations currently under consideration.

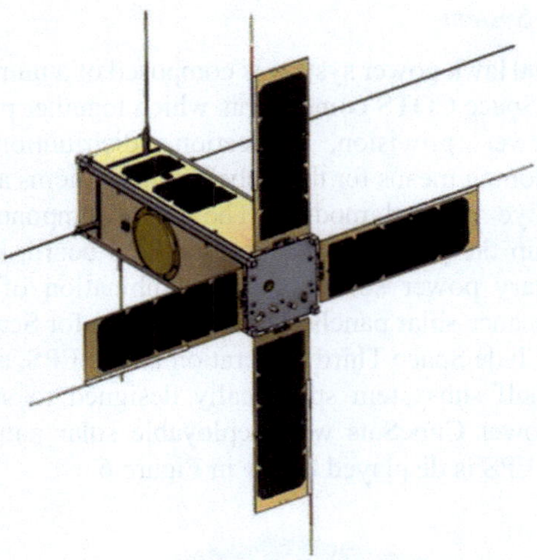

Figure 7 – CAD rendering of the PICASSO CubeSat currently in development by Clyde Space, ESA and the Belgian Institute for Space Aeronomy.

It is expected the HawkEye payload, the most power intensive system on-board, will be operated in sunlight with science acquisitions taking place between 10 AM to 2 PM local time, but never closer than 15 degrees to or from the terminator. This, along with power consumption, is a significant consideration in the design of the solar arrays in order to either maximise the amount of power available at the highest point of power consumption, or ensure that the batteries are fully charged before entering the eclipse period.

Attitude Determination and Control System (ADCS)

There are a number of challenges associated with the attitude determination and control of the SeaHawk CubeSats. Previously, Earth Observation missions using CubeSats have often been dismissed as it was thought that they would not be capable of providing the strict pointing accuracy and control required for earth observation applications. For Seahawk, the priority lies in obtaining the pointing knowledge and fine control, rather than an intensely-high pointing accuracy. It is expected the platform CubeSat must obtain a pointing accuracy of at least +/- 3° in all axes whilst maintaining adequate stability and minimising any jitter throughout the system, with a target of +/- 1° in order to reduce the load on the payload target correction capability.

A particular challenge for Ocean Colour Monitoring missions such as SeaHawk is the need to avoid sun glint off the ocean while observing. This places further constraints on the attitude control system, requiring the capability to pitch the spacecraft at intervals of 5° between 0 and +/- 20°.

SeaHawk's ADCS will consist of a range of sensors and actuators listed below,

- Clyde Space ADCS Motherboard
- Clyde Space 3-Axis Reaction Wheels
- Single Axis Coarse sun sensors
- 2-Axis Fine Sun Sensors
- Three Axis Magnetorquers
- Magnetometers
- Rate Gyro Sensors

The ADCS motherboard, displayed in , benefits from the heritage of the ADCS board that has successfully flown on UKube-1. An FPGA-based processing architecture has been specifically selected to ensure a system that is more robust to radiation events. The central Actel FPGA interfaces to the sensors and actuators, while a secondary processor acts as a watchdog, can place the spacecraft into a safe mode and can also be used to provide emergency detumbling of the spacecraft should the need arise.

Figure 8 - Clyde Space ADCS Motherboard

The ADCS will utilise the standard Clyde Space three-axis CubeSat reaction wheel system, depicted in Figure 9. Each reaction wheel is capable of providing a torque of up to 2mNm. The wheels will provide a total angular momentum of 3.53 mNms, with an angular velocity range of ± 7500RPM.

Figure 9 – Clyde Space 3-Axis Reaction Wheels

As previously mentioned, the CubeSat may be required to perform a lunar calibration to account for dark current effects in the HawkEye CCD arrays. It is expected that the lunar calibration could be performed by spinning the reaction wheels to send the CubeSat into a full 360° pitch rotation, which at some point the moon will be captured within the payload view. The important thing is to maintain the correct pitch rate to ensure the image captured from the CCD readout is not overly skewed. It is expected this will be achieved through the use of the 3-Axis Reaction Wheels.

The ADCS is validated during ground testing using Clyde Space's Hardware-In-the-Loop (HIL) simulator. The complete HIL set-up is a high fidelity, six degrees of freedom, spacecraft dynamical model interfaced directly with the ADCS hardware on which the attitude control algorithms run. The set-up allows validation of the autonomous attitude control software and hardware for all phases of the mission.

Voice coils embedded into the solar panels represent the magnetorquers (MTQ). These devices generate a magnetic dipole that interacts with the Earth's magnetic field, generating a mechanical torque. These actuators are used to de-tumble the spacecraft, to provide coarse pointing acquisition, and to manage the RWS angular momentum.

Critically, real data from the sensors and actuators will be used to simulate the entire mission. This allows many of the un-modelled dynamics that, because of the presence of unknown parameters, do not have clear mathematical formulations. These can include interference on the magnetometers from magnetorquer output and magnetometer reading; reaction wheel velocity and gyros output; sensor noise and pointing accuracy; and others. Consequently, the use of this system level HIL test vastly reduces the impact of 'non-ideal' operation of system hardware on the performance of the ADCS control algorithms, and therefore de-risks the potential for attitude control problems on-orbit.

Communications System

The communications system consists of two hardware components, a transceiver for TM/TC communication and a high data-rate transmitter specifically dedicated to downlink payload data. Telemetry and Telecommand communications are performed by a dedicated UHF/VHF transceiver, the CPUT VUTRX CubeSat transceiver. The VUTRX implements half-duplex GMSK and AFSK modulation schemes with data rates of 9600 and 1200 baud respectively using modified CCSDS packets. The transceiver also features a Beacon mode to broadcast identification and basic health data for tracking, particularly useful when the CubeSat enters the initial separation, de-tumbling or standby mission modes. The VU Transceiver interfaces to a combination of dual UHF and VHF omnidirectional Whip Antennas deployed from the Antenna Deployment Module (ADM).

Figure 10 – CMC VUTRX CubeSat UHF/VHF Transceiver

The science data downlink communications system has yet to be finalised. Initially an S-Band downlink solution was baselined; however, it soon became apparent that this mission would generate a large amount of data, meaning other options must be considered in order to maximise the downlink capabilities. Currently the team is investigating a number of options and trade-offs including data aggregation techniques, ground segment configurations and various operational considerations to determine a way to work around the possible issue. The team is also examining a possible move to an X-Band downlink solution. However, due to the higher power consumption and larger physical characteristics of an X-Band transmitter, this would put a large demand on the system design. These trade-offs will be analysed intensely over the coming months and it is highly expected this decision will be resolved following discussion at the PDR.

Command & Data Handling

SeaHawk shall have two separate data flows, one for science data and the other for Telemetry and Telecommand data. Both data streams will be routed through two different interfaces. The science data must be transferred from the payload to the On-Board Computer (OBC) within a matter of minutes to ensure the high-power payload can be powered down as soon as possible. This calls for a very high data rate interface; the team is currently investigating the available options before finalising their choice. For simplicity, and also due to power requirements, it has been decided to proceed without a fully dedicated payload computer to store and transfer payload data, instead all data will be transferred and routed via the OBC.

As discussed earlier in this paper the payload has an extremely large data generation. The raw data will be saved, and then aggregation performed to reduce the amount of data required for downlink whilst still maintaining good image quality. The team is investigating implementing a binned mode for the payload data collection, such as 2x2 binning, to further reduce the data generation if required.

On Board Software

The platform OBC runs the Bright Ascension Generation 1 On-board Software. This component-based software has an underlying framework including OS and hardware abstract libraries, as well as support for FreeRTOS and POSIX/Linux. The software components support CS platform subsystems including integrated EPS, Battery, solar panels, ADCS, VUTRX and the standard CS payload protocol. Activities including telemetry sampling, pooling, monitoring, logging, etc., as well as automated activities that are event-, time- or orbit-triggered are also supported.

Structure

The structure used will be the Clyde Space standard 3U CubeSat structure featuring custom cut-outs for payload apertures and mounting requirements. The standard structure has been designed specifically with design flexibility in mind, allowing a range of CubeSat stack and payload configurations to be easily designed and implemented within the confines of the structural components. A Clyde Space 3U structure is depicted in Figure 11.

Figure 11 – Clyde Space 3U Structure

An interesting structural requirement is the provision of a one-time door, something more typical of larger earth observation spacecraft. The door will be used to cover the sensitive optics during launch and the initial tumbling phase, ensuring the optics will never face the RAM direction or direct sunlight while unprotected. A solution currently being investigated is simply to use a deployable PCB panel as the door, utilising Clyde Space's existing solar panel deployment technology which has successfully flown on multiple missions previously. This will cover the payload face when stowed and deploy following successful de-tumbling. While stowed, the payload face will not be sealed, there may be a few millimetre of clearance between the panel face and the payload aperture, meaning the payload aperture would still be susceptible to the space vacuum environment. Therefore another purpose of the one-time door would be to allow any substances, such as oil, which may have condensed onto the lenses during the launch procedure to outgas before the payload face is subjected to any light.

5. SUMMARY

Ocean Colour Monitoring provides valuable scientific data which can be used worldwide, not only by scientists to fulfil important research but also by commercial businesses who operate on and rely on the health of our marine biosphere. Currently there is a need to develop a capability to produce high resolution ocean colour data on a global scale, something only possible by satellite observations. The SOCON project aims to bridge this gap using miniaturised push-broom CCD imagers integrated within two 3U CubeSats.

The HawkEye sensor will generate data in 8 wavelengths, very similar to the SeaWiFS observation bands, with a spatial resolution between 95-150 m. The amount of data generated by HawkEye is expected to be around 0.65 GB after aggregation, which places a large demand on the communications and data handling systems of the SeaHawk platform. In order to obtain the highest quality data, there are specific manoeuvrability requirements placed on SeaHawk. This includes the need to perform lunar calibration, accounting for dark current in the HawkEye CCD arrays, and discretised pitch manoeuvres to avoid sun glint off the ocean surface. The SeaHawk ADCS design is tailored, in both hardware and software capabilities, to perform these manoeuvres.

Successful demonstration of this mission will be ground-breaking for both the nanosatellite and ocean colour communities. Proving that the required high resolution optical performance can be achieved using a sensor designed to such tight physical and electrical limitations in itself is a huge success. If it can then also be proven this sensor can be successfully operated and integrated to such a low cost platform, this will open up a vast range of opportunities for ocean colour monitoring using smaller satellites and hopefully lead to an increase in the amount and type of ocean colour data available. A key goal of the SOCON project is to open the door to constellations of dozens of satellites utilising the SeaHawk design, providing more detailed information on our oceans than has previously been possible.

More generally, the lessons learned from this nanosatellite design will contribute to future platforms and will be yet another advancing step in nanosatellite capabilities. Showing that nanosatellites can compete with larger satellites and can provide opportunities which don't currently exist will push the reputation of CubeSats as Earth Observation platforms forward.

6. ACKNOWLEDGEMENTS

The authors thank the Gordon and Betty Moore Foundation for providing funding for the design and build of the HawkEye sensors and SeaHawk CubeSats.

7. REFERENCES

[1] National Research Council, "Assessing The Requirements for Sustained Ocean Colour Research and Operations," The National Academies Press, Washington, D.C, 2011.

[2] National Ocean Council, "National Ocean Policy Implementation Plan," 2013.

[3] National Research Council, "Ocean Research Priorities Plan and Implementation Strategy," National Academies Press, Washington, D.C, 2007.

[4] International Ocean Colour Coordinating Group, "Why Ocean Colour? The Societal Benefits of Ocean-Colour Radiometry," [Online]. Available: http://www.ioccg.org/reports/WOC_brochure.pdf. [Accessed 10 October 2015].

Cost Disruptive Reflector Surface for Large Deployable Antennas

Mike Lawton
Oxford Space Systems
Satellite Applications Catapult, Harwell OX11 OQR; 07740937935
mike.lawton@oxfordspacesystems.com

Juan R Reveles
Oxford Space Systems
Satellite Applications Catapult, Harwell OX11 OQR; 07827441738
juan.reveles@oxfordspacesystems.com

Zhong You
University of Oxford
Department of Engineering Science, OX1 3PJ; 01865283301
zhong.you@eng.ox.ac.uk

Ashley Dove-Jay, Amjad Khan, Vincent Fraux
Oxford Space Systems
Satellite Applications Catapult, Harwell OX11 OQR

ABSTRACT

Current flight solutions for Large Deployable Antennas (LDA) most commonly employ the use of highly flexible metal wire knitted meshes as a reflector surface material. These surfaces, also known as 'tricot meshes', are normally realized using gold plated molybdenum and tungsten wires and require tensioning in the range of 5-10g/cm to obtain sufficient electrical contact between wires and to reduce surface RMS deformation to an acceptable level. The cell or facet size of a knitted mesh is determined by the operating frequency; the higher the frequency the higher the cell density required. Meshes are well known for lower frequency (L and S-band) applications but the challenges of operating LDAs at desired higher frequencies (Ka-band) means metal mesh surfaces require increasingly complex cable tensioning nets and a deviation away from proven mesh materials. High frequency operation of metal meshes also increases the likelihood of Passive Intermodulation (PIM) occurring and thus a degradation in the performance of the antenna.

Due to the increasing difficulties and shortcomings of metal meshes encountered when attempting to operate at ever higher frequencies, Oxford Space Systems (OSS) is developing a flexible reflector surface based upon the use of flexible carbon fibre composite material. Preliminary results suggest the surface material lends itself to being folded in such a manner as to be compatible with the stowage requirements of common LDA architectures. A key property of the material is its ability to be pre-molded and thus it effectively seeks to achieve the shape of its original mold. The result is that, under micro-gravity, the surface will naturally tend toward the desired reflector surface profile thus minimizing the need of surface pre-tensioning. This reduces the need for cable tensioning net systems and thus the design complexity and cost of the reflector surface backing structure.

This paper describes the use of a proprietary outer ring deployment mechanism for the OSS surface material in order to achieve a complete LDA solution. OSS has successfully undertaken the development of a novel, mobility-1 TRL3 Reflector Deployable Structure (RDS) suitable for LDA applications. By overcoming the deficiencies of so called over-constrained mechanisms, OSS will present a cost disruptive, scalable LDA concept targeted at operation at frequencies up to Ka-band. Being mobility-1, the motion structure can be actuated from a single point by means of a novel linear actuator.

KEYWORDS: motion structure, large deployable antenna, foldable reflector, unfurlable reflector

INTRODUCTION

A high maturity European Large Deployable Antenna (LDA) capability in the diameter range of 4 to 9 metres is seen as strategically important[1] due to their wide applicability to broadband satellite telecommunication services.

S. Hatton (ed.), *Proceedings of the 13th Reinventing Space Conference*,
https://doi.org/10.1007/978-3-319-32817-1_10

Currently the LDA market is dominated by US corporate organisations with AstroMesh arguably regarded as the industry state-of-the-art LDA[1]. Developing a European supplier of LDAs to meet the demands of European customers in the first instance is seen as paramount by European industry primes and the European Space Agency (ESA).

The future market requirement for LDAs is expected to be driven by demands from commercial, military and scientific sectors. The commercial demand from fixed & broadcasting satellite services together with multi-media networks requiring satellites as an integral part of network infrastructure represent the greatest demand segment.

In Reference 1, it is reported that customer primary requirements for LDAs are driven by:

- High reliability
- Cost competitive
- Mass efficient
- Maintain high reflector surface integrity
- High stowage efficiency
- Side attachment capability
- Acceptable deployment time

Taking these requirements as their basis for their design, the Oxford Space Systems (OSS) team undertook the development of a novel outer ring mechanism suitable for a Large Deployable Antenna (LDA) capable of supporting both a novel flexible carbon fibre reflector surface as well as a more conventional metal mesh. This led to the realisation of a 4 metre diameter breadboard (BB) demonstrator of the deployable backing structure[4]. The purpose of the breadboard was to validate and demonstrate the kinematics of the chosen concept together with some of the key benefits of the selected deployable ring structure. The concept selected for the structure is an adaptation of the Sarrus mechanism with a pantograph adjusted to maintain a constant angle between the neighbouring vertical bars and therefore maintain a constant pyramid angle. The breadboard structure weighs 7.0kg and deploys from a minimum diameter of 0.23m to 4m. When the structure is stowed, an internal cylindrical volume of 0.18m diameter is achieved to allow the storage of the reflector surface material and supporting cable network. Several tests were conducted on the breadboard to determine the viability of the concept with encouraging results[5].

With regard to the flexible reflector surface materials, OSS' current development approach consists of formulating and manufacturing a range of reflector surface composite coupons from an array of space compatible carbon fibre structures, silicone compounds, and metallic nanoparticle materials. These coupons are tested for their radio frequency (RF) performance in terms of reflection efficiency and transmission losses at Ka-band, under the European Space Agency (ESA) definition of 26.5 – 40 GHz and over a range of temperatures (-180C to +100C). The best performing of these coupons has been taken forward for mechanical testing in the form of minimum bend radius testing to explore the folding feasibility and therefore stowage efficiency capabilities. Using this information, initial finite element (FE) studies into folding configurations and associated bend radii are presented. The preliminary results for the reflector surface are encouraging but it is acknowledged that the current development is still in its infancy and a rigorous qualification campaign is thus required.

DESCRIPTION OF KEY BUILDING BLOCKS OF THE LDA THE TECHNOLOGY

In this section the RDS architecture (the Sarrus-Pantograph based architecture) developed by OSS is described; this concept has been progressed to TLR3 and has undergone basic testing consistent with its technology level. Subsequently, the flexible reflector surface material is described and preliminary performance results presented.

Sarrus-Pantograph Based RDS Architecture

The Sarrus mechanism is a special case of a spatial closed chain in which the Kutzbach criterion for mobility becomes zero (or negative in the case of the Sarrus mechanism with a pantograph element)[3]:

$m = 6n - 5j - 6$ (1)

In Eq. 1, *m* refers to the mechanism mobility; *n* refers to the number of linkages and *j* to the numbers of joints. This holds for a closed chain where only lower pair joints (e.g. revolute, prismatic, screw, cylindrical, spherical or planar joints) are involved and where each joint has one degree of freedom. Evidently, for a spatial closed chain to have mobility 1 (*m*=1), seven links (*n*=7) and seven joints (*j*=7) are needed; any kinematic chain with fewer links and joints is to be either immobile or overconstrained. Immobile structures find no applicability in LDAs for obvious reasons. As for overconstrained mechanisms, they can be traced back to Pierre Frederic Sarrus who in 1853 reported a six bar mechanism capable or rectilinear motion. The OSS architecture is a closed kinematic chain which, following the accepted mobility criterion, can be described as a modified overconstrained 6-bar Sarrus linkage mechanism with 6 one-degree-of-freedom (revolute) joints which displays mobility one.

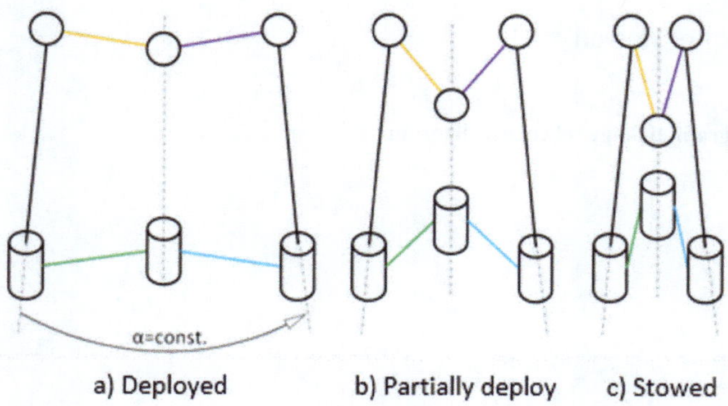

Figure 1 Sarrus linkage schematic diagram

This linkage arrangement allows the chosen angle α between the vertical bars (black bars in Figure 1 above) to remain constant (Figures 1 and 2). This characteristic is also displayed by other deployable structure concepts such as the one developed and patented by ESA[6]. By carefully arranging the facet presented in Figure 1 a hexagonal pyramid of constant angle can be obtained (Figure 3).

The pantograph is introduced into the Pure Sarrus mechanism as a simple means of synchronisation of all the elements of the close kinematic chain.

For the pantograph sliding joints to follow the right path while deploying and maintain a constant angle α between the vertical bars, the following dimensioning rule has to be respected (Figure 2)[3]:

$l1 = l2$ (2)

$\theta = 180 - \alpha/2$ (3)

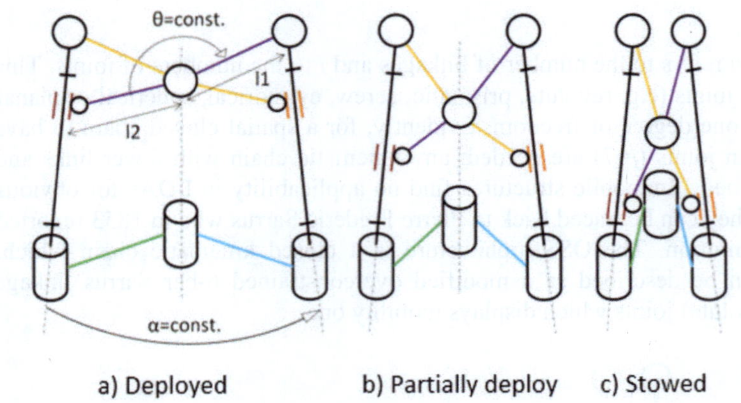

Figure 2 Sarrus-Pantograph linkage schematic diagram

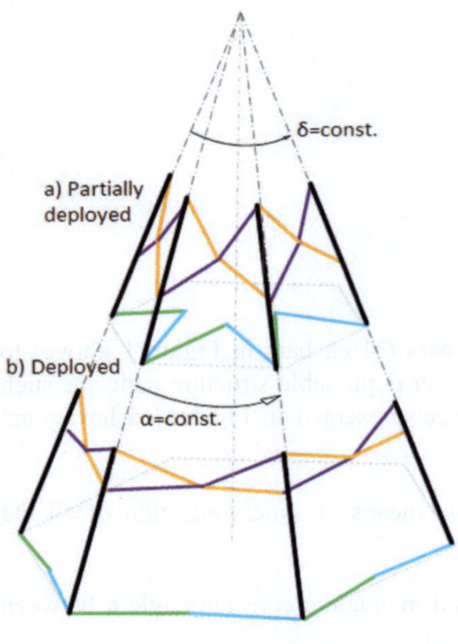

Figure 3 Sarrus-Pantograph hexagonal ring constant cone angle δ and α schematic

The concept presented above was successfully developed into a TRL3 demonstrator which to date has undergone more than 50 deployments under a 1g environment without a single failure. The physical demonstrator is depicted in Figure 4. Although the mobility criteria presented above reveals that the mechanism can be actuated from a single point, three actuation mechanisms are present in the structure to avoid single point failures and to evenly distribute deployment forces in the structural elements. Stored energy elements are strategically position in the mechanism to achieve its "blooming" immediately after the hold down and release mechanism is actuated. This allows a transition of the mechanism from a fully stowed geometry to a more advantageous configuration before motorized actuation kicks-in and realizes full deployment of the motion structure.

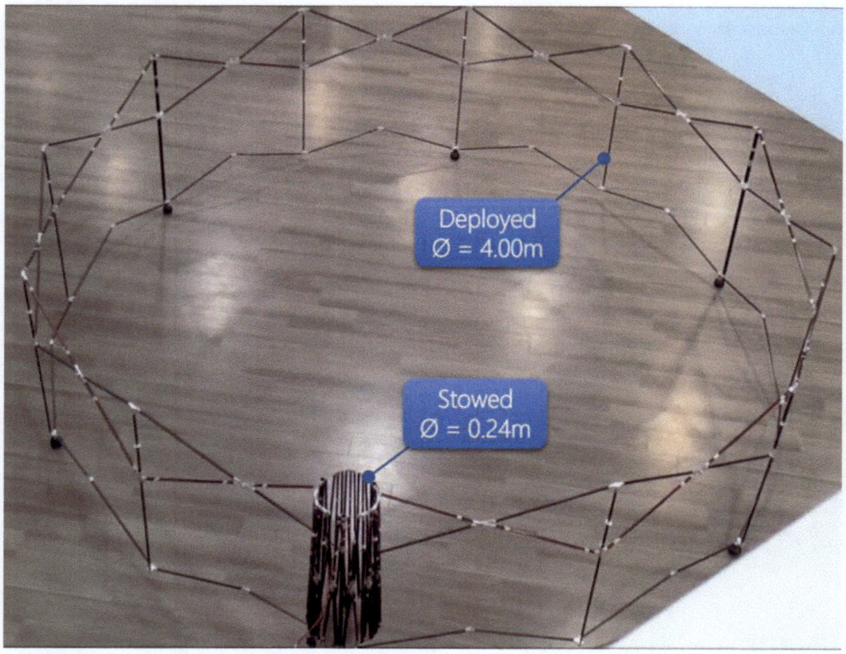

Figure 4 TRL3 demonstrator of the Sarrus-Pantograph RDS in its stowed and fully deployed configurations

Table 1 shows the budgets measured on the TRL3 demonstrator. The deployed diameter shown is the average diameter measure from 5 deployments.

Table 1: Main RDS budgets

Budget	Dimension measured or (estimated)	Margin	Dimension including margin
LDA Structure (incl. actuators) mass	7.0 kg	-	7.0 kg
4m Shell-membrane	(6.300 kg)	20%	7.560 kg
Total mass	-	-	**14.56 kg**
Deployed top diameter	3.953 m	-	3.953 m
Stowed top diameter	0.23 m	-	0.23 m
Stowed bottom diameter	0.33 m	-	0.33 m
Stowed internal diameter	0.18 m	-	0.18 m
Height	1.07 m	-	1.07 m
Deployment time	321 s	-	321 s

The power budget is a critical requirement that needs close monitoring and which limits the type and number of actuation points in all motion structures. Power demand during deployment in a 1 g environment was monitored during three deployment tests. The current was recorded for each actuator (unit 1, 2 and 3) every 10s during the said deployments and the calculated power was plotted against time in Figure 5.

The actuator power appears consistent over the 3 deployment events. Unit 2 power consumption (in red) is slightly higher than unit 1 (in blue) and 3 (in green); nevertheless all follow the same profile. The total power for each deployment (in orange, sum of 3 units) shows consistency. The maximum total power recorded was 8.40W. The thick light blue line on Fig 4 is the average total power over the 3 deployments.

The TRL3 demonstrator and the trade-off studies conducted to date show the proposed architecture offers a promising novel RDS which should serve as the foundation in future flight programmes.

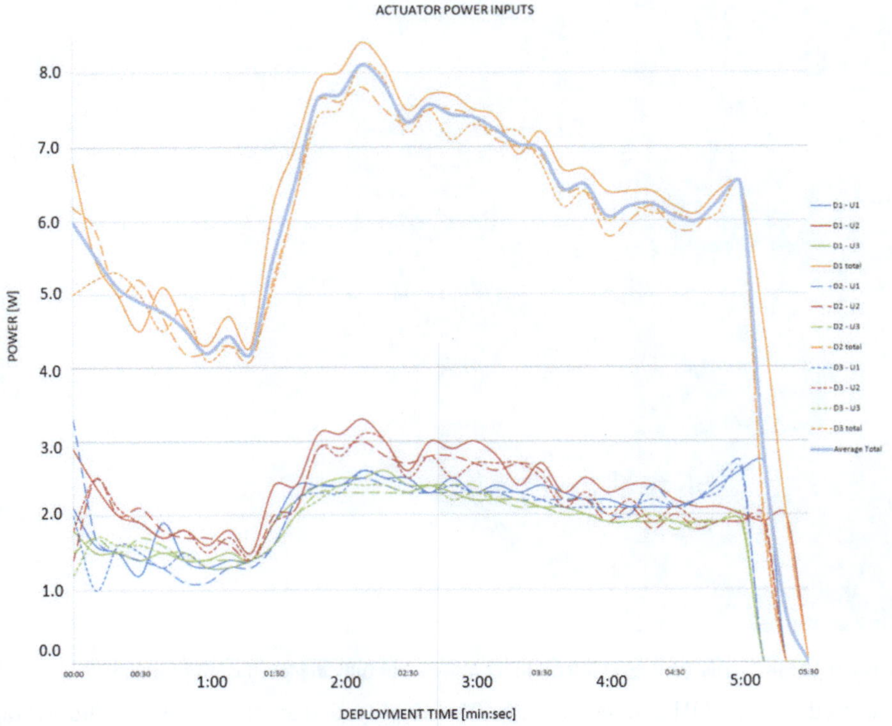

Figure 5 Power demand monitored in each actuator of the OSS RDS

Reflector Surface Technology Status

To meet the forecast emergence in demand and growth of a large market for high frequency (X- to Ka-band) deployable antenna reflectors over the coming decade, OSS are developing the necessary subcomponents to deliver such a product, including antenna deployment boom (currently at TRL6, part funded through InnovateUK), deployable reflector dish outer ring structure (currently at TRL3 as described in the previous section, funded through ESA) and the flexible reflector surface itself. This section highlights the initial research conducted by OSS into flexible reflector surface materials and gives an indication of planned future work, part funded through NATEP.

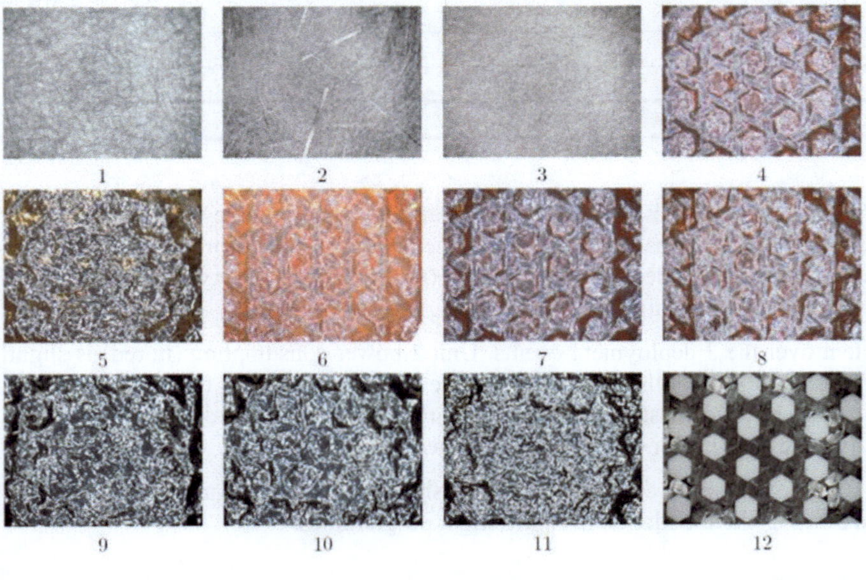

Figure 6 All 12 reflector surface materials tested. Each sample is composed from a unique combination of space compatible carbon fibre structures, silicone compounds, and metallic nanoparticle materials.

Reflector Surface RF Studies

All measurements on the 12 reflector surface material coupons in Figure 6 were obtained at ambient temperature with a Rohde&Schwarz ZVA-40 Vector Network Analyser (VNA). For the free space measurements, the samples were held at the aperture of a Ka-band corrugated feed horn. For that purpose they were placed on a 10mm thick sheet of low density Styrofoam which was in direct contact with the feed aperture. The reflections were determined by a S11 measurement at the waveguide port of the feed. An additional open ended waveguide probe was placed at a distance of 0.4m above the feed for simultaneous S21 measurements which were used to determine the transmission through the samples. The measurements were calibrated with a flat Aluminium plate reflector in the same setup as the samples under test.

The mismatch of the feed and the waveguide transition was determined by a free space absorber between the feed and waveguide probe acting as a matched load standard. Figure 7(a) shows this free space measurement setup without the supporting Styrofoam sheet. Though effective for lossy samples, the highly reflective samples under test and the reference reflector the S11 results are affected by sharp resonant features. These are caused by higher order modes that are trapped in the high-Q cavity which is formed by the feed horn and the sample. In order to mitigate this problem, the data was recorded with high frequency resolution (i.e. 5MHz for the 22–40GHz range or 3600 points), and only data points outside of the resonances were taken into account. The results were further binned using the median value to reduce the impact of outliers.

An alternative approach would be time gating of the signal in the VNA. The results of the freespace measurements depend significantly on the alignment and flatness of the samples. Because of this, it was difficult to determine the reflectivity with a resolution better than 0.1dB with the setup used.

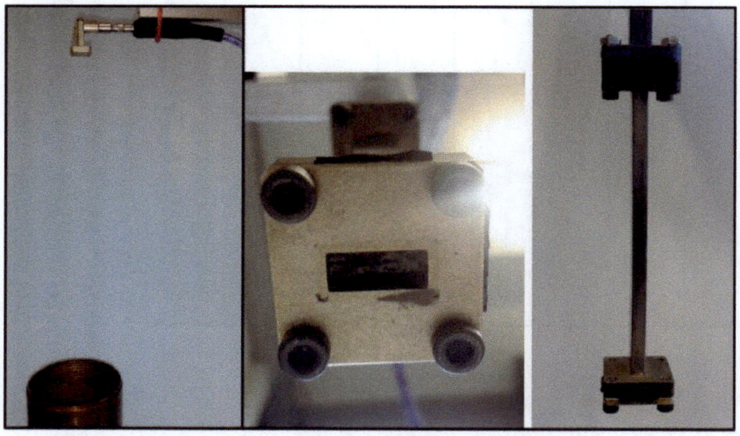

a.) Freespace b.) Waveguide

Figure 7 Test setup for ambient (a.) freespace and (b.) waveguide RF testing

In addition to the free space measurements, the reflectivity was also measured by S11 measurements: where the samples were placed in direct contact with an open ended waveguide port. The data of this test series was calibrated by a full 1-port calibration of the VNA using a matched load and two offset shorts for WR-28. Since the waveguide is singlemoded, the measurements are not degraded by any sharp resonances. However, this geometry is not fully representative for the free-space performance of the samples.

The results depend significantly on the physical contact between the sample and the waveguide interface. The isolating silicone layer at the surface of the composite will lead to a lateral power leakage at the waveguide flange. It appears as an apparent loss in this test, which would not be present in a free space setup. For this reason, the setup has overestimated the losses.

The dimensions of the waveguide are not much larger than the scale length of woven carbon fabric, meaning the S11 results should depend on the alignment between fabric and waveguide. This effect should be most pronounced for plain woven carbon sample #12. The measurements of sample #12 were repeated four times with different systematic alignments, resulting in differences up to 0.2dB. However, it was quite apparent that for samples #1 - #11 this variation is greatly suppressed due to an inherently different configuration of composite.

Given the uncertainties of both methods, the reflectivity results with the free space and waveguide setup are in reasonably good agreement with each other. On average, measurements of reflection efficiency using the two methodologies varied by 0.02dB / 0.5%.

In terms of reflection efficiency, sample #4 presented the most promising results. Not only had it achieved the highest reflection efficiency, averaging approximately -0.27dB / 94%, it demonstrated relatively high consistency in reflection efficiency over the full Ka-band spectrum, unlike say sample #12 that dropped in performance by approximately 0.4dB / 7% at 40 GHz compared to performance at 26.5 GHz. A great deal of this loss can be attributed with the relatively open weave of the structure. At higher frequencies, more Ka-band waves simply pass through these open apertures within the reflector surface. On average, 5.9% of waves directed at #12 simply pass through. Samples #1 - #11, however, have an additional component within the composite to reduce this particular form of transmission loss. This component is quite successful, reducing losses to between 0.0014 and 0.0389%, and increasing consistency of reflection efficiency through the full sweep of Ka-band.

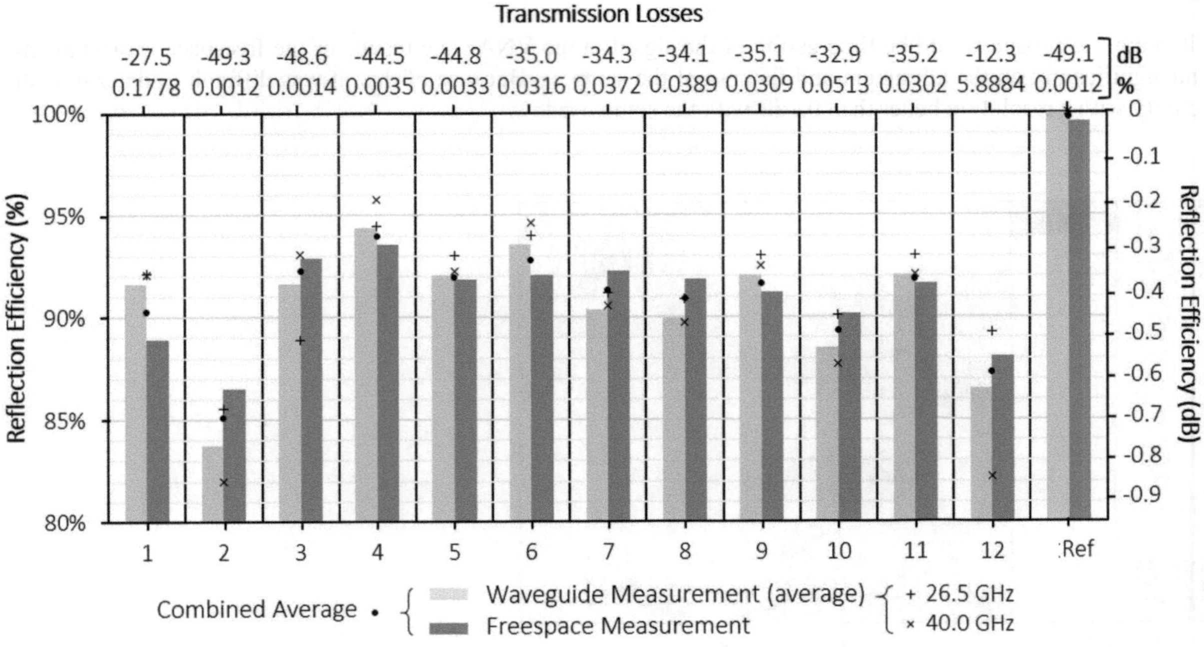

Figure 8 Summary of results for ambient reflection efficiency and transmission losses of the 12 reflector surface materials selected and developed compared to a reference aluminium plate

Considering sample #12 represents the current state-of-the-art in high frequency flexible carbon reflector surfaces globally, at least for this initial stage of RF testing, OSS seems to have formulated a more competitive solution – deeming further exploration of sample #4 at least. Compared to the performance of the AstroMesh 30 OPI metal mesh, the best commercially available flexible reflector surface considered usable for high frequency applications, sample #4 is comparable. Reflection efficiency for sample #4 at 26.5 GHz is -0.25dB / 94.5% and at 40 GHz is -0.19 dB / 95.8%. Reflection efficiency of the metal mesh at 26.5 GHz is -0.17dB / 96% and at 40 GHz is -0.36 dB / 92% . However, at the dish level, there can be expected to be additional complications inherently in the design of a metal mesh reflector that would further reduced reflection efficiency and overall system competitiveness compared to a moulded carbon fibre based reflector. These including pillowing, significant pre-tensioning, and thermal distortions. Quantification of the magnitude of the advantage a moulded carbon fibre based solution would have over a metal mesh solution in this sense is yet to be determined.

Reflector Surface Mechanical Studies

A significant mechanical aspect of the flexible reflector surface material developed is its minimum bend radius. This radius determines how tightly the material can be folded before the onset of damage within the material occurs. Damage to the reflector surface material will vary the mechanical properties in an uncontrolled way and will lead to plastic geometric distortions. These two variables inherently reduce the RMS accuracy achievable at the reflector dish level. Finding the minimum bend radius of the material developed by OSS will allow this constraint to be incorporated into the stowage technique adopted, if a practical one exists.

A Platen-type test was conducted with sample #4 to find approximately at what point damage occurred over a range of bend radii. This damage was determined on visual inspection using a CT scanner. Each coupon was CT scanned before testing to determine whether or not the material was already damaged, they were then bent to a given radii (between 10mm – 1mm), see Figure 9, and were finally CT scanned again to determine if any damage had occurred (and the nature of that damage). When CT scanning, each coupon was imaged at 6 equally spaced regions through-thickness, see Figure 10(a-c), and was inspected. An example of damage found through creasing is highlighted, see Fig 10(d).

Though far from conclusive, the onset of damage begins to appear at a bend radii of approximately 3mm. Damage was never more than very small, low in distribution and size, affecting never more than ~1% of the fibres bundles inspected. Though some coupons were bent to 1mm radius without damage, other coupons began showing damage at approximately 3mm. The conclusion from this testing was that the reflector surface material developed has a very small minimum bend radius, well within the requirements for the practical stowage of a flexible parabolic reflector.

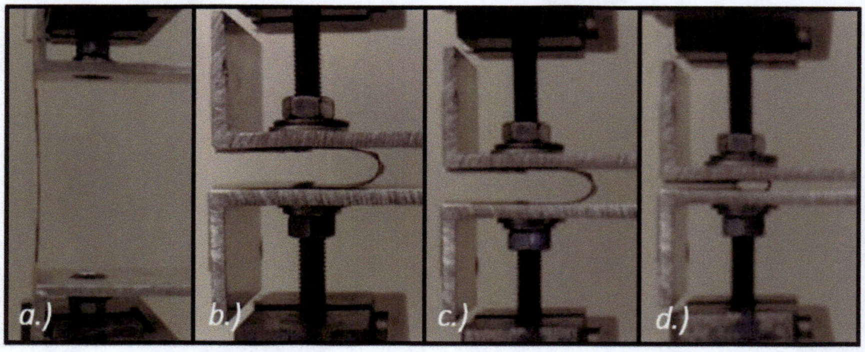

Figure 9 Platen-type testing using position controlled Instron, reducing bend radii of individual coupons of sample #4 from a.) ∞ to b.) 3.75mm, c.) 2.5mm, d.) 1mm.

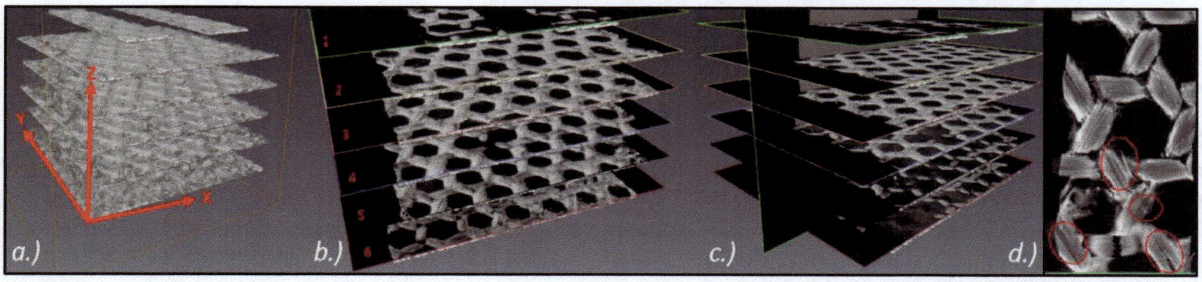

Figure 10 CT scan layers, a.) orientation of planes, b.) through-thickness slices of sample, c.) XZ and XY planes, d.) example of damage to material caused through a hard crease

Reflector Surface Folding Studies

Folding techniques for the parabolic reflector dish are being explored using FE simulations and the mechanical properties garnered thus far from sample #4. This line of study is being used to minimise the stowage volume of the folded reflector surface without causing damage to the material.

Creases in the material would reduce the level of control in the deployed shape of the reflector surface due to localised variation in mechanical properties and plastic geometric distortions. These introduced variables would subsequently reduce the RMS accuracy of the deployed shape achieved. This contribution to a reduction in RMS surface accuracy would reduce the frequency at which the structure is deemed capable of operation.

Creases, and the associated onset of damage, are determined by the minimum bend radius of the reflector surface material. FE studies allow for folding techniques to be explored whilst stresses, strains, and bend radii are relatively easily monitored. Non-linear FE techniques have been applied to model the mechanical and kinematic behaviour of the flexible reflector surface, this is an on-going study and conclusive results will be presented in a subsequent paper. Some membrane deployment simulation results are presented in Figure 11.

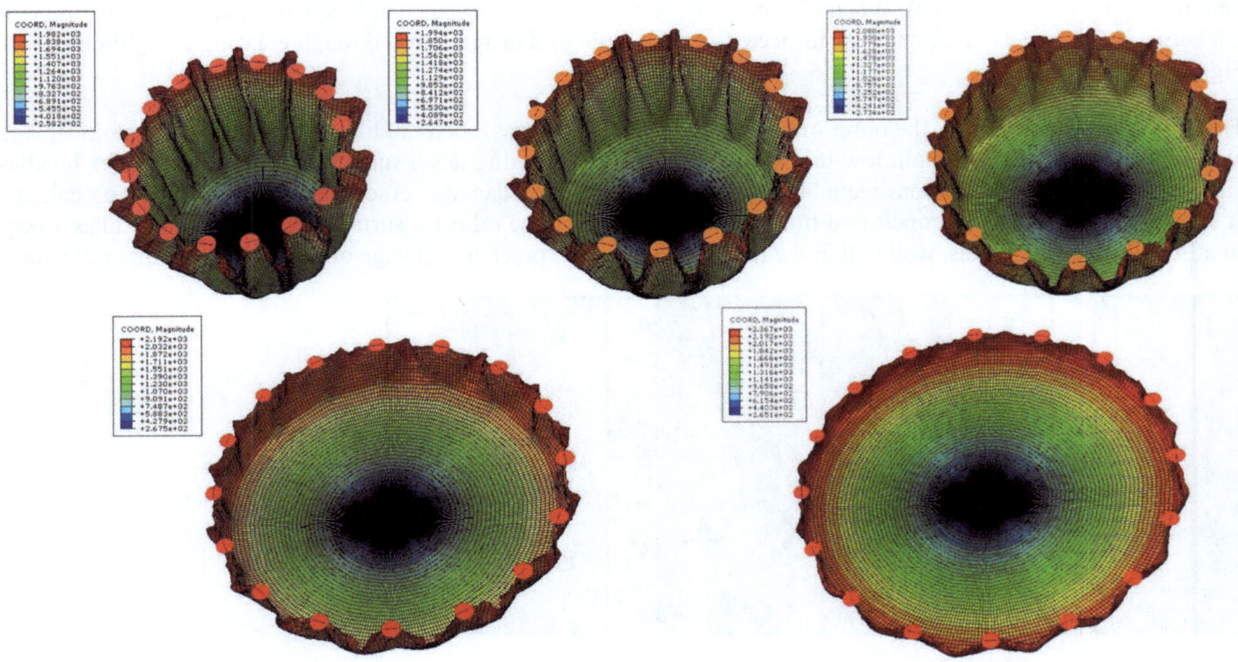

Figure 11 Finite Element analysis of the reflector surface deployment sequence

FUTURE WORK

OSS are in the process of conducting thermal tests on the reflector surface material at the composite and sub-component level in order to develop a deeper understanding of the thermal properties of the surface material, with particular focus on its thermal expansion coefficient and profile. This information will feed into FE studies to help indicate the geometric shape accuracy capabilities of the material, in terms of root-mean square (RMS) accuracy, at the reflector dish level and in the space environment.

OSS and project partners are currently at the detailed design level in the development of a generic deployment rig for a full-scale parabolic dish breadboard of the reflector surface material and support structure. A stiff and generic rig has been decided upon to test the deployment and deployed RMS accuracy of a range of deployable reflector surfaces (≤5m diameter) independent from the induced errors associated with the rest of a flight-worthy RDS. This mechanism and testing will not only allow for the validation of FE studies, it will enable iterative optimisation of the reflector surface material and support structure configuration through this coupled FE analysis and experimentation to maximise RMS accuracy. The generic deployment rig is expected to be operational by early Q2 2016.

Transition from coupon level to full-scale parabolic dish of the reflector surface composite material in terms of manufacturing technique is a significant challenge particularly when considering the geometric tolerances necessary for RF operations at Ka-band; <0.25mm RMS accuracy. Manufacturing techniques of such complex membrane structures will be the focus of subsequent papers.

ACKNOWLEDGEMENTS

OSS would like to thank ESA for their financial and technical support in the development of the RDS TRL3 demonstrator, the UK National Aerospace Technology Exploitation Programme (NATEP) for part-funding this research; without their backing, the research into foldable reflector materials would not have been conducted. OSS would also like to thank Dr Axel Murk from IAP Bern for conducting the RF testing presented in this paper the team under Prof. Michele Meo at the University of Bath carrying out the mechanical testing and inspection on the reflector surface material shown here.

REFERENCES

1. Mangenot, C., Saniago-Prowald, J, Van T'Klooster, K, et al (2010). Large Reflector Antenna Working Group - Final Report, ESTEC, 8 September 2010.

2. Magliorelli, L., Scialino, L., Breunig, E. *et al* (2011). Reflector Technology Trade-off for LDA on Telecom and Earth Observation Applications, *Proceedings of the 33rd ESA antenna workshop on challenges for space antenna systems.,* ESTEC Noordwijk, The Netherlands, 18-21 October 2011.

3. Z. You & Y. Chen (2012). Planar double chain linkages. In: *Motion Structure. Deployable Structural Assemblies of Mechanisms.* Spon Press 2012.

4. V. Fraux, J. R. Reveles, M. Lawton, Z. You, Novel Large Deployable Antenna Backing Structure Concepts for Foldable Reflectors, *CEAS Space Journal,* Vol. 5, Issue 3-4, pp 195-201, (2013).

5. V. Fraux, J.R. Reveles, M. Lawton, Novel Large Deployable Antenna Backing Structure Test Campaign, *Proceedings of the 2nd International Conference in Advanced Lightweight Structures and Reflector Antennas*, 1-3 October 2014, Tbilisi Georgia.

6. J. Santiago Prowald & M. Such Taboada. Innovative Deployable Reflector Design, *Proceedings of the 33rd ESA antenna workshop on challenges for space antenna systems.,* ESTEC Noordwijk, The Netherlands, 18-21 October 2011.

BIS-RS-2015-45

AN ANALYTICAL, LOW-COST DEPLOYMENT STRATEGY FOR SATELLITE CONSTELLATIONS

Ciara McGrath, Emma Kerr and Malcolm Macdonald
University of Strathclyde, United Kingdom

This work proposes a novel method for the deployment of a constellation of nano-satellites into Low Earth Orbit by using carrier vehicles to deliver the nano-satellites into the required orbit positions. The analytical solution presented allows for rapid exploration of the design space and a direct optimisation of the deployment strategy to minimise the time for complete constellation deployment. Traditionally, the deployment of satellite constellations requires numerous launches – at least one per orbital plane – which can be costly. Launching as a secondary payload may offer significant cost reductions, but this comes at the price of decreased control over the launch schedule and final orbit parameters. The analytical method presented here allows for the optimal positioning of the orbit planes of the constellation to be determined and the minimum time for deployment determined as a function of the manoeuvre ΔV. The effect of atmospheric drag on the manoeuvre propellant cost is also considered to ensure a realistic deployment scenario. A case study considering three constellation designs is presented which compares the cost of deployment using traditional launch methods with that of deploying the constellation using carrier vehicles. The results of this study show a significant reduction in cost when using the carrier vehicles on a dedicated launch, compared with launching the satellites individually. Most significantly, the launch cost when using carrier vehicles is primarily determined by the total number of satellites in the constellation, rather than the number of orbital planes. Thus, the carrier vehicle deployment strategy would allow for constellations with a large number of planes to be deployed for a fraction of the equivalent cost if traditional launch methods were used.

I. INTRODUCTION

Nano-satellites in general are becoming increasingly common with almost 150 nano-satellites currently operational and more than 400 launched in total since 1998 [1]. Nano-satellites are satellites with a mass of 1-10kg and include satellites conforming to the popular CubeSat standard [2]. The increasing flight heritage associated with their increased use means that they are no longer confined to Universities and educational institutions. Larger space organisations such as NASA, Boeing and The Aerospace Corporation are also building and launching their own nano-satellites either for technology demonstration or scientific research [3]. With the rapidly increasing capabilities of nano-satellites, their performance has now reached a point where they are capable of supporting Earth Observation (EO) missions. In particular, a large constellation of nano-satellites could prove valuable in supporting existing Earth Observation systems by reducing the burden on current EO satellites and providing data with a high temporal resolution that cannot be achieved by existing systems [4-7].

In line with these developments, the Advanced Space Concepts Laboratory at the University of Strathclyde has carried out a preliminary mission design study considering a constellation of nano-satellites capable of rapidly performing measurements of tropospheric properties to support real-time 'nowcasting' of severe weather [8]. The constellation proposed would be required to provide high temporal resolution and low data latency, while still remaining low cost. To fulfil these mission requirements the study proposed the use of CubeSats deployed in a constellation and performs a multi-attribute utility-cost trade-off analysis to identify the best value for money constellation architecture. One of the key costs identified is the launch cost which, in the case of a dedicated launch, increases as the number of satellite planes increases, and as the number of satellites per plane decreases. Rideshare launches are also considered, in which the satellites would be launched as secondary payloads alongside a primary customer, but the lack of control over the final orbit makes the achievable constellation performance unpredictable and reliance on their services undesirable.

As demonstrated by the FORMOSAT-3/COSMIC mission in 2006, an alternative method of constellation deployment is to launch a number of satellites into a single orbit plane and then separate the orbital planes of the satellites to achieve the required separation of the Right-Ascension of the Ascending Node (RAAN) and argument of latitude [9, 10]. The propellant mass associated with such a deployment manoeuvre can be reduced by making use of low-thrust propulsion and utilising the natural perturbations of the Earth's J_2 effect to produce the desired RAAN change, at the cost of a longer manoeuvre time [11].

This method of deployment has the potential to reduce the number of launches required to populate

S. Hatton (ed.), *Proceedings of the 13th Reinventing Space Conference*,
https://doi.org/10.1007/978-3-319-32817-1_11

a constellation and thus reduce overall mission cost. Traditionally, the design of constellation reconfiguration manoeuvres has been handled using numerical methods, often requiring the use of complex optimisation techniques [12, 13]. A semi-analytical method has also been proposed, but it requires full knowledge of the satellite orbit parameters before and after reconfiguration, meaning it is not ideal for performing a trade-space exploration [14].

A fully analytical solution describing satellite manoeuvres which could be used to reconfigure a constellation through Right Ascension of the Ascending Node (RAAN) and Argument of Latitude (AoL) has previously been presented by the authors [15]. This method is extended here to optimise the satellite deployment manoeuvres for a number of constellations designs and ultimately provide a comparison of the designs in terms of deployment time and overall cost.

II. GENERAL METHOD

Analysis of the deployment of a constellation of nano-satellites is done using the fully analytical method previously described by the authors [15]. In this method, two manoeuvres are considered independently; one manoeuvre to change the RAAN of a satellite, and one to change the AoL. Both manoeuvres are performed by varying the altitude of the satellites relative to each other, creating a variation in the rate of change of RAAN and AoL between the satellites. The most general case of this is considered here in which the satellite performs an initial spiral thrusting manoeuvre to either increase or decrease its semi-major axis. It then drifts at this altitude for a given time to achieve the required separation, before performing a final spiral manoeuvre to reach the desired final altitude. The resultant change in RAAN or AoL is considered with respect to a non-manoeuvring reference satellite as illustrated in Fig. 1.

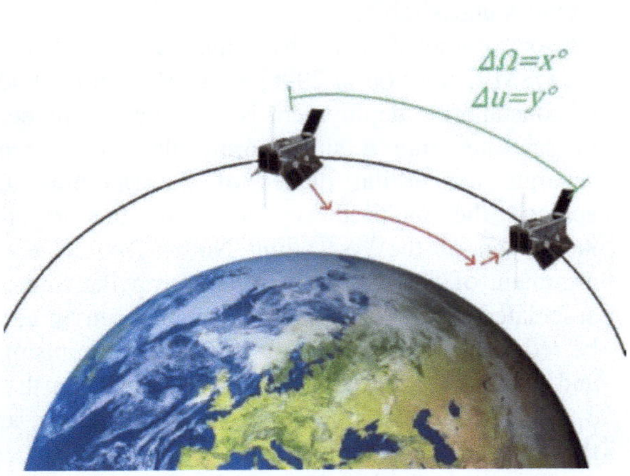

Fig. 1: Altitude lowering manoeuvre to separate through RAAN or AoL

While a change in either RAAN or AoL cannot in reality be performed independently of the other, the results previously presented by the authors show that, due to the relatively long manoeuvre time required to change a satellite's RAAN compared to the time required to change the AoL, the manoeuvres can be considered independently and the results later adapted to combine both [15].

The analytical method used is derived from the Gauss version of the Lagrange planetary equations [11] and considers a low-thrust manoeuvre with constant acceleration and no perturbations from drag or solar radiation pressure. It is assumed that the satellites maintain circular orbits throughout the entirety of the manoeuvre. These simplifications allow for the problem to be fully described and solved analytically.

The most general expression for the achievable change in RAAN, $\Delta\Omega$, is given by

$$\Delta\Omega = \frac{3}{256}\sqrt{\mu}\cos i\, J_2 R_e^2 \left(\frac{\sqrt{\mu}}{16Aa_0^4}\left(\frac{(\mu+\beta a_0)^4}{\mu^4} - 256\right) - \frac{16\sqrt{\mu}}{A}\left(\frac{1}{a_3^4} - \frac{\left(\beta + \frac{\mu}{a_0}\right)^4}{256\mu^4}\right) + \left(\frac{\sqrt{\mu}}{A}\left(\frac{1}{\sqrt{a_0}} + \frac{1}{\sqrt{a_3}} - \frac{1}{\sqrt{\frac{\mu a_0}{\mu+\beta a_0}}}\right) - t_t\right)\left(\frac{\mu a_0}{\mu+\beta a_0}\right)^{-7/2} + \frac{128 t_t}{a_{\mathrm{ref}}^{7/2}} \right) \quad [1]$$

where

$$\beta = \left(\sqrt{\frac{\mu}{a_3}} \pm \Delta V_t\right)\left(2\sqrt{\frac{\mu}{a_0}} + \sqrt{\frac{\mu}{a_3}} \pm \Delta V_t\right) \quad [2]$$

and μ is the standard gravitational parameter, i is the inclination of the satellite's orbit, J_2 is the central body's second dynamic form factor, R_e is the radius of the central body, A is the acceleration produced by the propulsion system, t_t is the total manoeuvre time, and ΔV_t is the total change in velocity required for the satellite to complete the full manoeuvre. a_0 is the semi-major axis of the satellite at the beginning of the manoeuvre, a_3 is the desired final semi-major axis of the satellite, and a_{ref} is the semi-major axis of a non-manoeuvring reference satellite against which the resultant change in RAAN is to be measured; in this case it is taken as a_3.

It should be noted that in equation 2, a '+' corresponds to the case where the satellite decreases its semi-major axis initially and increases its semi-major axis to reach its final orbit, and a '-' corresponds to the case where the satellite increases its semi-major axis initially and then decreases its semi-major axis to reach its final orbit. A positive A value corresponds to an increase in semi-major axis, while a negative A value corresponds to a reduction in semi-major axis.

The achievable change in AoL, Δu, is given by

$$\Delta u = \frac{1}{8}\left(\frac{8\gamma\mu}{Aa_0^{3/2}} - \frac{2\gamma^4\mu}{A} - \frac{12\gamma^2\mu}{Aa_0} + \frac{8\gamma^3\mu}{A\sqrt{a_0}} - \frac{2\mu}{A}\left(\frac{(\mu+\beta a_0)^2}{16\mu^2 a_0^2} - \frac{1}{a_3^2}\right) - 8\sqrt{\frac{\mu}{a_{\mathrm{ref}}^3}}t_t + \sqrt{\frac{(\mu+\beta a_0)^3}{\mu^2 a_0^3}}\left(t_t - \frac{\sqrt{\mu}}{A}\left(\gamma + \frac{1}{\sqrt{a_3}} - \frac{1}{2\sqrt{\frac{\mu a_0}{\mu+\beta a_0}}}\right)\right)\right) \quad [3]$$

where, as before,

$$\beta = \left(\sqrt{\frac{\mu}{a_3}} \pm \Delta V_t\right)\left(2\sqrt{\frac{\mu}{a_0}} + \sqrt{\frac{\mu}{a_3}} \pm \Delta V_t\right) \quad [4]$$

and

$$\gamma = \left(\frac{1}{\sqrt{a_0}} - \frac{1}{2\sqrt{\frac{\mu a_0}{\mu+\beta a_0}}}\right) \quad [5]$$

with all symbols as previously defined, and the use of '+' and '-' as in the case of the RAAN separation manoeuvre.

In both cases, these general solutions can be reduced to represent specific simple manoeuvres by applying the relevant boundary conditions.

III. MODEL VALIDATION

In order to validate the model, the FORMOSAT-3/COSMIC constellation deployment was analysed using the analytical method described in Section II and the results compared with existing mission data. This constellation consists of six satellites which were initially launched into an approximately circular orbit with an altitude of 522km. The altitude of each satellite was then raised to 800km with the manoeuvres timed to achieve a -30° RAAN separation between the satellites [9].

The six satellite manoeuvres were carried out over an 18 month period in 2006 and 2007. One of the satellites (FM3) experienced a solar array deployment failure and could not complete the orbit-raising manoeuvre. The other five satellites all reached the required final altitude and achieved the desired RAAN separation [10].

With knowledge of the initial and final semi-major axes of each satellite, and with the assumption of circular orbits and ignoring atmospheric drag, the required ΔV can be calculated as 152.494m/s per satellite manoeuvre. This allows the achievable RAAN separation to be described as a function of the transfer time only, as shown in Fig. 2. Here, the total transfer time consists of the time spent in the initial orbit as well as the time required to complete the orbit-raising manoeuvre. The lines on the graph indicate the desired RAAN separations to be achieved and the corresponding total time as calculated using the analytical method.

Using the two-line element (TLE) data of the FORMOSAT-3/COSMIC satellites it is possible to track the satellites through their manoeuvres, as shown in Fig. 3, and thus to approximate the true time required to achieve the desired RAAN separation.

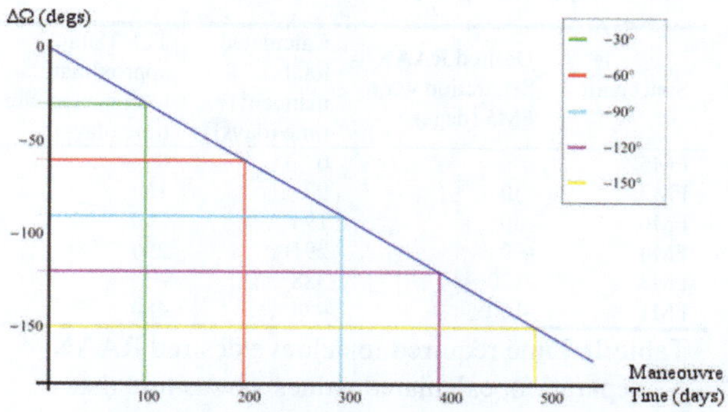

Fig. 2: Time required to achieve the desired separation between the FORMOSAT-3/COSMIC constellation satellites

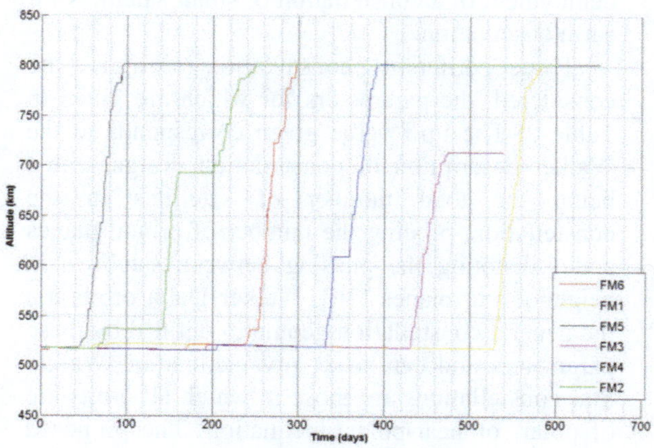

Fig. 3: TLE data from FORMOSAT-3/COSMIC constellation showing RAAN phasing manoeuvre

To compare the calculated results with the actual results, the desired RAAN separation is defined in reference to the first satellite to be manoeuvred (FM5), which for the purposes of analysis is assumed to be the reference satellite. This means that the initial time, t_0, is taken as the time at which FM5 reaches its final orbit. This gives the time required to achieve the desired RAAN separations compared with the true manoeuvre time as shown in Table 1. These results show that for FM6, FM4 and FM1 the proposed analytical method accurately predicts the time required to achieve the given RAAN separation with less than 5% error. FM3 cannot be used for comparison as it never reached the desired final orbit altitude, and the error in the prediction of the time for FM2 can be explained by the approximately 40 day pause at 700km altitude during the first manoeuvre. While the consideration of drag and other influences will likely give improved results, the current solution is considered to be sufficiently accurate to predict the required time and ΔV for constellation deployment in the case of approximately circular orbits.

Spacecraft	Desired RAAN Separation w.r.t. FM5 (degs)	Calculated total manoeuvre time (days)	TLE Data approximate total manoeuvre time (days)
FM5	0	0	0
FM2	-30	97	150
FM6	-60	194	200
FM4	-90	291	290
FM3	-120	388	-
FM1	-150	484	480

Table 1: Time required to achieve desired RAAN separation, calculated values versus true data

IV. CASE STUDY

Having validated the analytical method, a case study was then carried out considering the deployment of a constellation of small satellites for earth observation.

Three different constellation designs are considered, the parameters for which are given in Table 2. The t/p/f value given corresponds to the Walker description of constellation design with t being the total number of satellites in the constellation, p being the number of orbital planes and f denoting the phasing between satellites in neighbouring planes [16]. Walker Delta orbits are the general constellation geometry defined by these parameters and can sit at any inclination; Walker Star constellations are those in which all orbits are of polar, or near-polar, inclination. The proposed constellation designs contain a number of orbit planes in each category.

Designs 1 and 2 are the designs selected from the previous University of Strathclyde study as the best balance of utility to cost constellation designs, while Design 3 is another option which was explored as part of the study [8].

Design No.	Altitude (km)	Inclination (degs)	Delta t/p/f	Star t/p/f
1	550	50	16/4/3	6/2/1
2	550	60	20/4/2	6/2/1
3	550	50	18/6/2	6/2/1

Table 2: Constellation Design Parameters

IV.I. Optimal Satellite Distribution

In order to consider the deployment of a satellite constellation it is necessary to define the final positions of each of the individual satellites. Generally, satellite positions within a constellation are described relative to each other, as in the case of a Walker Delta or Walker Star constellation [16-18]. However, it is also necessary to define the position of each satellite with respect to the launch injection point and, due to the lengthy manoeuvre times involved in changing the RAAN of the satellites, the positioning selected may have a large influence on the overall manoeuvre time and propellant cost. It has also been shown that for a given orbit, achieving a change in RAAN or AoL can be done more efficiently in one direction than in the other [15]. This means that evenly distributing the satellites from the launch injection point in both directions is unlikely to be the most efficient deployment method.

To find the ideal satellite distribution with regards to the launch injection point it is necessary to first define the spacing of the satellites relative to each other, again considering RAAN and AoL separately. If the satellites, or satellite planes, are evenly distributed this can be simply described by

$$\Delta\Omega_i = \Delta\Omega_1 + (i-1)\left(\frac{2\pi}{n}\right) \qquad [6]$$

and

$$\Delta u_i = \Delta u_1 + (i-1)\left(\frac{2\pi}{n}\right) \qquad [7]$$

for $i: 1 \rightarrow n$ where i is the satellite number and n is the total number of planes when considering RAAN distribution, or the number of satellites within a plane when considering AoL distribution. The two satellites positioned furthest from the manoeuvre starting point in this case will be satellite 1 and satellite n. By describing the change of RAAN or AoL of these two satellites analytically using equations 1 and 3 respectively, it is then possible to solve for the shortest time manoeuvre by setting the requirement that both satellites must reach their final position at the same time.

Note that the method described can be applied even if the satellites are not evenly distributed, but the position of the satellites relative to each other would need to be explicitly defined.

IV.II. Drag

Accounting for atmospheric drag in an analytical solution is not straightforward as the effective drag force does not vary linearly with altitude. However, the general perturbations method provided by Kerr and Macdonald [19, 20] can be used with some simplification to determine if a satellite in the constellation will deorbit during deployment.

This method provides orbit lifetime predictions contingent on the launch date of a satellite as it includes an analytical atmospheric density model incorporating solar flux. As this study is a theoretical deployment strategy no launch date is known and therefore the solar flux is assumed to be constant at an average rate over the entire deployment time period. In reality some satellites in the constellation may deorbit more quickly than others depending on the solar flux conditions during the drift period, and this should be taken into consideration before applying this method to a proposed constellation design.

In order to account for atmospheric drag in the analytical deployment method presented, the satellite distribution and manoeuvre is considered excluding drag, and the drift phase is assumed to occur at a constant altitude. By making use of the analytical orbit lifetime prediction method it is then possible to calculate the true final altitude of each satellite at the end of the drift phase. Whilst this does not account for the variation in the rate of change of RAAN due to the change in altitude throughout the drift phase, it does ensure that none of the satellites deorbit during deployment and therefore that the constellation deployment strategy is feasible. In addition, the total ΔV required for the manoeuvre is calculated using the post-drift altitude with drag taken into account.

IV.III. Costing

While there are CubeSat propulsion systems in development, it is currently unlikely that the necessary plane change manoeuvres described in Section II could be carried out by individual CubeSats due to the required ΔV cost [6]. However, it would be possible to stow individual satellites on a larger carrier satellite which could deliver the satellites to the required orbit plane. From here, the satellites could be distributed within the plane using their own on-board propulsion or by using springs of varying strengths to control deployment [21].

Design, development and manufacture costs have already been considered as part of the previous University of Strathclyde study [8] and are assumed to be consistent regardless of whether traditional launch methods or the use of the proposed in-orbit deployment strategy is employed. As such, the costing done here focusses on the launch costs associated with both methods.

Two different launch providers are considered and the most applicable of their available launches selected to meet the mission requirements. These launch providers are Spaceflight Industries Inc. [22] and Firefly Space Systems [23]. Spaceflight Industries Inc. currently provide rideshare launch opportunities for small satellites; this means that the satellites would be considered secondary payloads and would have inexact knowledge of the final orbit and no control over the launch itself. Firefly Space Systems are in the process of developing a dedicated small satellite launch vehicle with a maximum payload of 400kg. This has the advantage of being able to provide dedicated launches, allowing the customer to choose their orbital parameters and launch schedule. However, as the cost in this case is per launch, rather than per satellite, the cost of the launch may be much higher than in the rideshare case unless the launch vehicle payload capacity is fully used or other satellites can be found to make use of the remaining payload capacity.

IV.III.I Traditional Launch Methods

Traditional launch methods here assume that no carrier vehicle is used and that the individual satellites have little to no manoeuvring capability. This means the satellites must be inserted at the correct altitude, inclination and RAAN by the launch vehicle. In this case, one launch will be required for each plane of the constellation. In the case of a rideshare launch the total launch cost is simply calculated as

$$C_{total} = C_{sat} \times n \times p \qquad [8]$$

where C_{sat} is the launch cost per satellite, n is the number of satellites in each plane and p is the total number of orbit planes. In the case of a dedicated launch the cost would be

$$C_{total} = C_{launch} \times p \qquad [9]$$

where C_{launch} is the cost of a single dedicated launch.

IV.III.II Carrier Vehicle Method

In the case of manoeuvrable carrier vehicles being used to deploy the constellation, the number of launches required to place all satellites into orbit will be dependent on the number of satellites to be launched and the maximum payload capabilities of the launch vehicle. In the case of a rideshare launch the total launch cost will be calculated as

$$C_{total} = C_{carrier} \times p \qquad [10]$$

where $C_{carrier}$ is the launch cost per carrier vehicle. In the case of a dedicated launch the cost would be

$$C_{total} = C_{launch} \quad [11]$$

where C_{launch} is the cost of a single dedicated launch.

The size of the carrier vehicle will be primarily dependent on the manoeuvre it is required to perform and the number of spacecraft it is required carry. As an initial estimate the dry mass of the carrier is estimated as

$$m_f = m_p \times 3.3 \quad [12]$$

where m_p is the mass of the satellites to be carried [24]. A low power Xenon resistojet propulsion system is considered as a baseline with a specific impulse of 48secs and the ability to deliver up to 100mN thrust [25].

From this, the maximum allowable propellant mass is calculated to make use of the full payload mass available on the dedicated launch vehicle, and the maximum allowable ΔV calculated from this using the rocket equation [26]. A margin of 20% is applied to both the spacecraft total mass and the ΔV calculation to ensure a conservative estimate.

It is assumed for these analyses that one carrier vehicle is used per orbital plane; while it would be possible to use one carrier to deliver satellites to multiple orbit planes, the length of time required to deploy the constellation using a single carrier vehicle would in all cases be longer than when using one carrier per plane and as such it is not considered in this study.

V. RESULTS

V.I. Optimal Satellite Distribution

V.I.I RAAN Separation

In the case used for analysis, four orbital planes are considered which are evenly distributed through 360° (i.e. 90° separation between each plane). The mission parameters are given in Table 3 and Table 4.

It is assumed that of the furthest two satellites to be placed, satellite 1 and satellite 4, satellite 1 will initially lower its semi-major axis, resulting in a negative ΔΩ, and satellite 4 will initially raise its semi-major axis above the final desired altitude resulting in a positive ΔΩ.

As the required separation of satellite 1 and satellite 4 is known to be 270°, by plotting the achievable ΔΩ of satellite 1 against the achievable ΔΩ of satellite 4 minus the required 270° separation as shown in Fig. 4, an intersection can be found along which both satellites will arrive at their required final position at the same time. The time at which this occurs is dependent on the total ΔV used for the manoeuvre.

Once a position for these first two satellites has been selected, the position of the other satellites will be decided relative to them. The time required to place the remaining satellites in position will be dependent on the ΔV selected, but in any case will be shorter than the time required for the first two satellites to reach their final positions.

V.I.II Argument of Latitude Separation

In considering the placement of the satellites with regard to argument of latitude, four satellites are considered for even distribution within each orbital plane, corresponding to 90° separation between each satellite. For an initial analysis it is assumed that of the furthest two satellites to be placed, satellite 1 and satellite 4, satellite 1 will initially lower its semi-major axis, resulting in a positive Δu, and satellite 4 will initially raise its semi-major axis above the final desired altitude resulting in a negative Δu.

Parameter	Symbol	Value	Units
Gravitational Parameter	μ	3.986E14	m^3/s^2
Radius of Earth	R_e	6.371E3	km
J_2 Parameter	J_2	1.0827E-3	-

Table 3: Orbital Constants

Parameter	Symbol	Value	Units
Propulsion acceleration	A	± 0.001	m/s^2
Inclination	i	50	degs
Initial semi-major axis	a_0	6771	km
Final semi-major axis	a_3	6921	km

Table 4: RAAN Analysis Mission Parameters

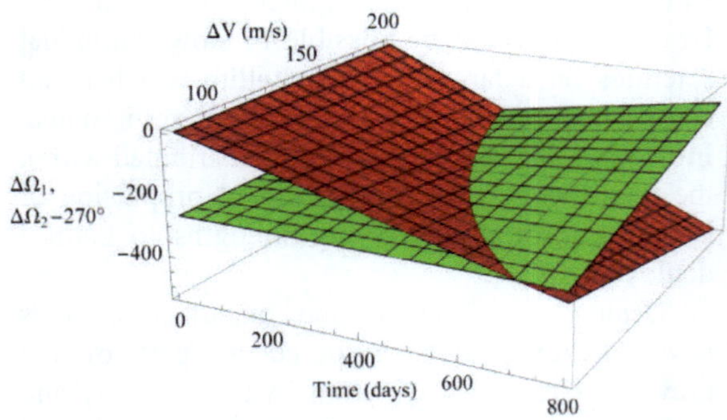

Fig. 4: Optimal ΔΩ of satellite 1 (red) and the relative ΔΩ of satellite 4 (green) as a function of ΔV and manoeuvre time

For this case it is assumed that the satellites begin their manoeuvres at the desired final altitude. This is in accordance with the recommendation made in the authors' previous work that the RAAN distribution manoeuvre be completed first and then the AoL manoeuvre carried out [15]. The orbital constants are as in the case of the RAAN separation and are given in in Table 3 and the mission parameters are as given in Table 5.

Similar to the case of the RAAN distribution the achievable Δu of satellite 1 is plotted against the achievable Δu of satellite 4 minus 270° as shown in Fig. 5. Here the results are only plotted for cases in which the total manoeuvre time is greater than the necessary thrust time in order to show only realistic scenarios.

In this case the graphs do not intersect indicating that when distributing the satellites within the plane for this constellation, it will always be more efficient to lower the altitude of the satellite initially and move all satellites in the same direction through a positve Δu. Thus the minimum time manoeuvre would correspond to a case in which one satellite remains at the initial location and the other satellites are moved relative to it.

Parameter	Symbol	Value	Units
Propulsion acceleration	A	± 0.001	m/s^2
Inclination	i	50	degs
Initial semi-major axis	a_0	6921	km
Final semi-major axis	a_3	6921	km

Table 5: AoL Analysis Mission Parameters

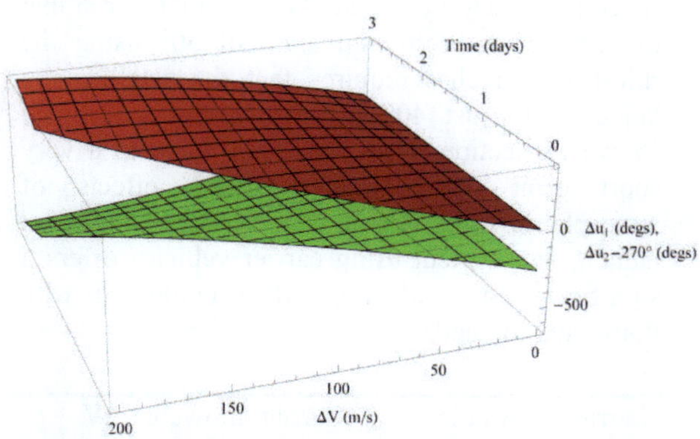

Fig. 5: Optimal Δu of satellite 1 (red) and the relative Δu of satellite 4 (green) as a function of ΔV and manoeuvre time

V.II. Drag

During the proposed satellite manoeuvres, the satellites will spend a relatively long time in the drift orbit before manoeuvring to reach the desired final altitude. For satellites lowering their altitude in this phase the effects of atmospheric drag must be considered to ensure they do not deorbit during this drift period. In addition, the ΔV required to reach the desired constellation altitude will be dependent on the altitude of the satellite at the end of the drift phase.

The results of a general analysis are shown in Fig. 6. From this it is clear that the lower the drift orbit and the longer the satellite spends in this drift phase, the greater the influence of atmospheric drag. As a result of this analysis, combined with the deployment scenario results from Section V.I.I, it is decided that for the case study considered the satellites should be launched to an initial altitude of 550km to prevent them from deorbiting before the full constellation can be deployed.

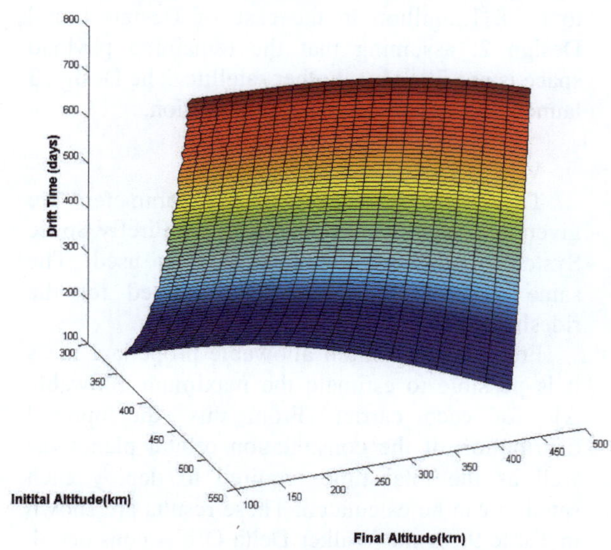

Fig. 6: Final altitude as a function of initial altitude and dift time

V.III. Costing

The launch costs of both launch service providers considered are given in Table 6 and Table 7, as well the most appropriate orbital parameters that can be provided by the launch vehicle for the three constellation designs considered. In the case of Spaceflight Industries Inc. the cost per kilogram is calculated for satellite launches carrying the closest mass to that of all carriers to be launched. The payload capacity of the Firefly Space Systems dedicated launch vehicle is dependent on the altitude and inclination of the launch injection orbit [27].

Spaceflight Industries Inc.	Launch Type	Cost per 6U CubeSat (£)	Cost per kg Delta/Star (£)	Launch Altitude Delta/Star (km)	Launch Inclination Delta/Star (degs)
Design 1 & 3	Rideshare	354,250	19,067 / 25,277	400 / 510	51.6 / SSO (≈97.41)
Design 2	Rideshare	354,250	18,200 / 25,277	500-600 / 600	63.4 / SSO (≈97.76)

Table 6: Spaceflight Industries Inc. Launch Vehicle Datasheet

Firefly Space Systems	Launch Type	Cost per launch (£)	Launch Altitude Delta/ Star (km)	Launch Inclination Delta (degs)	Max Payload Mass Delta / Star (kg)
Design 1 & 3	Dedicated	5,200,000	550 / 550	50 / 90	315 / 215
Design 2	Dedicated	5,200,000	550 / 550	60 / 90	280 / 215

Table 7: Firefly Space Systems Launch Vehicle Datasheet

V.III.I Traditional Launch Methods

If using the Spaceflight Industries Inc. rideshare launches, the total cost for launching the entire constellation is calculated to be £7.79million for Design 1, £9.21million for Design 2 and £8.5million for Design 3 when using traditional launch methods.

Using the dedicated launch vehicle provided by Firefly Space Services the launch cost is calculated to be £31.2million in the case of Design 1 and Design 2, assuming that the remaining payload space is not filled by another satellite. The Design 3 launch cost is calculated as £41.2million.

V.III.II Carrier Vehicle Method

The calculated carrier vehicle parameters are given in Table 8 for the case in which Firefly Space Systems dedicated launch vehicle is used. The same carrier vehicle mass is assumed for the rideshare launch.

From the maximum allowable propellant mass it is possible to estimate the maximum allowable ΔV for each carrier. From this, the optimal distribution of the constellation orbital planes, as well as the total time required to deploy each satellite can be calculated. These results are shown in Table 9 for the Walker Delta Orbits considered. The actual manoeuvre ΔV value listed in Table 9 differs from the allowable value shown in Table 8 in some cases. This is because the actual value is the maximum value which can be used without the satellite deorbiting during the deployment manoeuvre as a result of atmospheric drag due to the low altitude of the drift orbit. In the case of the Walker Star orbits, which are to be placed at 90° inclination, the time required to separate the orbital planes by the required amount is such that the satellites would deorbit before the manoeuvre could be completed. Thus it is assumed that one launch would be required to populate each Walker Star orbit plane individually.

These results show that while each carrier in all three constellation designs uses a similar ΔV value, the time to deployment varies greatly. In the case of Design 2 this is because the RAAN change manoeuvre is naturally slower at the higher inclination [11]. In Design 3, the greater number of orbital planes means the satellites must travel further to reach their final position.

Using the Spaceflight Industries Inc. rideshare launches, the total cost for launching the carrier vehicles is calculated to be £22.9million for Design 1 and 3, and £31.2million for Design 2.

The cost of deployment using the Firefly Space Systems dedicated launch is calculated as £20.8million for Design 1 and Design 3 and £31.2million for Design 2.

The costs of all methods considered are summarised in Table 10 from which it can be seen that traditional launch methods utilising rideshare opportunities offer the most economical means of constellation deployment. However this has the disadvantage of allowing the customer minimal control over the orbit parameters and launch schedule. In addition it can be seen in Table 6 that to achieve an orbit inclination of 50° using the rideshare launches requires that the satellites be launched to just 400km altitude. As previously shown in Section V.II this would result in a very short orbit lifetime due to the effects of atmospheric drag. When considering a dedicated launch, deployment using carrier vehicles offers a significant cost reduction when compared with traditional methods.

Firefly Space Systems	Carriers per Launch	Carrier dry mass (kg)	Carrier allowable propellant mass (kg)	Carrier allowable ΔV (m/s)
Design 1	2	79.2	52.05	198
Design 2	1	99	134.3	336
Design 3	3	59.4	28.1	152
Star Orbits 1, 2 & 3	1	118.8	60.4	161

Table 8: Carrier Vehicle Parameters for Various Constellation Designs

	RAAN of Orbital Planes (degs)						Manoeuvre time for carrier (days)						Actual ΔV (m/s)
Design 1	-139	-49	41	131			439	155	139	439			187
Design 2	-138	-48	42	132			740	257	236	740			160
Design 3	-153	-93	-33	26	86	146	616	375	134	113	365	616	152

Table 9: Deployment Results for Various Constellation Designs

Cost (million £)	Traditional Launch Method		Carrier Vehicle Method	
	Rideshare	Dedicated	Rideshare	Dedicated
Design 1	7.79	31.2	22.9	20.8
Design 2	9.21	31.2	31.2	31.2
Design 3	8.5	41.2	22.9	20.8

Table 10: Cost for various launch methods and constellation designs

VI. CONCLUSION

Using carrier vehicles to deploy a constellation of CubeSats is shown to be a practical alternative to traditional launch methods. Optimising the position of each satellite to be deployed allows for the total deployment time to be minimised and considering the effect of atmospheric drag ensures that the manoeuvres and associated propellant costs are realistic. While the time for deployment can be lengthy depending on the ΔV used, planning for this could allow a limited service to start once some satellites are in place. This deployment strategy would also enhance system responsiveness by allowing for the deployment to be adapted in the face of changing mission requirements and removing the reliance of the mission on uncertain launch schedules.

The case study considered demonstrates that a constellation of CubeSats for earth observation could be deployed by carrier vehicles using existing propulsion systems and launch vehicles. While the cost of launching individual CubeSats using rideshare launches is identified as the lowest cost scenario, it reduces the usefulness of the constellation significantly due to the lack of control over each satellite's final position as well as the launch schedule. Launching carrier vehicles on rideshare launches would partially combat this by allowing the satellites to manoeuvre after launch, but the increase in cost is significant.

Launching the carrier vehicles using a dedicated launcher costs less than launching them via rideshare, and also costs significantly less than launching the individual satellites using a dedicated launch vehicle. This would give the customer full control over their launch injection and schedule and so is recommended for deploying a constellation of this kind.

The greatest reduction in cost when using the carrier vehicles comes when considering constellations with a large number of planes, as shown by the significant reduction in cost when comparing the launch of Design 3 using carrier vehicles versus traditional launch methods. The use of the carriers could allow for constellation designs with a large number of planes to be implemented at little or no increase in cost compared to those with fewer planes, and as such may allow for increased system performance and greater mission flexibility.

VII. REFERENCES

[1] Radius Space (2015). *Nanosatellite Database*. Available: http://www.nanosats.eu/

[2] J. Tristancho Martínez, "Implementation of a femto-satellite and a mini-launcher for the N Prize," 2010.

[3] J. Bouwmeester and J. Guo, "Survey of worldwide pico-and nanosatellite missions, distributions and subsystem technology," *Acta Astronautica,* vol. 67, pp. 854-862, 2010.

[4] D. Selva and D. Krejci, "A survey and assessment of the capabilities of Cubesats for Earth observation," *Acta Astronautica,* vol. 74, pp. 50-68, 2012.

[5] European Space Agency. *Announcement of Opportunity #2 'Nanosatellite constellations for Earth Observation'*. Available: http://gsp.esa.int/web/guest/ao2_description

[6] K. Woellert, P. Ehrenfreund, A. J. Ricco, and H. Hertzfeld, "Cubesats: Cost-effective science and technology platforms for emerging and developing nations," *Advances in Space Research,* vol. 47, pp. 663-684, 2011.

[7] S. Müncheberg, M. Krischke, and N. Lemke, "Nanosatellites and micro systems

technology—capabilities, limitations and applications," *Acta Astronautica,* vol. 39, pp. 799-808, 1996.

[8] C. Lowe, M. Macdonald, C. Rogers, R. Thompson, M. Jessup, S. Wakelam, *et al.*, "SysN.TN.004 Mission Design Report," University of Strathclyde, Scisys, Chemring Technolgy Solutions 2015.

[9] C.-J. Fong, C.-Y. Huang, V. Chu, N. Yen, Y.-H. Kuo, Y.-A. Liou, *et al.*, "Mission results from FORMOSAT-3/COSMIC constellation system," *Journal of Spacecraft and Rockets,* vol. 45, pp. 1293-1302, 2008.

[10] C.-J. Fong, W.-T. Shiau, C.-T. Lin, T.-C. Kuo, C.-H. Chu, S.-K. Yang, *et al.*, "Constellation deployment for the FORMOSAT-3/COSMIC mission," *Geoscience and Remote Sensing, IEEE Transactions on,* vol. 46, pp. 3367-3379, 2008.

[11] D. A. Vallado and W. D. McClain, *Fundamentals of astrodynamics and applications* vol. 12: Springer Science & Business Media, 2001.

[12] O. L. De Weck, U. Scialom, and A. Siddiqi, "Optimal reconfiguration of satellite constellations with the auction algorithm," *Acta Astronautica,* vol. 62, pp. 112-130, 2008.

[13] L. Appel, M. Guelman, and D. Mishne, "Optimization of satellite constellation reconfiguration maneuvers," *Acta Astronautica,* vol. 99, pp. 166-174, 2014.

[14] N. Crisp, K. Smith, and P. Hollingsworth, "Launch and deployment of distributed small satellite systems," *Acta Astronautica,* vol. 114, pp. 65-78, 2015.

[15] C. McGrath and M. Macdonald, "Design of a Reconfigurable Satellite Constellation," presented at the 66th International Astronautical Congress, Jerusalem, Israel, 2015.

[16] P. Fortescue, G. Swinerd, and J. Stark, *Spacecraft systems engineering*, 3rd ed.: John Wiley & Sons, 2011.

[17] J. G. Walker, "Circular orbit patterns providing continuous whole earth coverage," DTIC Document1970.

[18] J. G. Walker, "Continuous whole-earth coverage by circular-orbit satellite patterns," DTIC Document1977.

[19] E. Kerr and M. Macdonald, "Incorporating Solar Activity into General Perturbations Analysis of Artificial Satellite's Orbit Lifetime," In Draft for: *Proceedings of the Royal Society A,* 2015.

[20] E. Kerr and M. Macdonald, "A general perturbations method for spacecraft lifetime analysis," in *25th AAS/AIAA Space Flight Mechanics Meeting*, 2015.

[21] J. A. King and N. J. Beidleman, "Method and apparatus for deploying a satellite network," ed: Google Patents, 1993.

[22] *Spaceflight Industries Inc.* (2015). Available: http://www.spaceflight.com/

[23] *Firefly Space Systems* (2015). Available: http://www.fireflyspace.com/

[24] W. J. Larson and J. R. Wertz, "Space mission analysis and design," Microcosm, Inc., Torrance, CA (US)1992.

[25] "Low Power Resistoject," vol. 0072017 v005, S. S. T. Ltd, Ed., ed, 2014.

[26] M. Macdonald and V. Badescu, *The International Handbook of Space Technology*: Springer, 2014.

[27] Firefly Alpha Payload User Guide (2015). Available: http://www.fireflyspace.com/downloads

Reinventing Space Conference
BIS-RS-2015-11

A New Era in Space Flight: The COTS Model of Commercial Partnerships at NASA

Dennis Stone

Commercial Space Capabilities Office
National Aeronautics and Space Administration

S. Hatton (ed.), *Proceedings of the 13th Reinventing Space Conference*,
https://doi.org/10.1007/978-3-319-32817-1_12

ABSTRACT

The commercial space industry can supply affordable and responsive space missions if government engages industry in a partnership. The National Aeronautics and Space Administration (NASA) demonstrated the success of the partnership approach in the Commercial Orbital Transportation Services (COTS) program. By providing milestone-based funding and technical support to commercial partners, NASA stimulated development of commercial capabilities which today are servicing the International Space Station. The COTS program was conducted by NASA using funded Space Act Agreements. It was followed by the purchase of cargo transportation services by NASA under the Commercial Resupply Service contracts. This paper reviews the key attributes and lessons learned from COTS, and how the model has stimulated new commercial partnerships within NASA and beyond.

EARLY COMMERCIALIZATION AT NASA

Throughout most of its history, NASA has performed missions primarily by developing and flying systems which are owned and operated by NASA. This has been the one and only *modus operandi* of human spaceflight: Mercury, Gemini, Apollo, Skylab, Space Shuttle, and the International Space Station have all been government-owned, government-operated systems. This was generally true of unmanned scientific spacecraft as well. In both human and unmanned space exploration, the demanding focus on mission success and the lack of commercial demand for what NASA needed allowed no other option. Only in NASA's aeronautics enterprise, which continued the legacy of the National Advisory Committee on Aeronautics, has the focus been on supporting commercially-operated systems. NASA also had application programs at various times which strived to support commercial industry, such as in communications and microgravity.

It was in communications where the first significant attempt at a public-private partnership (PPP) in space was made by NASA. In the 1970s, because of the successful commercial communications satellite industry, NASA decided to use a commercial approach to fulfill its need for a Tracking and Data Relay Satellite System (TDRSS). However, commercial TDRSS was not successfully implemented, and NASA fell back to a government-owned-and-operated solution.

In the early days of the Space Shuttle in the 1980s, there were various space commercialization initiatives which NASA supported. The Shuttle flew commercially owned and operated systems, such as SPACEHAB modules and the McDonnell Douglas Payload Assist Module which launched satellites from the Shuttle's cargo bay. These provided augmented Shuttle capabilities as a commercial service. It also began flying commercially-funded experiments, such as the Continuous Flow Electrophoresis capability funded by McDonnell Douglas with Ortho Pharmaceuticals. NASA established an Office of Commercial Programs in 1984 to encourage such commercial use of space.

NASA's early engagement with commercial space was not always smooth. As NASA began developing the Space Station, a commercial Industrial Space Facility (ISF) was proposed. In 1988, ISF was defeated in Congress because it was considered a threat to the Station. In 1996, Lockheed Martin began work on the X-33 reusable launch vehicle in partnership with NASA, but NASA withdraw five years later after a $900M investment. In 1995, a task force headed by former NASA Johnson Space Center (JSC) Director Dr. Christopher C. Kraft recommended consolidating Space Shuttle operations under a single operator to, in part, give space commercialization a chance. Yet his successors opposed efforts by the new operator, United Space Alliance, to privatize an orbiter. Commercial space was indeed both an opportunity and a challenge for NASA.

FOUNDATIONS OF COTS

Throughout the Space Shuttle era, numerous commercial launch companies, such as Space Services, Rotary Rocket, and Kistler Aerospace, tried to lower the cost of space launch. In 2000, NASA's Alternative Access to Space initiative funded studies of contingency resupply of the International Space Station (ISS) by commercial launch vehicles. Transformational Space Corporation promoted a design for a low-cost commercial system to transport crew and cargo to and from ISS.

The market for cargo and crew transportation to ISS was indeed an appealing one to the commercial space industry. It offered sizable, predictable demand which could attract commercial financing of innovative solutions. Importantly, new launch systems would have a market other than NASA, namely satellite launch, which could enable sharing of costs between NASA and other customers. The cost of using Shuttle for routine ISS transportation to post-assembly, and NASA assigned its new exploration spacecraft, Orion, the task of transporting crew and cargo to ISS after the Shuttle program ends. The emerging commercial industry sought an opportunity to show how it could do this routine task and save NASA money which could be reallocated to new space exploration challenges.

These factors aligned to support a new attempt by the United States for a public-private partnership in human spaceflight. In 2004, President Bush announced the Vision for Space Exploration which included a provision to pursue commercial solutions for transportation to the Space Station. NASA Administrator Griffin fenced $500M for what would soon be called Commercial Orbital Transportation Services (COTS). The Commercial Crew & Cargo Program Office (C3PO) was then formed at NASA JSC in 2005 with these goals:

- Implement U.S. Space Exploration policy with investments to stimulate the commercial space industry,
- Facilitate U.S. private industry demonstration of cargo and crew space transportation capabilities with the goal of achieving reliable, cost-effective access to LEO, and
- Create a market environment in which commercial space transportation services are available to Government and private sector customers.

THE COTS APPOACH

In January 2006, C3PO issued the COTS Announcement, which defined two distinct phases:

> **Phase 1** – Development and demonstration by private industry of space transportation capabilities to and from LEO. NASA has a role similar to a lead investor. Space Act Agreements (SAAs) are used in lieu of traditional cost contracts.
>
> **Phase 2** – Purchase of commercial transportation services to and from ISS. NASA is a customer. Traditional commercial services contracts are used.

The announcement gave commercial industry an opportunity to compete for COTS funded SAAs. Unlike a typical procurement, COTS did not impose a set of requirements. Instead it established goals, and allowed industry to choose which goals it met. This allowed COTS partners to optimize their business plans around systems with both government and commercial customers. To enable this trade space, NASA defined four broad sets of capabilities of interest to the Government:

> A – External (unpressurized) cargo to a LEO test bed and safely dispose of external cargo.
>
> B – Internal (pressurized) cargo to a LEO test bed and safely dispose of internal cargo.
>
> C – Internal (pressurized) cargo to a LEO test bed and safely return internal cargo to Earth.
>
> D – Crew to a LEO test bed and safely return crew to Earth.

NASA offered use of ISS as the LEO test bed. If COTS partners opted for use of ISS, they would have to meet ISS interface requirements. Otherwise the partner could provide its own LEO test bed, provide it accounted for the additional work needed to integrate their COTS spacecraft with ISS. NASA would provide up to $500M to fund demonstration of cargo capabilities A, B, and C. Crew capability D was an option that would provide additional funds if exercised. NASA offered to pay these funds only as chosen partners completed pre-negotiated milestones, which could reflect technical or business achievements.

The COTS strategy was unique compared to anything NASA had done before. Key characteristics included:

Use of Funded SAA – NASA did not use a contract for the COTS development and demonstrations, because it was not procuring anything. Rather, NASA was supporting a public purpose. This resulted in the first use of a funded Space Act Agreement.

Investor Mindset – Because C3PO considering COTS an investment in commercial capabilities, it hired a venture capitalist to assist NASA. To bid on COTS, companies had to furnish business plans as well as traditional technical proposals. NASA would then conduct "due diligence" as in a traditional investment.

Limited Termination Rights – NASA voluntarily limited its ability to terminate the SAAs to specific situations, such as failure to get appropriated funds or failure of a COTS partner to meet its milestones. This was intended to improve chances of COTS partners attracting private financing.

Limited Government Investment – NASA's contributions were fixed to the milestone payments negotiated in the COTS SAAs. Any cost overruns were the partner's responsibility.

NASA Goals, Not Requirements – Partners could choose which capabilities to pursue based on NASA's future projected needs and their knowledge of other markets. Partners then manage their own requirements.

Pick a Portfolio -- To maximize coverage of all NASA goals and to reduce risk, NASA would invest in a portfolio of partners.

Buy a Ticket, Not a Vehicle – COTS would help industry develop capabilities which NASA and others might later buy as a service. NASA thus would not own COTS vehicles. They would be company-owned and company-operated.

FAA Licensing – COTS partners would be required to obtain commercial launch licensing from the Federal Aviation Administration.

NASA as an Advisor – Traditionally NASA technical personnel oversee all aspects of a contractor's work. On COTS, NASA was an advisor, making expertise available upon partner request. NASA relied on insight not oversight of its partners, except when evaluating the partner's milestone performance.

COTS PARTNERS

In 2006, NASA announced two winners of COTS funded SAAs: Space Exploration Technologies (SpaceX) and Rocketplane Kistler (RpK). SpaceX, founded by Elon Musk, proposed the Falcon 9 launch vehicle and the Dragon spacecraft to carry cargo and crew to and from LEO, addressing all COTS capabilities. RpK, a merger of Rocketplane and Kistler Aerospace, proposed the fully-reusable K-1 launch vehicle for the same capabilities.

Proposing to use traditional aerospace companies as its contractors, RpK required significant private investment. Understanding this risk, NASA required that RpK's financing rounds be considered formal milestones in the SAA so that NASA had an off-ramp if financing failed to materialize. When market conditions contributed to repeated delays in financing, NASA terminated the RpK SAA. Only the first three of 15 milestones had been paid by NASA, allowing retention of sufficient funds to find a new COTS partner.

NASA begin a competition for a replacement COTS partner in October 2007. In February 2008, NASA announced selection of Orbital Sciences Corp. as that partner. Orbital committed to providing the necessary non-NASA financing from internal corporate funds. Orbital proposed the Taurus II (later renamed Antares) launch vehicle and Cygnus cargo spacecraft. It eventually focused on Capability B, pressurized cargo up and destructive return.

COTS partners SpaceX and Orbital Sciences made significant early progress. To reduce risk, NASA negotiated with the partners $300M of additional testing and demonstration milestones. As is common in launch vehicle development programs, each partner experienced approximately two years of schedule delays. However both partners were fully successful technically. Each partner opted to demonstrate cargo missions to ISS. SpaceX completed its flight demonstrations in May 2012, and Orbital Sciences completed its in September 2013.

In keeping with the announced two-phase strategy, NASA's ISS Program Office began to procure commercial cargo transportation services in 2008. Through an independent competitive process, it awarded Commercial Resupply Services (CRS) contracts to both SpaceX ($1.6B for 12 deliveries) and Orbital Sciences ($1.9B for 8 deliveries). Flights to ISS under CRS began after each contractor completed its COTS demonstrations. Each partner has to date had one CRS mission failure.

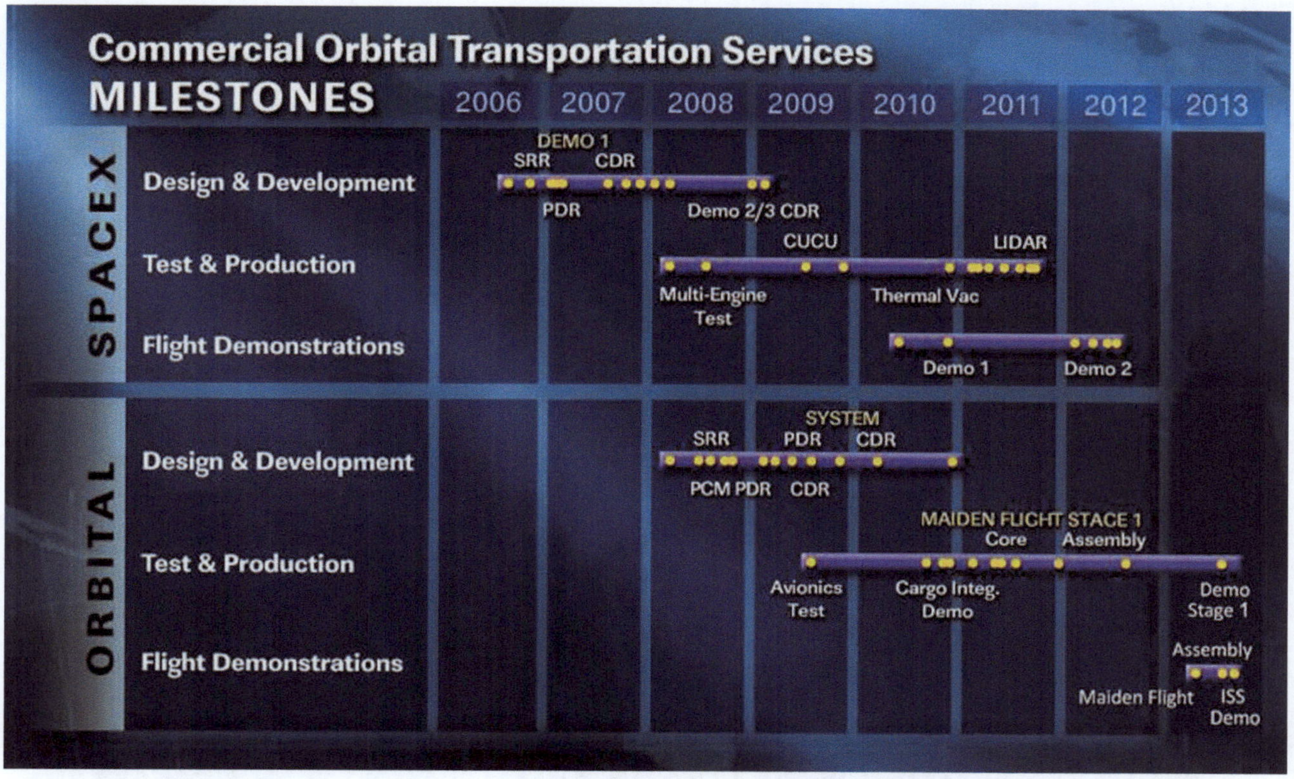

Figure 1-- COTS Milestone Summary

Figure 2- Spacex Dragon captured by ISS.

Figure 3 – Orbital Sciences Cygus captured by ISS

THE LESSONS AND LEGACY OF COTS

NASA's unique approach to Commercial Orbital Transportation Services proved highly successful:

Economy – In seven years, with an investment of $800M, the United States obtained two new medium-class launch vehicles and two automated cargo vehicles which could rendezvous and berth with the Space Station.

Contingency to Dependency – When COTS began, it was a side-bet; NASA was depending on Orion for Earth-to-ISS transportation. By the time COTS ended, NASA was depending on COTS for this service.

Official Policy – President Obama in 2013 made it national policy to use public-private partnerships in space exploration.

A New Model – The success of COTS gave NASA a new model to fulfill mission needs, beginning with PPP to invest in commercial capabilities, followed by a procurement for services.

Economic Development – COTS led to not only $1 billion in investment in the space industry, but capture by the United States of a greater share of the global launch market.

Lower Launch Costs – The success of COTS partners has stimulated efforts across the launch industry to lower launch costs, which may open new markets in space in the future.

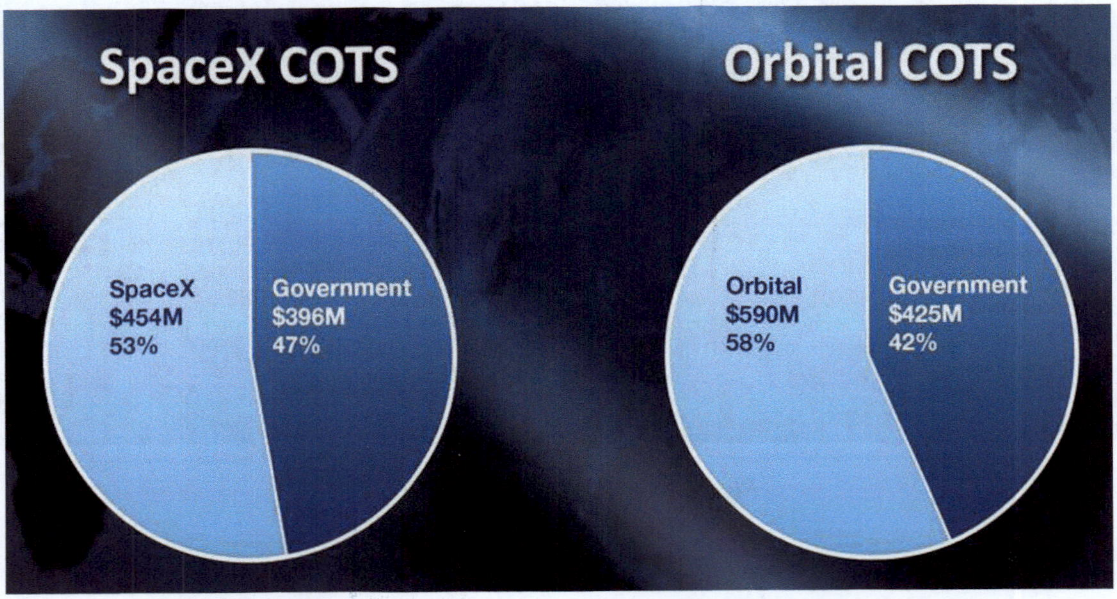

Figure 4 – NASA paid less than half of COTS development and demonstration costs.

COTS opened the door for renewed commercialization efforts at NASA. These included:

- **Commercial Crew Program** – Development and demonstration of crew transportation to and from ISS
- **Partnership Council** – Provides guidance and review of NASA public-private partnerships
- **Collaborations for Commercial Space Capabilities** – Provides technical expertise to commercial space companies on a variety of capabilities relevant to NASA human exploration, including the United Launch Alliance Vulcan launch vehicle and the SpaceX Red Dragon unmanned Mars lander
- **Lunar CATALYST** – Provides technical support to companies to land small payloads on the Moon

The US Air Force is also using aspects of the COTS model for its EELV Rocket Propulsion System development.

SUMMARY AND CONCLUSIONS

NASA's Commercial Orbital Transportation System initiative funded the development and demonstration of privately owned and operated systems to carry cargo to and from low Earth orbit. COTS has been widely considered a success from both the technical and financial standpoints. It demonstrated that public-private partnerships, if properly formulated and executed, can develop capabilities to fill government mission needs in space reliably and efficiently.

COTS had moved commercial spaceflight closer to the mainstream, allowing governments to more routinely consider public-private partnerships when government needs intersect with commercial markets. With greater use, commercial partnerships may open a new era in spaceflight.

REFERENCE

1. Commercial Orbital Transportation Services: A New Era in Spaceflight, NASA/SP-2015-617, National Aeronautics and Space Administration, 2014.

Reinventing Space Conference
BIS-RS-2015-31

IODISPLay: Capturing European needs and capabilities for in-orbit demonstration of space technologies

Giovanni Binet
GMV Aerospace and Defence

Gabriele Novelli
GMV Aerospace and Defence

Celestino Gomez Cid
GMV Innovating Solutions ltd

Marco Bolchi
Valdani Vicari e Associati

13th Reinventing Space Conference
9-12 November 2015
Oxford, UK

S. Hatton (ed.), *Proceedings of the 13th Reinventing Space Conference*,
https://doi.org/10.1007/978-3-319-32817-1_13

IODISPLay: Capturing European Needs and Capabilities for In-Orbit Demonstration of Space Technologies

Giovanni Binet, Gabriele Novelli
GMV Aerospace and Defence
Isaac Newton 11, PTM, Tres Cantos, Madrid-Spain; +34 91 806 16 94
gbinet@gmv.com, gnovelli@gmv.com

Celestino Cid Gomez
GMV Innovating Solutions ltd. (GMV-UK)
Harwell Innovation Centre, Building 173 Curie Avenue; OX 11 0QG Harwell, Oxford (UK); +44 1235 838536
cgomez@gmv.com

Marco Bolchi
Valdani Vicari & Associati Srl
Via Torino 61; Milano (italy) +39 (0)2 72 733 1
m.bolchi@vva.it

ABSTRACT

Current In-Orbit Demonstration (IOD) possibilities are restricted to either the identification of carriers of opportunity (where IOD has to fulfill with fixed requirements and interfaces, limited to a top-down approach) or to dedicated missions where a satellite is designed as a compromise among the needs of a number of identified technologies to be demonstrated in orbit. Moreover, often political choices drive the selection of the technologies to be validated in orbit, sometimes at the expenses of more interesting technologies in terms of innovation, time-to-market and future mission or industrial application. On the one hand, this approach strongly limits the maximum available potential of IOD. On the other hand, current trends in modular satellites, the now dynamic panorama of space launchers and innovative concepts certainly offer new and extended possibilities for IOD.

We believe that the optimum approach to build IOD missions shall consistently investigate and merge both the bottom-up and the top-down directions, i.e. on the one hand there must be a clear and extensive assessment exercise of the current and future technologies candidate for IOD and, on the other hand, a thorough identification exercise of IOD carriers and launcher services. This shall then drive the selection of IOD missions to be implemented at European level.

This paper will then present the results of a European-wide survey on current needs and capabilities for IOD, as well as the software tool that has been prepared in order to use such information on technologies, launchers and carriers in order to identify IOD missions. Also, this paper will describe the results of an analysis of the potential market of a commercial IOD service at European level, identifying the supply and the demand for such service (including the willingness to pay). The activities described in this paper are part of the IODISPLay project which has received funding from the European Union's H2020 research and innovation program under grant agreement No 640253.

KEYWORDS: IODISPLay, in-orbit demonstration, IOD, in-orbit validation, IOV, Technology Readiness Level, Technology Domain, H2020.

1. , INTRODUCTION

Current In Orbit Demonstration/Validation (IOD/IOV) possibilities are restricted to either the identification of carriers of opportunity (where IOD/IOV has to fulfil with fixed requirements and interfaces, limited to a top-down approach) or to dedicated missions where a satellite is designed as a compromise among the needs of a number of identified technologies to be demonstrated in orbit. Moreover, often political choices drive the selection of the technologies to be validated in orbit, sometimes at the expenses of more interesting technologies in terms of innovation, time-to-market and future mission or industrial application.

On the one hand, this approach strongly limits the maximum available potential of IOD/IOV. On the other hand, current trends in modular satellites, the now dynamic panorama of space launchers and innovative concepts certainly offers new and extended possibilities for IOD/IOV.

In 2014, the European Commission requested proposals within its Horizon 2020 framework programme for Research and Innovation for studies aimed at helping define the envelope and the requirements for the implementation of affordable missions of IOD/IOV (in combination with the launching system to be selected) within the Horizon 2020 timing and development contexts

In this context, the IODISPLay Consortium (led by GMV and formed by INTA, VVA and GAUSS) answer that an optimum approach shall consistently investigate and merge both the bottom-up and the top-down directions, i.e. on one hand there must be a clear and extensive assessment exercise of the current and future technologies candidate for IOD/IOV and, on the other hand, a thorough identification exercise of IOD/IOV carriers and launcher services.

Moreover, the Consortium believes that the final goal that the European space community shall pursue is to have a European commercial IOD/IOV service. In this respect, the Commission should then be in position to co-finance, within H2020 horizon, only those IOD/IOV mission concept/s that have the real potential of maturing into a commercial IOD/IOV service.

In this view, the IODISPLay project (www.iodisplay.eu) has the following objectives:

- To assess current IOD/IOV needs (in terms of current and future European space technologies) and capabilities, as well as current IOD/IOV service market. This will be done with a dedicated survey, and will include proposal from European developers.

- To analyse current and future available/existing IOD/IOV carriers concepts and also ad-hoc modifications to enhance the IOD/IOV capabilities of already existing concepts. The cost-to-benefit ratio shall be the main criterion when analysing those ad-hoc modifications.

- To provide an IOD/IOV mission configuration tool (IOD MITO) which can intelligently analyse a database of carriers and technologies in order to intuitively provide a number of IOD/IOV missions.

- To identify a portfolio of IOD/IOV missions and concepts achievable and affordable within H2020 timeframe and to down-select a number of most interesting IOD/IOV missions.

- To identify, of those down-selected IOD/IOV missions, the business case for an IOD/IOV service.

- To carry out a preliminary design of those missions, including programmatic aspects.

The project has been named IODISPLay (IOD/IOV Service mission PortfoLio), as its output will be a display of potential IOD/IOV missions, as well as a display of European space technologies that could benefit for IOD/IOV. This will be enabled by a dedicated infrastructure (the IOD MIssionization TOol, IOD MITO in short) built and provided as output of the project to the EC. This paper will present the main preliminary results of our study, with particular highlight on the identification of a database of technologies and analysis of market perspectives

2. CAPTURING IOD CAPABILITIES AND NEEDS IN EUROPE

While our project has carried out an extensive survey of all potential carriers (satellite platforms or modified upper stages that can host an IOD/IOV payload) and launchers, this paper will focus on the identification of the European IOD/IOV technologies as the most important part any space mission is its payload. In an IOD mission, the list of technologies to be demonstrated in orbit forms the mission payload.

Our approach to identify the list of European innovative space technologies that would benefit from an in-orbit demonstration has been to carry out a survey, getting in contact with all technology developers (companies, research centres and universities) across Europe. In a large number of emails and phone calls (a dedicated register has been used to coordinate all contacts), we have communicated our project goals and asked to voluntarily propose technologies to be included in the database. The benefit in providing such data would be in the inclusion of a database to the EC.

To this extent, we have designed a data collection process composed of two different questionnaires to be filled out by companies/institutes that wanted to propose a technology:

A first questionnaire, where some basic data about the technology to be proposed is requested, is contained.in the project website: www.iodisplay.eu . This relatively light document to fill out allows the IODISPLay team to have a first contact and identification of the technologies, as well as building a working relationship with the technology owner. The questionnaire contains information about the innovation content, the targeted application, the TRL status and other generic information

A second questionnaire where more detailed information about the technology is requested. The purpose of this questionnaire is to gather details about what types of requirements a specific technology would impose on a carrier for its in-orbit validation.

We have therefore followed this 2-step approach with all technology actors that we have been in contact with. This has allowed us to gather and consolidate the information of European Technologies in a single structured database.

3. IDENTIFIED TECHNOLOGIES FOR IOD

The total amount of technologies that has been submitted to IODISPlay is of 152 technologies (gathered through the first questionnaire). The number of technologies that we can include in the IOD missionization tool is of 99, as it refers to the received second questionnaires. The main reason for this decrease in the second step of the survey is due to the effort that has to be devoted in filling out the second questionnaire.

Generally, it can be said that the IODISPLay survey has succeeded in collecting a very good amount of information on technologies that could profit from IOD: GMV has been therefore considered as a valid counterpart by the European space technology owners community. The main outcomes of the technology survey are reported here below by mean of charts that have been produced using aggregated information contained in the database.

Figure 1 depicts the distribution of the technology owners per country. Note that the same entity (i.e. technology owner) can submit more than one technology.

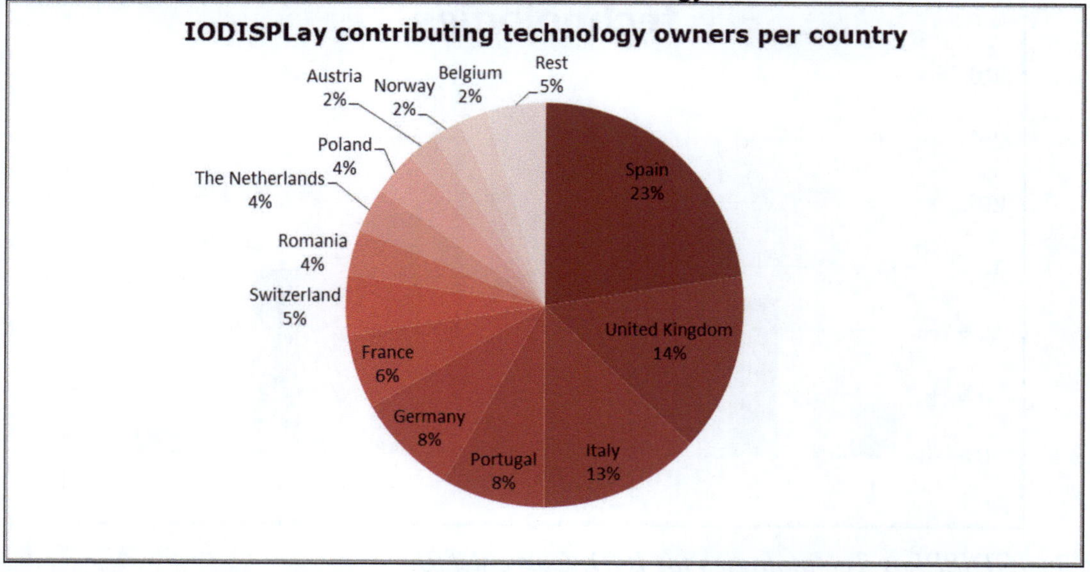

Figure 1. IODISPLay Technology Owners, Geographical Coverage

The graph shows a fairly good coverage of European countries represented in our database. Still, apart from the prominence of Spanish contributors, mainly due to the Consortium connections and the open support to our initiative from CDTI, UK's interest in the project and IOD activities is still striking if compared to other countries with strong space heritage. The case of France and Germany is quite interesting, as their low contribution to the database could be related to the fact that they have national programs and therefore technology owners don't see any special opportunity through the EC.

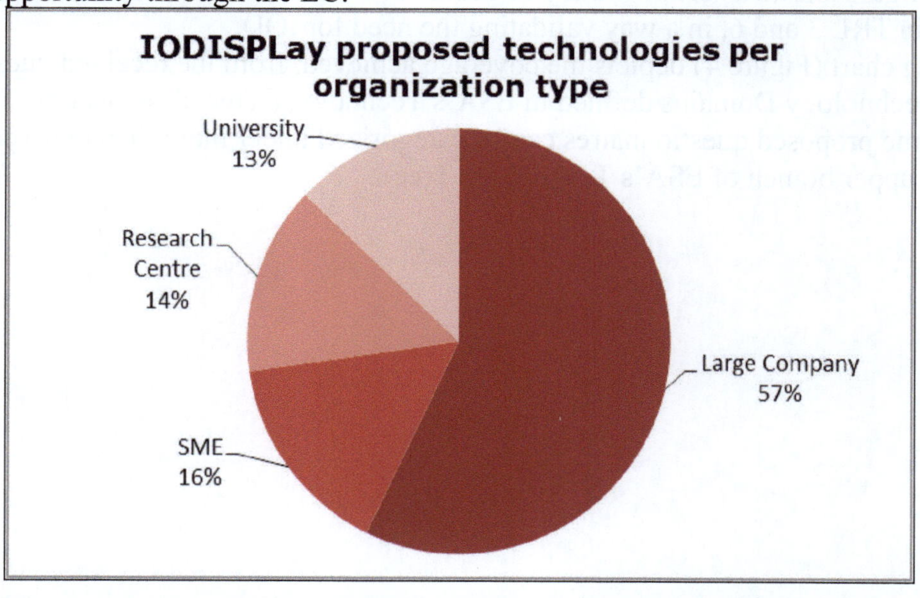

Figure 2 - IODISPLay Technologies for IOD, Organization Type Coverage

Looking at the distribution of the types of organization (Figure 2), we see that more than half of the contributions come from large companies (i.e., not SMEs). More than a quarter of the submitted technologies come from research centres or universities.

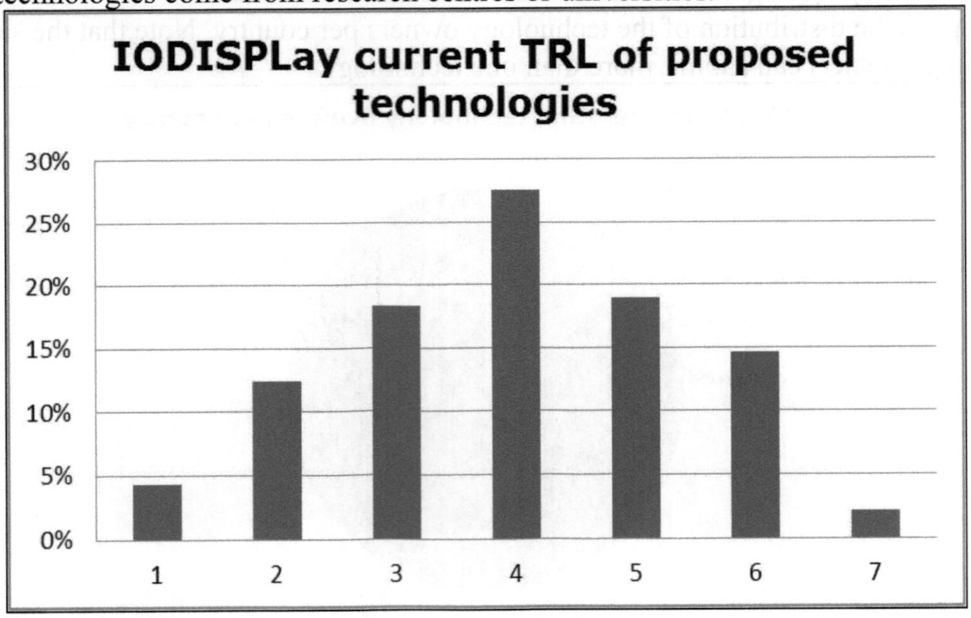

Figure 3 - IODISPLay Technologies for IOD, Current TRL

Concerning the maturity level of the presented technologies, we can see from the picture above (Figure 3) that most of the technologies presented are at least TRL 3. Actually, more than half of the technologies have been stated with at least TRL 4 (Component or breadboard validation in relevant environment): this somehow validates the fact that there is indeed a huge need for IOD of technologies that can mature rapidly on ground, but that have difficulties in finding a way to be flight proven. The effort associated to bring a TRL to 6, which is the needed level in order to be embarked on an IOD mission, varies largely on the technology and on the type of validation needed. The figure shows a sort of "valley of death" of a high number of lots of technologies trapped within TRL 3 and 6, in a way validating the need for IOD.

The following chart (Figure 4) depicts the coverage achieved, from the received questionnaires, in terms of Technology Domains defined in ESA's Technology Tree classification. It has to be noted that some proposed questionnaires can be categorized under more than one Technology Domain (the upper branch of ESA's Technology Tree).

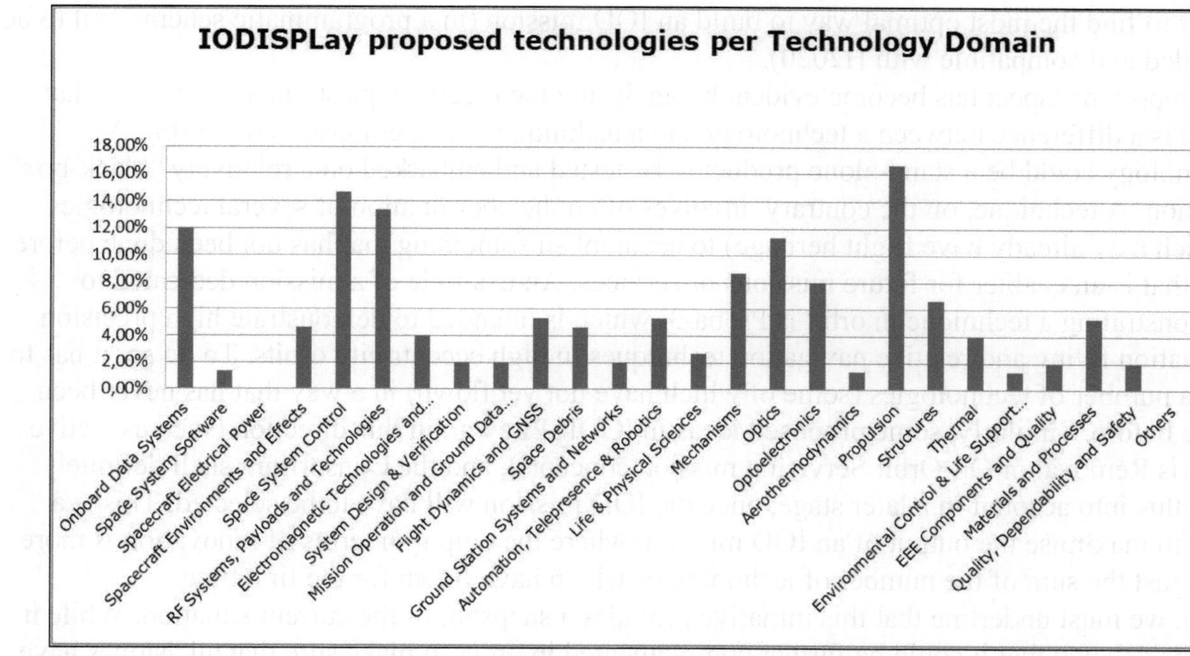

Figure 4 - IODISPLay technologies for IOD, ESA Technology Tree coverage

Analysing the above figure, it can be said that there is in general a good coverage of the whole technology domains that are part of a space mission. In particular, the following aspects should be highlighted:

- The highest interest in IOD resulted for propulsion technologies and "Space Systems Control", "Onboard data systems" and "RF Systems, payloads and Technologies", .

- Very good participation from technologies on "Optics" and "Optoelectronics", "Mechanisms" and "Structures".

- Limited (or absent) proposed IOD has been received on "Spacecraft Electrical Power", "System Design and Verification", "Aerothermodynamics" and "Ground Station Systems and Networks" (TD 26 is an "Others" category which has not been used)

The peak corresponding to propulsion technologies is mostly linked to the current trend in enabling new missions for cube, mini and small satellites, which often do not embark a propulsion system because of the high costs and complications that current solutions impose on the spacecraft. The market is therefore rapidly addressing this issue (as it currently prevents the use of such compact and economic platforms for a number of applications) and in this context we can observe a high need in IOD for propulsion systems (both chemical and electrical) for small satellites. Also, propulsion is a field where the concept of the "TRL valley of death" is particularly strong: many technology owners have told us that the reason for an IOD for their technology is often more to convince people with a space heritage of their technology rather than carrying out specific technical validation in space environment.
We have also seen how disperse the scenario of the received technologies with IODISPLay is, as shown by the Technology Domain coverage. This, together with the fact that many technologies are mature (or about to be) to be demonstrated in orbit, validates the need for an initiative like IODISPLay that can identify the needs for IOD and coordinate these disperse technologies in

order to find the most optimal way to build an IOD mission (in a programmatic scheme still to be detailed and compatible with H2020).
An important aspect has become evident by analysing the received questionnaires: that is, that there is a difference between a technology and a technique to be demonstrated in orbit. A technology could be a stand-alone product to be tested and embarked on a relatively "black-box" mission. A technique, on the contrary, involves often the coordination of several technologies (which may already have flight heritage) to accomplish something that has not been done before and that is an enabler for future missions or services. An example of a mission dedicated to demonstrating a technique in orbit is Proba-3, which is intended to demonstrate high precision formation flying and relative navigation techniques in high eccentricity orbits. To do so, it has to use a number of technologies (some of which have not yet flown) in a way that has never been done before. Similarly, some proposed ideas in IODISPLay go in this direction (such as Active Debris Removal or On-Orbit Servicing mission concepts), and the Consortium shall definitely take this into account in a later stage, once the IOD mission will have to be selected. This is a way to maximise the output of an IOD mission, where the output in terms of innovation is more than just the sum of the number of technologies which have flown for the first time.
Also, we must underline that this initiative provides a snapshot of the current situation. While it can be debated that it could be further complemented in order to make sure that all "stones have been turned" in our search, it would definitely make sense to update such snapshot on a periodic basis (every one or two years, for instance) in order to have a consistent and updated vision of the IOD needs in Europe. Also, it would be very interesting to follow the evolution of each single technology that is contained in the database, i.e. if the technology has found an IOD opportunity and, otherwise, if there is still interest for an IOD.

4. THE MARKET FOR AN IOD SERVICE

Up to now, IOD has been mainly linked to institutional missions. While the hosted payload approach has offered some opportunities, and so are also doing Cubesats for small technologies, at the time being there is no such thing as a commercial service for demonstrating technologies in-orbit. In our study, we will look into the possibility that the missions which will be identified in the project can be implemented within H2020 and be the first step of a commercial service of IOD. In this context, in our study we have first carried out a market survey for an IOD service. The demand for IOD is composed by technology suppliers in the manufacturing sector, which can be grouped in two main categories: private companies and research organisations, though the vast majority of the needs arise from the companies. The need for IOD is expressed for systems, equipment and components. Differences in requirements for IOD concern the aim of the demonstration, which can be:

- Demonstrating that a technology performs in space the way it was designed for;
- Demonstrating that a technology performs in space for a certain amount of time.

The entities representing the potential demand for IOD services obviously share the same characteristics of the overall EU space industry, composed of a small number of very large players and a large number of smaller companies. Aside the differences in size between these two categories of players, there are also differences in the average financial outlook, with large system integrators tending to enjoy, on average, higher profitability margins, whereas most smaller companies are under pressure, recording low, zero or negative profitability.

As mentioned above, the companies contacted within the project expressed a strong interest in IOD, with more than 150 technologies being submitted. After removing outliers, the average budget indicated as necessary for IOD has been indicated to be around 720 EUR thousand (estimated as the cost needed to produce the flight model of the technology for its IOD). Further analysis, carried out through in-depth interviews with selected technology developers, revealed that companies, in particular small ones, do not necessarily set aside this budget within the overall resources allocated to technology development. Some companies do not even consider an IOD budget. Reasons behind can be different:

- The technology is driven by institutional needs rather than by commercial objectives. In this case, an institutional player requires the technology and funds its development through contracting. Thus, technology developers focus on achieving the TRL steps required in the contract and expect the institutional player to fund the IOD activity, if and when they deem necessary;

- The technology is a component, meant to be tested as a part of a whole system. In this case, it is the system integrator, not the technology developer that will move to the IOD activity, only after finalising the development of the system;

- The technology developer does not have direct access to the players offering (the quite limited number of) IOD opportunities, and can rely only on its network of contacts, which often encompasses only other component manufacturers and the system integrators, or the technology developer does not have the necessary budget for IOD – which could often be the case considering the overall profitability figures of medium and small space companies.

Developing an IOD service is particularly important to solve this third issue. Some technologies might have a high commercial potential; however the absence of contacts, relationships with IOD providers and/or money could hamper this potential. The upcoming activities of the project will include a definition of a business model and a business plan for the provision of an IOD service, which is expected to be able to:

- Provide a cost-effective opportunity for technology developers which are currently missing opportunities for IOD; and

- Solve the "market failure" faced by developers of technologies with promising commercial potential, which are stopped by the lack of IOD budget. In this case, the business model and plan will evaluate the need and the extent of a possible public intervention to close this gap.

5. IDENTIFYING A PORTFOLIO OF IOD MISSIONS

Once the information about the technologies, carrier and launchers has been captured, the next step has been to consolidate all this information into a single database and to develop a tool that can analyse it and assess the feasibility of IOD missions. In this context, we have developed a tool that checks a number of rules and filters in order to confirm to the user that its selected mission configuration (composed by at least one technology, one carrier and one launcher) is feasible. High-level filters can assess, among other checks, if the orbital regime of the different technologies is compatible and if their mass and volume is consistent with the available payload room in the carrier. Upon confirmation that its configuration is deemed as compatible by the tool, a report is generated that presents to the user also a critical assessment of such mission configuration. This assessment contains a number of criteria derived from the included technologies (overall potential, estimated cost, maturity of involved technologies, need for in orbit demonstration and confidence level based on available information) that can help the user in comparing several selected mission configurations. The mission configuration report also provides to the user a number of interesting information, such as the coverage of the Technology Tree or of the involved countries.

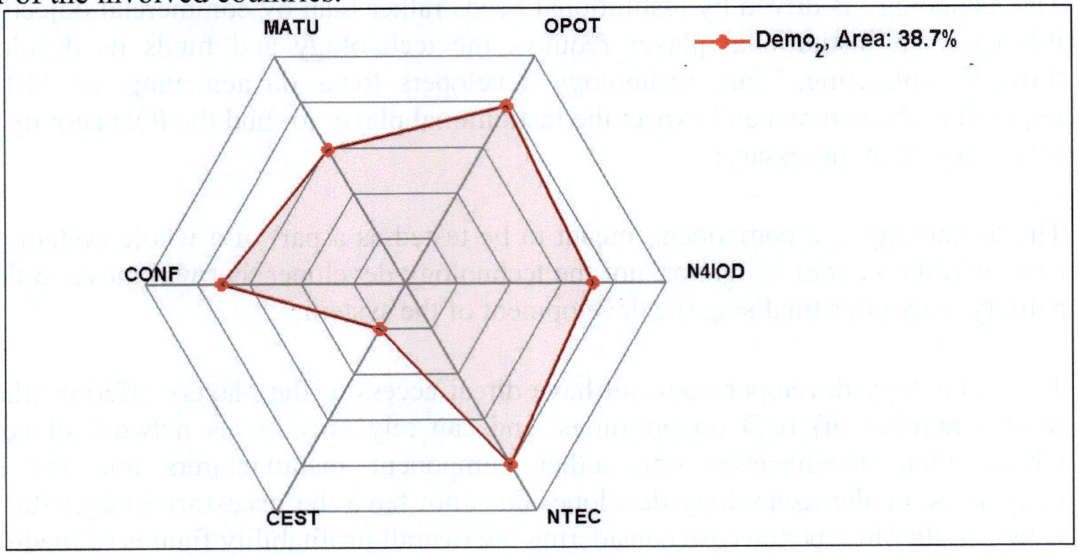

Figure 5. Assessment of a Mission Configuration as Calculated by the Tool

The tool also provides hints on additional technologies that could be added to the user selected mission configuration, based on available margins in the carrier and on couplings and compatibilities among different technologies in the database. The MITO has to be considered as a useful tool for browsing and selecting a first design point for an IOD mission design, which has to be obviously followed by a classical rigorous system engineering approach.

The tool can also be used for navigating within the database, as it can presents summary reports of the technologies included, as well as comparing different technologies.

6. WAY FORWARD

The following activities of the project will focus on the utilization of the data which has been collected on technologies, carriers and launchers in order to identify (thanks to the developed missionization tool) of a number of IOD missions that are implementable within H2020. For two down-selected missions, we shall also carry out an exercise of business case analysis for implementing them as an IOD service.

The mission identification will be based on the information contained in the tool about technologies and on the main conclusion of the "Workshop on IOD opportunities and priorities in Europe" that will be organized by the IODISPLay project and will include also the other 3 parallel H2020 projects on IOD (GOTOFLY!, INVEST and PLUGIN). The workshop will be held at ESA/ESTEC and will discuss, including major IOD institutional and private stakeholders, about what are the main priorities of In-Orbit Demonstration.

Once the missions have been identified, a business case for their implementation as an IOD commercial service will be carried out. This will be followed by a preliminary requirements and design of those missions, including programmatic (schedule and costs) implementation aspects within H2020.

7. CONCLUSIONS

In this paper, we have presented the activities carried out so far in the H2020 IODISPLay project devoted to the identification of a portfolio of IOD missions. The first activity that we have carried out is an in-depth survey of all potential technologies, at European level, that could benefit from an IOD. Such survey has been conducted by GMV and has identified more than 150 technologies across all Europe and covering most Technology Domains. Such information of technologies has been included in a structured database, which includes detailed information about each technology scope and main requirements (e.g. size, mass, power) and needs for in-orbit demonstration. The database and its associated tool are a powerful tool especially for policy makers and analysis.

This exercise of technologies identification has allowed us to take a snapshot of the current situation, and it would be interesting to update its status on a yearly basis in order to draw conclusions about which technologies have found its way into a demonstrations and which other ones have remained trapped in the "TRL valley of death".

Also, we have studied the potential of introducing a commercial service that would offer In-Orbit demonstration possibilities to technologies developers. In the following phases of the study, we shall propose a service scheme for IOD within Europe, as well as identifying a number of missions to be implemented within H2020 dedicated to IOD.

The activities described in this paper are part of the IODISPLay project (**www.iodisplay.eu**) that has received funding from the European Union's H2020 research and innovation program under grant agreement No 640253.

Acknowledgments

The activities described in this paper are part of the IODISPLay project (**www.iodisplay.eu**) that has received funding from the European Union's H2020 research and innovation program under grant agreement No 640253. The consortium of IODISPLay is formed by GMV, INTA, GAUSS and VVA.

Optimal (not opportunity) Orbits for Rideshare Payloads

Chris Pearson, Marissa Stender
Moog Integrated Systems – Golden, Colorado

Chris Loghry
Moog Integrated Systems – Chatsworth, California

Joe Maly
Moog CSA – Mountain View, California

13th Reinventing Space Conference

9-12 November 2015
Oxford, UK

S. Hatton (ed.), *Proceedings of the 13th Reinventing Space Conference*,
https://doi.org/10.1007/978-3-319-32817-1_14

Optimal (not opportunity) Orbits for Rideshare Payloads

Chris Pearson & Marissa Stender, Moog Integrated Systems
1113 Washington Avenue Golden, CO 80401; 303-216-9777 x204, cpearson2@moog.com

Chris Loghry, Moog Integrated Systems
21339 Nordhoff Street Chatsworth, CA 91311; 720-289-7041, cloghry@moog.com

Joe Maly, Moog CSA
2581 Leghorn Street Mountain View, CA 94043; 650-960-4226, jmaly@moog.com

ABSTRACT

As rideshare launches become more commonplace, secondary payloads continue to be challenged by the limited choice of orbits, upper stage restart capability and risk-averse nature of primary payloads to allow for flexibility in the deployment sequence. The result is that a secondary payload's final orbit is limited by its host and the propulsion capability of the individual spacecraft, particularly so for cubesat class passengers. Many of these challenges can be met through the use of a propulsive rideshare adapter or Orbital Maneuvering Vehicle (OMV). In addition, the use of an OMV to act as both a long-term host for payloads, as well as a traditional tug, has enabled novel mission concepts such as hub and spoke architectures for communications and data processing functionality for distributed sensing systems.

Moog has analyzed, developed and supported numerous missions employing OMV functionality. In this paper a number of case studies are described to illustrate the utility, value and flexibility of the OMV as a mission enabling technology. Cases examined include:

- Accelerated deployment of a smallsat constellation, which results in the optimal satellite placement (revenue generation) in less than half the time of a traditional (passive) deployment.
- A joint operational heliophysics and technology demonstration mission at the Earth-Sun L1 point piggybacking from a rideshare GTO launch opportunity.
- A scenario to ferry multiple small payloads to Low Lunar Orbit (LLO) from earth orbit.

The performance of small satellite technology continues to improve at an exponential pace but, if small satellites continue to compromise optimal orbit for general space access, true potential cannot be fulfilled. In each of the scenarios identified, the particular use of an OMV gives rise to a number of shared launch opportunities that would not have previously been considered and improves the overall access to space for rideshare passengers.

KEYWORDS: Orbital Maneuvering Vehicle, Orbit, Rideshare, Secondary Payload, Small Satellite, Payload, Space Tug

INTRODUCTION

The increased utility of small satellites and CubeSats over the last decade has led to a boom in the number of spacecraft produced and has highlighted the challenges associated with launching secondary payloads to an orbit they desire, rather than an orbit they will settle for.

This problem has only come to the forefront of discussions now that the initial launch bottleneck is starting to loosen. The role of secondary payload brokers has opened up a much larger quantity of payload slots than many small satellites have had access to previously. The limitation still stands, however, that a majority of the launches are delivering primary spacecraft to sun-synchronous orbit and offer little to no ability to deploy secondaries at a different inclination and/or altitude.

To fill this gap in capability with minimal effect on the launch stack and no changes to the design of the small spacecraft, a propulsive launch adapter, such as the Orbital Maneuvering Vehicle (OMV), should be considered. With an independent propulsion system and avionics, OMV upgrades a launch adapter from a structural entity alone to a fully capable spacecraft platform or final insertion stage. This configuration provides the capability for multiple, discrete payloads and spacecraft to achieve their optimal orbit and maximize their value, rather than settle for "opportunity" orbits.

SECONDARY PAYLOAD ACCOMMODATION

A key factor in the launch and deployment of secondary payloads is the ability to prove that they will do no harm to the primary spacecraft. They must be able to launch without adding additional risk to the overall mission. Therefore, a variety of adapters and dispensers have been designed to meet these stringent requirements and safely secure both smallsats and CubeSats until it is safe to release them.[i] One such adapter is the Evolved Expendable Launch Vehicle (EELV) Secondary Payload Adapter ring, or ESPA ring, which typically sits beneath a primary payload in the launch stack to provide the additional volume and attachment points for rideshare passengers. Due to the large volume available within the ring for avionics and other subsystems, this adapter is particularly well suited for use as a final stage or spacecraft structure for longer duration missions.

EELV Secondary Payload Adapter

The ESPA Ring was designed to use excess launch capacity on EELV medium-class launch vehicles. The ring is a multi-payload adapter for large primary spacecraft (up to 9072 kg/20,000 lb) and six auxiliary spacecraft (up to 180 kg/400 lb). A ring variant called ESPA Grande can accommodate up to five auxiliary spacecraft up to 318 kg/700 lb. The ESPA mounts directly to the launch vehicle upper stage, below the primary spacecraft. Stacked ESPA configurations are also possible.

The maiden flight of the ESPA ring was in March 2007 for the STP-1 mission, pictured in Figure 1. Further ESPA options have been developed to offer varying port configurations, ring heights, and increased auxiliary spacecraft carrying capability. The first NASA mission to utilize an ESPA, Lunar Reconnaissance Orbiter (LRO)/Lunar Crater Observation and Sensing Satellite (LCROSS), launched in June 2009. The first ESPA Grande mission occurred in July 2014, with the launch of the first group of satellites for the ORBCOMM Generation 2 constellation. In 2016, Spaceflight Inc. will launch SHERPA, a five-port ESPA Grande ring, with 87 satellites.[ii]

Figure 1: STP-1 Payload Stack

Building on these successes, the OMV can offer the same flexibility in launch that a standard ESPA ring mission provides, with the added benefit of propulsive capabilities for orbit optimization.

ESPA-based Missions

Several ESPA-based missions have shown the feasibility of ESPA as the structure for an independent mission.

LCROSS—The Lunar Crater Observation and Sensing Satellite flew as a secondary mission with the LRO spacecraft. The mission was also the first use of the ESPA Ring with on-board propulsion. The spacecraft acted as a "shepherding" craft for the Centaur stage that impacted the lunar surface; it imaged the impact with on-board sensors, then followed it into the lunar surface as a secondary impactor.

DSX—The AFRL Demonstration and Science Experiments (DSX) mission will use a four-port ESPA as the hub of a free flyer in Medium Earth Orbit. The platform is currently ready for flight at Kirtland AFB and awaiting launch on the STP-2 Mission on Falcon Heavy.

EAGLE—The U.S. Air Force is developing the ESPA Augmented Geostationary Laboratory Experiment (EAGLE) based on a modified six-port ESPA. This propulsive ESPA platform will be capable of operating in GEO, GTO, and LEO.

SoftRide Vibration Isolation

The launch of an ESPA can also be coordinated with the use of SoftRide vibration isolation to provide individual satellites or the entire launch stack with lower environmental loads. The ORBCOMM Generation 2 (OG2) integrated payload stack, with satellites by Sierra Nevada Corporation (SNC), employed a SoftRide system at the base of the double-ESPA dispenser, as shown in Figure 2.[iii] Six OG2 satellites were launched using this dispenser in July 2014 from Cape Canaveral; eleven more OG2s will launch on a dispenser consisting of three ESPA Grande rings with SoftRide in late 2015.

Figure 2: Stacked ESPAs with SoftRide for OG2 Mission 1 (Photo credit: ORBCOMM)

The OG2 Dispenser SoftRide system has design heritage from the SoftRide family of whole-spacecraft vibration isolation systems. Whole-spacecraft isolation has been developed to attenuate dynamic loads for launch vehicles ranging from Minotaur 1 to Delta IV Heavy, and it has been successfully flown on over 35 missions.

Vibration isolation systems work by adding compliance between a base structure such as a launch vehicle and the payload,[4] as shown in Figure 3. The isolator has low relative stiffness compared to the base structure and payload, and a precise amount of structural damping. SoftRide isolator stiffness is designed to result in a payload isolation frequency that attenuates dynamic loads from the launch vehicle interface. Isolator damping reduces payload response at the isolation frequency.

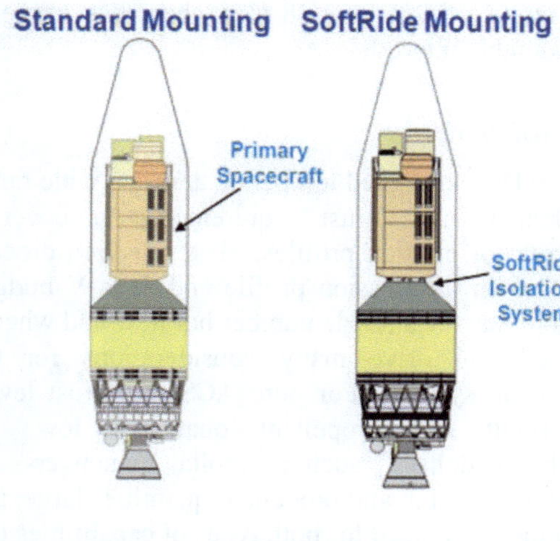

Figure 3: SoftRide Installation at Spacecraft Interface

The isolator must allow relative motion between the vibrating base structure and the payload at the isolation frequency, which is referred to as the isolator stroke. The isolation system effectively works as a low-pass filter and attenuates vibration energy above the isolation frequency. The reduction in the time domain can be substantial, as shown in the flight data reproduced in Figure 4.

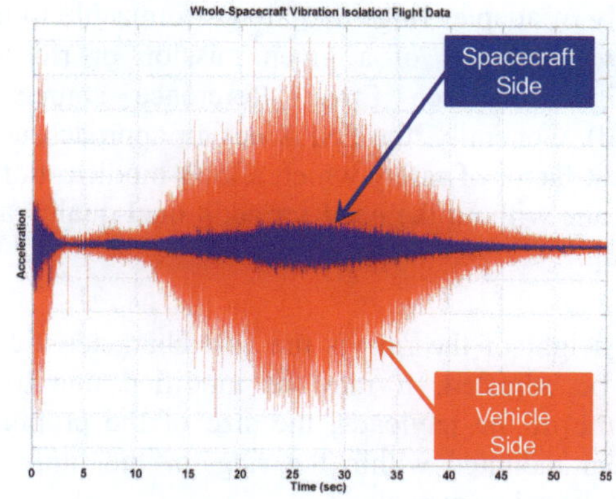

Figure 4: SoftRide Flight Acceleration Data

MODULARITY AND SCALABILITY OF THE OMV

One of the most valuable features of the OMV is the scalable nature of the system. The structure, power and propulsion subsystems can be expanded to meet the needs of more demanding mission profiles while maintaining the same core avionics.

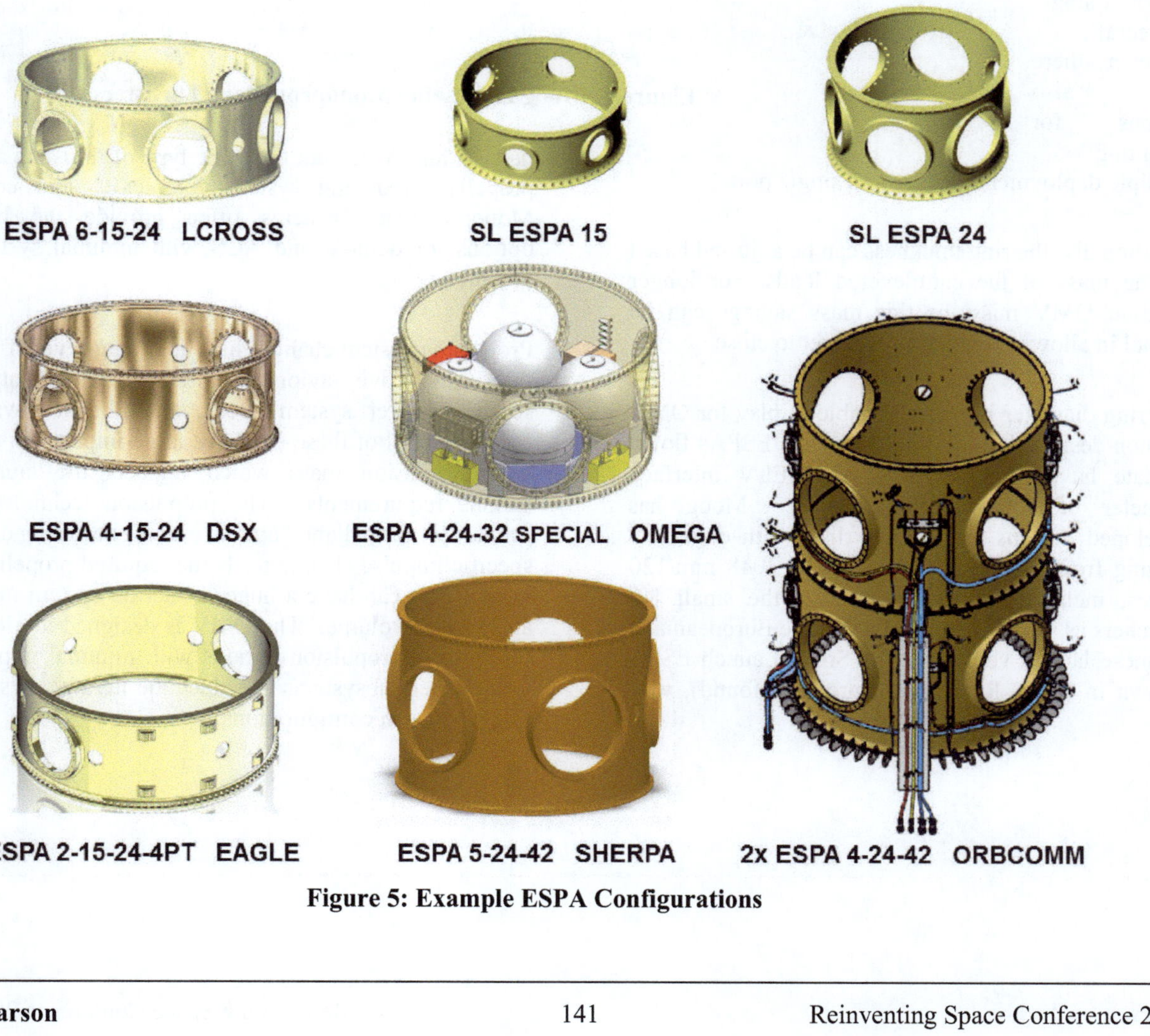

Figure 5: Example ESPA Configurations

Structure Options

The ESPA product line has been developed into a family of adapter rings that are customizable to meet the requirements of a given mission or rideshare configuration (see Error: Reference source not found). Generally, the ESPA design approach uses a no-test factor of safety which allows modifications to be made without the need for additional qualification testing.

The height of the ring is the first characteristic that can be adjusted. It may be modified due to the volume of the payloads, the size of the propulsion system mounted within the ring, or the limitations from the primary payload.

The option to use standard or custom ESPA ports, external brackets that are configurable for a specific mission design, and/or internal mounting features is another feature that contributes to the flexibility of this structure. Typically, each port is used to launch and deploy a single spacecraft, however there are also options for mounting multiple deployment devices to a single port.

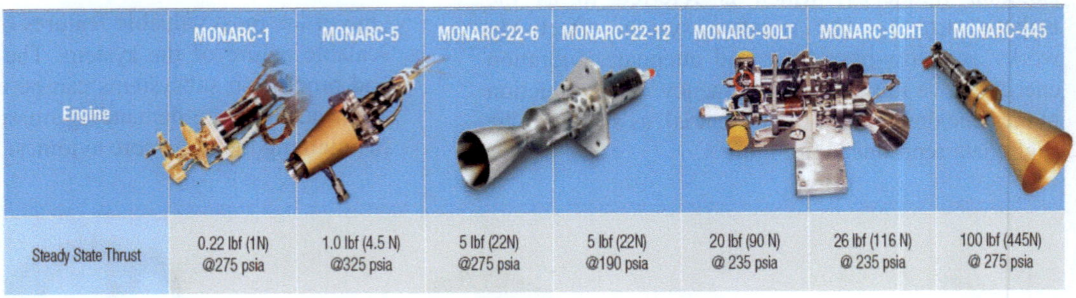

Figure 6: Moog Hydrazine Monopropellant Thrusters

Additionally, the ring thickness can be adjusted based on the mass of the cantilevered loads. For longer duration OMV missions this mass savings can be crucial in allowing the delta-v budget to close.

The ring diameter is also a variable in play for OMV mission design, despite the fact that all ESPAs flown to date have bolt circles at the EELV interface diameter of 1575 mm/62 inches. Moog has developed designs for adapter rings with diameters ranging from 800 mm/31.5 inches to 3048 mm/120 inches, including configurations for the small US launchers as well as the small and large European and Japanese launch vehicles. The Small Launch ESPA (shown in Error: Reference source not found), with bolt circle of 986 mm/38.81 inches, has attracted considerable interest.

Propulsion System

The OMV is designed to operate across a wide range of delta-V and thrust requirements to cover a spectrum of mission profiles. The selection process begins with the mission profile and delta-V budget. Delta-V may be a single number but how and when it is used can drive many considerations for the propulsion system. For pure ACS the thrust levels are small and propellant quantities low, for translation delta-V, such as orbital maneuvers, the thrust can be high and propellant quantities large, and often there is a need for both types of capabilities that may drive other propellant considerations. Once the translational delta-V is established and sized, the ACS requirements are included. In some instances a system can be sized to handle both the translational delta-V and ACS with the same type of thrusters and propellant reducing system complexity and cost. Monopropellant systems often provide the best options for delta-V and ACS with minimal system complexity.

Propulsion system changes have a system-level effect that can drive avionics, attitude determination sensors, power system needs, and almost always structure. All of these can have large impacts on the overall mission mass which impacts the launch vehicle requirements. The propulsion technology, primarily propellant options and corresponding specific impulse (I_{sp}), impacts the required propellant mass. This can have a huge impact on system mass and system volume. The OMV is designed to allow for flexible propulsion options with minimal impact to the other subsystems to reduce the iterative design effort between configurations.

For any appreciable delta-V maneuvers, the overall thrust size and efficiency requirements drive the system towards chemical propulsion over a simple cold gas system. Even with chemical propulsion there are varying levels of complexity and risk such as Hydrazine monopropellant, Green monopropellants, and storable bipropellants. In relatively extreme cases the efficiency requirements are such that high efficiency electric propulsion is required. Hybrid chemical/electric propulsion systems are sometimes used to benefit from both technologies.

Many OMV applications have been baselined with a Hydrazine monopropellant system that can easily be used in blowdown or pressurized mode with the simple addition of the pressurization system. The available thrusters provide a wide range of thrust levels from 1 N to 445 N (see Error: Reference source not found) with the larger thrusters used for translational delta-V and the smaller ones for ACS. A cluster of larger thrusters can be used to provide large thrust, small thrust, or some ACS through off-pulsing of engines. The components and architecture are selected to be converted to the green monopropellant LMP-103S if required by the mission. This propellant launched on the PRISMA mission in 2010 and to date is the only one of the new class of green monopropellants to have flight heritage. Moog worked closely with ECAPS providing many of the components and building the complete Hydrazine system that was used to validate the LMP-103S performance in flight. ECAPS has a similar selection of thrusters that provide relatively simple conversion without major GN&C changes. The only system change is the thrust chamber itself as ECAPS and Moog use the same valves and similar catalyst bed heaters. The titanium and stainless steel components are compatible with newer green monopropellants in development that offer higher Isp and a more dense liquid allowing for a greater mass in the same size tanks. This ensures future growth capability with the OMV design.

Figure 5 shows a typical propulsion system block diagram using six 60 liter 19" titanium tanks, six RCS thrusters (for 3-DOF), four Translational thrusters (for primary delta-V), and a pressurization system to maximize the system propellant capacity and performance.

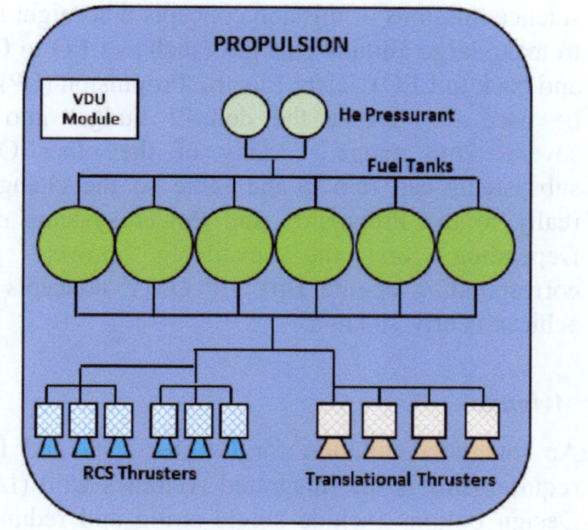

Figure 5: OMV Propulsion System Block Diagram

The pressurization system can be easily removed depending on the mission requirements. To demonstrate flexibility, the 19" titanium tanks can be replaced with 18" aluminum rolling metal tanks that provide a lower cost option that are also demisable [for low-earth orbit (LEO) missions] or other tank options such as four 22" tanks fit within the standard ESPA dimensions. The tanks can also be replaced with four larger 23" diameter cylindrical tanks increasing the gross propellant capacity by 30% with only a modification to the composite support deck and a taller ESPA (such as the ESPA Grande). The translational thrusters can easily have the quantity adjusted (or even removed) and the thrust level adjusted depending on the mission needs (the baseline configuration can deliver 625 N thrust which is greater than large GEO spacecraft typically require). The propellant tank capacity and propellant type allow for a wide range of performance characteristics.

Monopropellant propulsion provides a capability for a wide range of mission scenarios, but depending on payload mass and required delta-V a practical limit is reached. The OMV is flexible enough to accommodate a higher performance "dual mode" propulsion system that increases the performance for the delta-V engine while maintaining the same ACS engines. This provides parts and design commonality with a wide variety of OMV options.

Chemical propulsion does becoming limiting for high delta-V missions, particularly any interplanetary science missions or mission concepts that might need to make large altitude changes (such as LEO to GEO and back to LEO). Here Electric Propulsion (EP) can be used to increase the delta-V budget into the several km/s range. Many of the other OMV subsystems can remain the same so the change is really to the Propulsion and Power systems only. Depending on the available power (and corresponding engine Isp), EP OMV concepts can achieve nearly 10 km/s!

Avionics

An integrated solution for C&DH, EPS and POD requirements is the Integrated Avionics Unit (IAU). Design options include single string and redundant systems. All avionics leverage heritage systems with tailoring and/or new designs as needed to meet mission requirements. An example IAU is shown in Figure 6.

Typical IAU functionality consists of:

- <5 kg
- <32W at 28V
- 750 W Solar Array Input
- 90 power switches
- 32-Ch RS-422
- MIL-STD-1553 BC/RT
- High Speed LVDS I/O
- 12-bit A/D
- 112 Analog Inputs
- 16 MB NVRAM
- 512 MB DDR RAM
- >266 MIPS/266 MFLOPS CPU
- CCSDS Cmd/Tlm Support

Figure 6: Integrated Avionics Unit

Redundant IAU options meet specific mission requirements by tailoring existing IAU design and board heritage. The tailored Redundancy Management Unit (RMU) design accommodates hot, warm, or cold sparing in an A/B side implementation. Features include scheduled "wake up" for verifying B-side functionality and various watch dog functions.

CASE STUDY 1: SMALL SATELLITE CONSTELLATION ACCELERATED DEPLOYMENT

The first OMV mission concept is based on the need for an accelerated deployment of a commercial small satellite constellation.

The example constellation consists of 12 spacecraft, each with a mass of approximately 60 kg (including a hydrazine propulsion system). The final orbit for the spacecraft is 800 km with a 72° inclination in six separate orbital planes (two spacecraft per plane). The right ascension of the orbital planes are evenly separated at 30° intervals.

The need for an OMV is derived from the following challenges:

- Complete deployment in the minimum time possible to accelerate revenue generation
- Leverage rideshare opportunities to minimize launch cost

Similar on-orbit small satellite constellations, such as COSMIC (aka FORMOSAT-3), have taken up to 20 months to maneuver and disperse satellites (with limited Delta-V) into the desired orbital planes with correct relative placement, often settling for less than optimal placement[5]. This OMV scenario is representative of one of the two orbits that are chosen for the follow-on COSMIC-2 constellation[6].

A typical method for deploying such a constellation is to deploy all of the spacecraft into a lower orbit, such as 500 km, and then individually raise the spacecraft to a higher orbit after the precession of the ascending node of the orbit has reached the desired spacing (in this example, increments of 30°). Due to the low differential rate of the ascending nodes with respect to the final orbit, this method can take over a year to achieve a fully deployed constellation and realize full revenue potential.

Use of two OMV can dramatically accelerate this deployment down to a 6-month period. The spacecraft are strategically deployed from the OMVs at different times, with some spacecraft immediately leaving the vehicle and others being "ferried" to the correct position over a 6-month period. A conceptual design of this carrier, with a solar array affixed to the top of the ESPA, is shown in Figure 7.

Figure 7: Multiple OMVs on a Single Launch, Carrying 6 Spacecraft Each

CONOPS

- Two OMVs are launched and deployed in the same orbit
- OMV 1 immediately deploys 2 spacecraft, which boost themselves to 800 km – these are the reference spacecraft in the final orbit
- OMV 2 deploys two spacecraft which remain in the 500 km orbit to precess with respect to the final orbit for 90 days
- OMV 1 deploys two additional spacecraft which remain in the 500 km orbit to precess with respect to the reference spacecraft for 183 days
- Both OMVs perform delta-v maneuvers to change inclinations from 72 degrees to 67 degrees and increase their orbital precession rate with respect to the final orbit
- After 96 days, OMV 2 maneuvers back to the initial orbit to deploy 4 additional spacecraft, which boost themselves to their 800km final orbits after 1 day and 90 days, respectively
- After 162 days, OMV 1 maneuvers back to the initial orbit and deploys 2 additional spacecraft, which boost themselves to the final orbit
- After 187 days all spacecraft are positioned in the 6 separate orbits.

Delta-V Budget

The specific details in Table 1 and show the delta-v for deploying this constellation of 12 spacecraft into six equally spaced orbital planes around the earth, starting at an inclination of 72 degrees.

Table 1: OMV #1 Maneuvers

Deployments & Maneuvers	OMV Delta-V	OMV Fuel Required	Duration after Launch
Deploy S/C 1 & 2 (Individually boost to 800 km)	0 m/s	0 kg	Day 1
Deploy S/C 3 & 4	0 m/s	0 kg	Day 2
Decrease OMV Inclination to 67°	664 m/s	224.7 kg	Day 3
Increase inclination to 72°	664 m/s	171.35 kg	Day 161
Deploy S/C 5 & 6 (Individually boost to 800 km)	0 m/s	0 kg	Day 162
S/C 3 & 4 Individually boost to 800 km	0 m/s	0 kg	Day 183
TOTAL	**1328 m/s**	**396 kg**	**183 days**

CASE STUDY 2: HELIOPHYSICS SCIENCE MISSION AND TECHNOLOGY DEMONSTRATION AT THE EARTH-SUN L1 POINT

Many recent missions have brought to our attention the usefulness of the Earth-Sun Lagrange points for science and space weather missions.

Although many of the large, high priority science missions, such as ACE and DSCOVR, have bought their own launch vehicle in order to minimize the propulsion requirements of the spacecraft, reaching a Lagrange point orbit is feasible from an initial launch into a Geo Transfer Orbit (GTO). Due to the large number of commercial GEO spacecraft launched each year (upwards of 25 spacecraft), there are many opportunities for secondary payloads to take advantage of in the future[7].

This second OMV concept describes a mission that Moog has previously teamed with multiple NASA field sites to develop and mature.[8] In this scenario the OMV is launched as a rideshare passenger on a launch to Geostationary Transfer Orbit (GTO) and serves as a ferry to the first earth-sun Lagrange point (L1) for a technology demonstration payload and as a platform for a 5-year sun-staring science mission.

Table 2: OMV #2 Maneuvers

Deployments & Maneuvers	OMV Delta-V	OMV Fuel Required	Duration after Launch
Deploy S/C 7 & 8	0 m/s	0 kg	Day 1
Decrease inclination to 67°	664 m/s	253.2kg	Day 2
S/C 7 & 8 Individually boost to 800 km	0 m/s	0 kg	Day 91
Increase inclination to 72°	664 m/s	193.1 kg	Day 96
Deploy S/C 9 & 10 (Individually boost to 800 km)	0 m/s	0 kg	Day 97
Deploy S/C 11 & 12	0 m/s	0 kg	Day 97
S/C 11 & 12 Individually boost to 800 km	0 m/s	0 kg	Day 187
TOTAL	**1328 m/s**	**446.28 kg**	**187 days**

CONOPS

- Secondary launch into GTO below a commercial GEO spacecraft
- Three large burns are completed to raise the orbit and send the OMV into the L1 transfer orbit.
- Halo orbit insertion occurs approximately 3 months later.
- The technology demonstration payload is deployed from the OMV.
- The operational, sun-staring mission payloads are commissioned to allow the science mission to begin.

OMV Tailoring

A longer duration mission outside of LEO required three primary modifications to the OMV baseline design. These included the communications system, the redundancy of the avionics, and the power system.

At the Earth-Sun L1 point, the Deep Space Network (DSN) was baselined for ground system operations. A system similar to the one flown on the LADEE mission was chosen, including a Spacemicro transponder, two low-gain S-band antennas and 1 medium-gain S-Band antenna.

A redundant avionics system was required to meet the 5-year mission duration

Unlike an OMV operating in LEO, this mission is sun-staring, and therefore the battery sizing was based upon the worst-case eclipse length and delta-V burn durations that occurred prior to reaching the final operational orbit. Ultimately, two deployed solar arrays were sized to meet the requirements and stow during launch.

Delta-V Budget

A summary of the maneuvers and corresponding propellant requirements are shown in Table 3.

Table 3: Delta-V Budget for L1 OMV Mission

Deployments & Maneuvers	OMV Delta-V	OMV Fuel Required	Duration after Launch
Perigee Burn 1	250 m/s	91.7 kg	Days 1-14
Perigee Burn 2	250 m/s	82.3 kg	
L1 Transfer Orbit Injection Maneuver	250 m/s	73.9 kg	
Trajectory Correction	10 m/s	2.8 kg	
L1 Halo Orbit Insertion	50 m/s	13.8 kg	3 months
Stationkeeping (15 m/s per year)	75 m/s	20.1 kg	5 years
TOTAL	**885 m/s**	**284.6 kg**	

CASE STUDY 3: SPACECRAFT TUG FOR DELIVERY OF SMALL SATELLITES TO LOW LUNAR ORBIT (LLO)

The final scenario outlines a mission to ferry multiple small payloads to the moon via an OMV deployed in GTO. The announcement of NASA's Centennial Challenges Program for "CubeSat Lunar Propulsion and Communications" and "CubeSat Deep Space Communications" demonstrates the interest from NASA's Space Technology Mission Directorate in seeing new technologies demonstrated in deep space.

There are many reasons an OMV provides a useful capability for smaller spacecraft to leverage for lunar missions:

- Limited launch opportunities for small spacecraft to find a ride to lunar orbit
- Large delta-V required to reach lunar orbit
- Communications systems capable of transmitting high data rates back to Earth can use a significant portion of a small satellite's power and mass budget.

  The baseline mission concept described here assumes a total payload carrying capability of 100 kg within a single FANTM-RiDE small spacecraft dispenser module (the gray box shown in Figure 8).

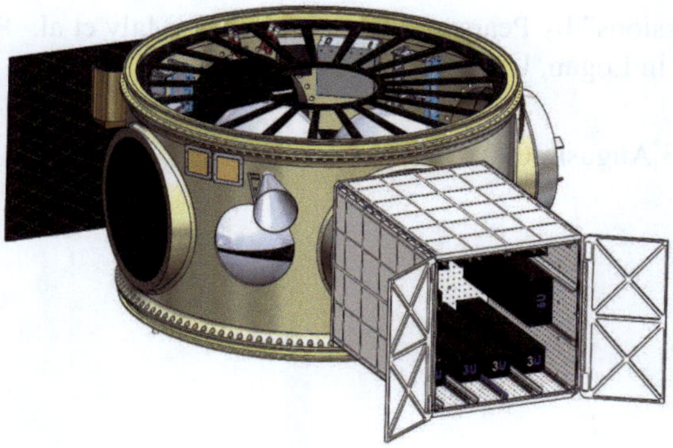

Figure 8: Lunar OMV Configuration

CONOPS

- Secondary launch into GTO below a commercial GEO spacecraft
- Three perigee burns are completed to raise the orbit apogee to the earth-moon distance
- A lunar capture burn is completed to achieve a 200 km lunar orbit
- Multiple small spacecraft (100 kg Payload capacity) are deployed in Low Lunar Orbit (LLO)
- OMV acts as a communication relay to Earth

OMV Tailoring

Similar to the requirements of the previous deep space mission, the OMV must have upgraded avionics and communications hardware to operate at the moon as compared to a LEO mission.

Delta-V Budget

The OMV lunar transportation scenario is based on a delta-v budget originally designed for a proposed ESA mission, MORO.[9] The total transfer time is 8 days, as shown in Table 4 .

Table 4: Lunar OMV Delta-V Budget

Deployments & Maneuvers	OMV Delta-V	OMV Fuel Required	Duration after Launch
Initial GTO Orbit	0 m/s	0 kg	
Perigee Burn (1)	240 m/s	102.2 kg	Day 1
Perigee Burn (2)	240 m/s	92.7 kg	Day 2
Perigee Burn (3)	240 m/s	84.1 kg	Day 4
Mid-course Maneuver	50 m/s	16.5 kg	
Lunar Capture (at 200 km)	810 m/s	225.3 kg	Day 8
TOTAL	**1580 m/s**	**520.8 kg**	**8 days**

CONCLUSION

The OMV provides small satellites the capability to reap full benefit from their rapidly improving performance via operation in optimal not opportunity orbit while still saving launch cost through rideshare. The OMV provides the capability to perform as a pure mechanical adapter and deployer as well as a tug or freeflying payload host. Importantly, these capabilities are not mutually exclusive that is opening up new potential low-cost mission architectures.

[i] Maly, J.R., J.M. Howat, "Launch Infrastructure for Small Satellite Rideshare," Proceedings of the 4S Symposium (Small Satellites Systems and Services), Porto Petro, Majorca, Spain, May, 2014.

[ii] Kelly, K., M. Elson, J. Andrews, "Deploying 87 Satellites in One Launch: Design Trades Completed for the 2015 SHERPA Flight Hardware," Paper SSC15-II-2 presented at the August 2015 Small Satellite Conference in Logan, Utah.

[iii] Maly, J.R., et al. "ESPA Satellite Dispenser for ORBCOMM Generation 2," Paper SSC13-V-3 presented at the August 2013 Small Satellite Conference in Logan, Utah.

[4] Johnson, C.D., P.S. Wilke, S.C. Pendleton, "SoftRide Vibration and Shock Isolation Systems that Protect Spacecraft from Launch Dynamic Environments," Proceedings of the 38th Aerospace Mechanisms Symposium, Langley Research Center, Virginia, May, 2006.

[5] Fong, Chen-Joe, et al. "Constellation Deployment for the FORMOSAT-3/COSMIC Mission." IEEE Transactions on Geoscience and Remote Sensing. November 2008.TBD

[6] Cook, Kendra. "The Future of GNSS-RO for Global Weather Monitoring and Prediction – A COSMIC-2/FORMOSAT-7 Program Status Update." Eighth FORMOSAT-3/COSMIC Data Users' Workshop. October 2, 2014.

[7] Silverstein, Sam. "Satellite Builders Innovate as Telecom Industry Rockets Ahead." Via Satellite. 17 July 2014.

[8] "Rideshare and the OMV: the Key to Low-cost Lagrange-point missions" by Pearson, Stender, Loghry, Maly et al. Paper SSC15-II-5 presented at the August 2015 Small Satellite Conference in Logan, Utah.

[9] Biesbroek, R. and Janin, G. "Ways to the Moon?" ESA Bulletin 103 August 2000.

RIS-2015-13

"Gateway Earth" – Low Cost Access to Interplanetary Space

Derek Webber, FBIS, Senior Member AIAA

Spaceport Associates

PO Box 614, Damariscotta, Maine 04543, USA 207-563-2660

DWspace@aol.com, www.SpaceportAssociates.com

INTRODUCTION

The paper describes some aspects of a proposed space development whose aim is to dramatically reduce the cost of missions, both crewed and robotic, to anywhere in the solar system. The concept uses a funding mechanism combining governmental and commercial financing working in new ways together. The aim is to enable interplanetary space transportation and exploration within realistic funding projections.

The basic architecture uses a combined governmental/commercial space station ("Gateway Earth") located in geostationary orbit (near the edge of Earth's gravity well) and a logistical supply chain of reusable tugs going regularly between LEO (Low Earth Orbit) and GEO (Geostationary Orbit) and back. Part of the commercial station would be a GEO space tourist hotel. Other commercial uses of the Gateway complex would include servicing of GEO communications satellites. "Gateway Earth", the station at the edge of interplanetary space, would consequently be partly funded by space tourism revenues, as would also be the operation of the tugs. It's a way of exploring the solar system by using space tourism revenues to augment government funding.

The paper addresses some design implications of including an industrial scale 3-D manufacturing facility at the Gateway complex, for building and assembling components of craft intended for ongoing interplanetary travel. Elements of a business case framework are presented, using market analogs for this next destination of space tourists beyond LEO, which supports the argument that the "Gateway Earth" architecture represents a way to achieve relatively low cost interplanetary travel.

SUMMARY OF RATIONALE

The case has been made many times, going back at least to von Braun and Tsiolkovsky, for why we need to develop our capabilities to access interplanetary space (one reason is that, eventually, life on Earth will become unsustainable due to a range of inevitable long term astronomical factors). The initial robotic explorers will, therefore, eventually need to be followed by craft carrying human occupants. The difficulty in practice has always been in finding ways to finance such ventures. There has been an implicit miss-match between the associated time frame of government funding for such long time-frame eventualities (as, eg, stellar evolution of our star, the Sun) in relation to the short term presidential and budget cycles. Apollo, funded at about 5% of the US GDP throughout the 1960's, was conceived and funded as a reaction to a perceived *imminent* national threat. Current NASA funding levels are nearer 0.5% of GDP, and likely to remain at that level for the foreseeable future (since that level reflects the public's reaction in opinion polls). The way forward, it would seem, is to use the notion of economic development in space to augment the public funding contribution. Are there enough commercial markets in space to make this feasible? The author believes the answer is in the affirmative, provided enough time is allowed for their development. The first such markets are going to be in space tourism. Then there could be satellite servicing and refueling markets. And ultimately asteroid mining and other uses of the unlimited resources of space. References 1-9 provide further documentation on this joint governmental/private funding concept.

S. Hatton (ed.), *Proceedings of the 13th Reinventing Space Conference*,
https://doi.org/10.1007/978-3-319-32817-1_15

Reduced to its simplest form, the issue is how to regularly, and at low cost, get from the surface of the Earth, out of Earth's gravity well, and across the relatively flat geopotential plateau to the edge of the gravity wells of neighboring planets. Reusability becomes an essential element of any proposed architectural solution. Figure 1 provides an overview of the proposed architecture. Steps 3, 4, 5 are in fact interrelated and carried out in parallel. Although the Figure represents the near-Earth neighborhood, the approach is equally applicable to any interplanetary destination at any point across the relatively flat geopotential plateau.

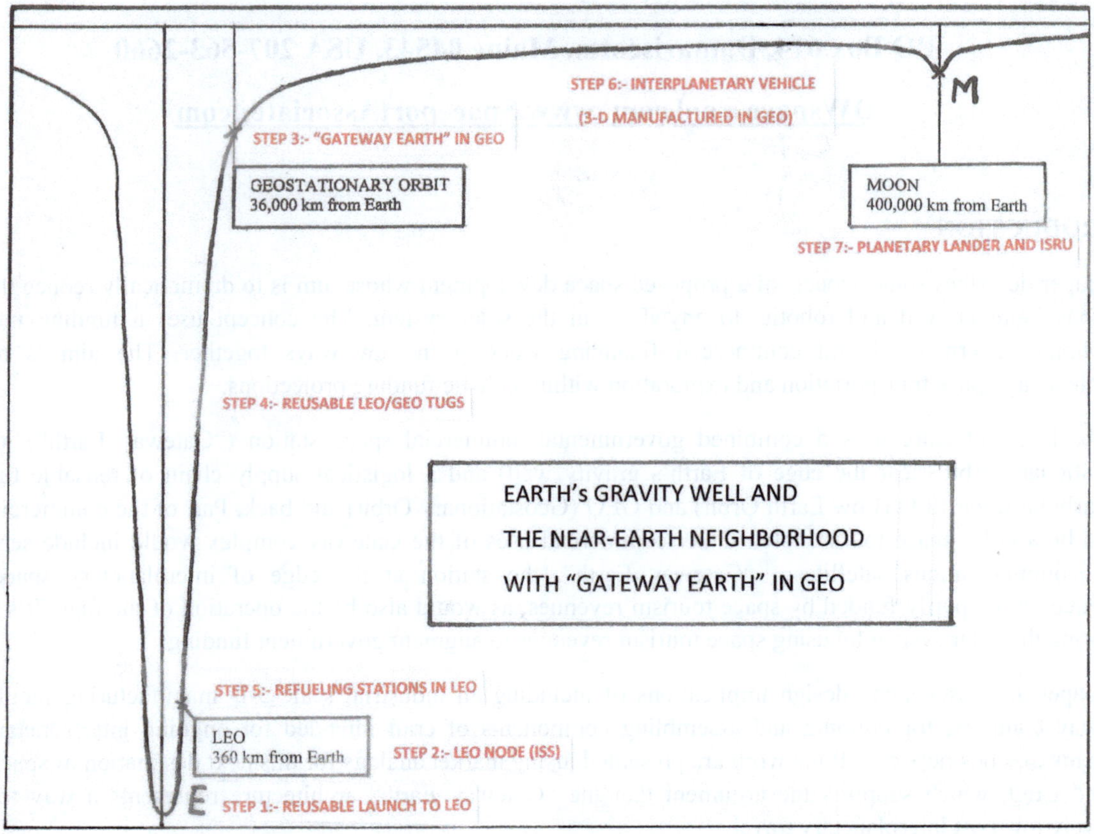

Fig 1. Overall 7-Step Process for Regular Interplanetary Travel (Credit: Author)

The proposed solution is a 7 stage infrastructure (Ref 9), a key focal point of which (Step 3) is to install "Gateway Earth" more or less near the rim of Earth's gravity well. This then becomes the effective start and end point for interplanetary missions. At the Gateway, the interplanetary vehicles are built and assembled (Step 6). The Gateway is also the place to where the vehicles return after their journeys. Therefore, the interplanetary vehicles themselves do not need to be built to withstand the rigors of a launch from Earth's surface, either from the point of view of structural strength or aerodynamics. And furthermore, they do not need to possess the weighty thermal control system associated with atmospheric re-entry, or the powerful motors needed to provide the energy for getting from the Earth to GEO. Another way of thinking from the "gravity well" perspective is to view the "getting to the edge" part of the operation as "powered flight", and traversing the vast interplanetary spaces as much more like "gliding". Much of the structure for the interplanetary "glider" craft can therefore be built using additive manufacturing techniques. The returning interplanetary space travelers of the future will consider "Gateway Earth" as "home" on coming back in an interplanetary Aldrin cycler from, say, the vicinity of Mars. Once they have docked at the Gateway they may return to Earth's surface using the regular tug service introduced as part of the proposed architecture (Step 4). How can the tugs be "regular'? This is because their main purpose is to shuttle back and forth between LEO and GEO to take space tourists, and their associated supplies, to their GEO space hotel. The tugs are in fact a business operation in their own right. They are refueled, and restocked, in LEO each time before they make their next trip to GEO. So a LEO refueling component is an important part of the logistical infrastructure (Step 5).

This might be provided as a separate commercial venture, or partly supported by government funding. There is a need for a continuing LEO station, such as the ISS, as the LEO node of the architecture (Step 2). Space travelers will take two effective shuttle rides to reach their interplanetary craft; first they ride up to the ISS node – then they transfer to the LEO-GEO tug. And of course the process is repeated on returning from their interplanetary trip; first they dock at "Gateway Earth", then they take the tug down to LEO, finally they take the last leg to the surface using the traditional aero-thermal braking approach for atmospheric re-entry. Only the craft designed for providing the service back and forward between Earth's surface and LEO (Step 1) needs to be designed to handle the demands of transferring each way through our atmosphere, and of course such vehicles already exist (eg Soyuz, Dragon, etc). Ongoing developments at this stage for the Earth to LEO craft are approaching the achievement of reusability, but are not discussed further in this paper.

In summary, therefore, what is proposed is a "Gateway Earth" station placed in the GEO orbit. The station itself is partly funded by space tourism and other commercial businesses; there is a regular tug service between LEO and GEO – again largely funded by commercial space businesses; and this Gateway may be used by governmental astronauts (NASA and other) both as a place to assemble interplanetary craft, and as a starting and end point for subsequent journeys across the solar system gravity plateau. Almost the entire system is reusable. The governmental space program budgets only need to provide funding for *part* of the "Gateway Earth" complex, and for paying the operators for "taxi rides" both from Earth to LEO and from LEO to GEO. The capital costs for government are therefore limited, and this, together with the consequential lightweight/low energy design of the interplanetary spacecraft, will thereby dramatically reduce the costs compared with missions performed in the "traditional" ways. Of course, a major challenge will be the coordination between the commercial and governmental aspects of the architecture, and this is discussed later in the paper. There is a need for a flexible, rather than a monolithic planning process which is perhaps more familiar in traditional space programs. In particular, for the approach to succeed, there must be an acceptance of a timeline that recognizes the need to allow the space tourism and other commercial businesses to develop and generate revenues.

This section has so far provided an overview of 6 separate elements of the proposed "Gateway Earth" – based architecture for interplanetary travel. The first two (Earth to LEO vehicles, and LEO space station node) exist already. The specific design features of the LEO orbiting refueling depots, and of the interplanetary craft, are *not* the subject of this paper. We here only concentrate on the LEO to GEO tugs, and the key multi-functional "Gateway Earth" complex itself. There is also a seventh element (Step 7), which would probably vary for each destination planetary surface, and that comprises the lander and associated ISRU technologies. This seventh step is also not addressed further in this paper, but is here simply noted for completeness. This overall 7-step architecture makes possible relatively low cost access to the Moon, the asteroids, Mars or any other interplanetary destination (Ref 9).

"GATEWAY EARTH" DESIGN ASPECTS

This complex has been described in earlier papers (eg Ref 2,3). Its main feature, apart from its location in GEO, is that it consists of a combination of government and commercial elements, and will *not* have a pre-defined shape. There will be a need to coordinate the chosen orbital slot for "Gateway Earth" with the ITU organization, which is the relevant international authorizing and regulating authority.

The ***commercial*** part will grow only as the demand for services grows. It is assumed that the commercial modules will be Bigelow-type, or maybe Galactic Suites - type structures. The Bigelow modules have already been demonstrated in space with orbiting prototypes, and one is soon to be added as part of a NASA contract to augment the operating volume of the LEO ISS facility (Ref 19). Some of the modules will be operated as a GEO space tourist hotel, possibly capable of supporting up to about 6 space tourists and a commercial astronaut "hotel keeper". Others will provide a base of operations for a potential GEO communications satellite servicing business. One might even be a kind of "general store" or eatery. The funding of the commercial part of the complex will be provided by the commercial operators, only as they are able, and as the markets develop. Tourists at this new space hotel in GEO will have views of an almost entire hemisphere of the Earth, while also being able to observe the operations of the government astronauts who will be constructing the interplanetary spacecraft on site (Fig 3, 4). Although not addressed further in this paper, it is a possibility that space tourists in GEO may wish to pay more to use a tug in which to travel around the complete geostationary orbit, (which requires very little fuel to achieve) and thereby would see all parts of the Earth below, both in day and night.

The ***government*** part of the complex is intended to be very small compared with the size of the existing ISS facility in LEO. It will be designed *a priori* only to provide a safe haven for astronauts going onward to, and returning from, interplanetary destinations, and will initially be used as a base for the construction crews building and assembling the interplanetary spacecraft. It will carry a manipulating arm to unload cargoes and also assist in construction. It is *not*, however, intended for the conduct of science experiments, which will remain the preserve of the ISS laboratory in LEO. It will, though, contain the solar arrays for power generation, waste management, docking ports, storage facilities for oxygen, water, fuel and printer supplies, and the 3-D manufacturing facility. It will effectively become "the 3-D printer at the edge of the universe". The traffic control inbound and outbound from the entire "Gateway Earth" complex could also be managed by the government astronauts in this part of the Gateway. This government part of the complex will be funded by the respective space agencies who are going to use it. The construction will need to be phased to align with national space budget cycles. Clearly, an initially minimalist government part of the complex will make its funding more probable. In any case, none of this "Gateway Earth" facility will ever be returning to Earth, and it will likely grow over time as the need arises, and as the budgets make possible. So it will make sense to initially consider only enough infrastructure to support a small number of government astronauts in preparing initial agency budgets.

3-D manufacturing, otherwise known as additive manufacturing, is a newly emerging technology with great potential for space operations. Some parts for lunar rovers being built to attempt the Google Lunar XPRIZE challenge have already been constructed in this way. With a 3-D printer in orbit, for example, it becomes possible to manufacture on site replacement parts for failed systems, and tools, by uploading the design details. This will obviate the need to store pre-constructed spares inventory. And an industrial-sized facility at the "Gateway Earth" complex will be capable of manufacturing the main elements of interplanetary craft. Such craft do not need to be as complex or robust as a similar vehicle manufactured on Earth. They may even in the case of cargo variants use a "lightsail" for propulsion (Ref 15, 16, 17). The technology has already been demonstrated that quite large and complex shapes can be produced out of a variety of materials, and the resulting parts have the requisite structural integrity and strength. There have already been demonstrations of this capability in orbit (Ref 15), particularly using the Made in Space, Inc. equipment installed and operated on the ISS in 2014. Engine parts have been constructed of copper by Aerojet Rocketdyne. A joint New Zealand/US firm named Rocket Lab has manufactured a prototype engine largely using additive manufacturing techniques. Lockheed Martin is working on building propellant tanks using 3-D printing technology. NASA designers at Marshall Space Flight Center are also working on using 3-D printing, and NASA is even exploring converting plastic waste into high quality 3D printer filament for construction purposes. Fig 2 shows an example of a large scale industrial 3-D manufacturing capability.

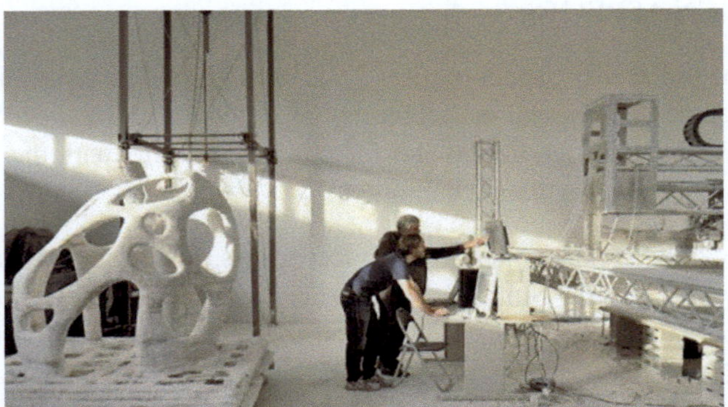

Fig 2. Large Scale Industrial 3-D printing of Building Elements (credit: Edyta Zwirecka)

For the purposes of the "Gateway Earth" arrangements, we need a printer setup which can operate using a range of different raw material supplies. The printer will need to be big enough to effectively extrude structural elements of an interplanetary craft. Industrial sized 3-D printers are emerging, and will need to be demonstrated in space as part of the checkout of the proposed architecture. With the availability of the on-orbit 3-D manufacturing equipment, the supply logistics from Earth are considerably simplified, and storage at the complex may thus largely be relegated to a matter of bulk raw material. Such material needs no special handling during launch from Earth. Fig 3 is an impression by artist Phil Smith, showing the dual nature of the "Gateway Earth" complex, with the commercial modules (including the GEO space hotel) nearest the viewer, and the governmental part including trusses and robotic arms for assembly work.

Fig 3. "Gateway Earth" in GEO, with Arrival of a LEO/GEO Tug (Credit: Phil Smith)

Fig 4 is another perspective, also by artist Phil Smith, of the "Gateway Earth" complex in GEO, with space tourists at the GEO space hotel observing the Earth behind the government modules, and the activities of governmental astronauts as they manufacture parts for an interplanetary craft using the Gateway's 3-D manufacturing facilities and then assemble them for onward exploration. A tug is arriving with more raw material for the manufacturing facility.

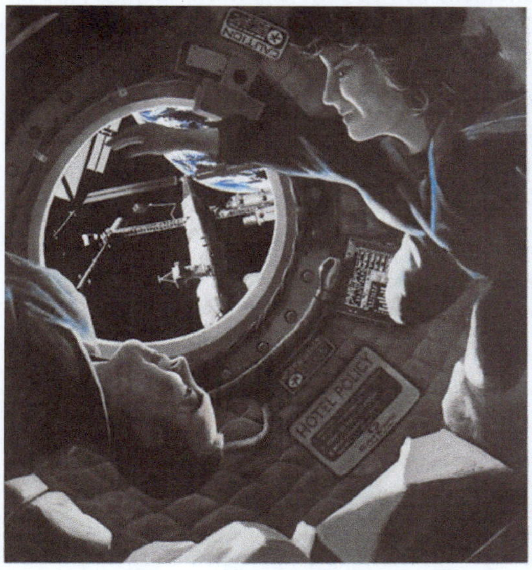

Fig 4. "Gateway Earth" Complex. Tourists Observe Construction Work in GEO (Credit: Phil Smith)

One of the main considerations in bringing a "Gateway Earth" facility into operation will be a joint working agreement between government space agencies and commercial operators. The main elements of such an agreement will resemble an interface agreement. This will be a heads of understanding about respective responsibilities, about the specification of interconnections, power levels, safety procedures, storage, training requirements, etc. The agreement will also need to specify the prices that the governmental users will pay for the ongoing use of the commercial LEO to GEO tugs.

DESIGN OF THE LEO-GEO TUG

The basic design of the tug has also been provided in earlier papers (eg Ref 2,3). The LEO to GEO tug does not need to ever return to Earth, and so does not need to carry any re-entry thermal controls or heatshield. Also, it will always be operating in the vacuum of space and does not therefore need to be aerodynamic (once it has been delivered into LEO). There may be more than one variant, perhaps one type for pressurized cargos and personnel transportation (including space tourists), and one for bulk cargo deliveries. For each type, the common features will be docking ports (for docking at the ISS serving as the LEO node, and at "Gateway Earth"), refueling ports (for linking up with the orbiting refueling stations in LEO (Ref 18)), the transportation space for the cargo and/or crews, propellant tanks, and a motor. For the cargo version, probably an ion motor will serve; for a crewed version a more conventional reusable in-space engine will be required. The tug could be based for example on an adaptation of the "trunk" section of the SpaceX Dragon spacecraft, or of the Orbital Cygnus transporter, or versions being developed by Lockheed Martin corporation.

What will be the cargoes that the tugs will be required to carry on a regular basis both up to the "Gateway Earth" complex, and back down again from the Gateway to the LEO node at the ISS? The human cargoes include the regular arrival and departures of the space tourists. Each group (of initially about 4 to 6) will probably be staying for one to two weeks at "Gateway Earth". The space tourists will also have a "hotel manager" commercial astronaut in support. Other human cargoes include the governmental astronauts from NASA, and other space agencies, who are arriving and departing from their tour of duty at "Gateway Earth", and who ultimately will include the crews of interplanetary missions.

There will be a mix of non-human cargoes. There will be oxygen, water, food and laundry for the humans. Also in the downward direction there will be trash. There will be a need to transport rocket and reaction control fuel. As the GEO satellite servicing business develops, then there will be spare or upgraded communications satellite modules. But probably the main bulk transport requirement will be for 3-D printer supplies, the raw materials for building the interplanetary craft. We will quite literally be "3-D printing our way to the planets". These supplies may come in pellet form or in rolls of raw material. The raw material itself might be plastic or metal or carbon. Fig 5 is a notional design by artist Phil Smith showing two versions of the GEO-LEO tug, one for crew/space tourists, and the other for cargo, including 3-D manufacturing supplies.

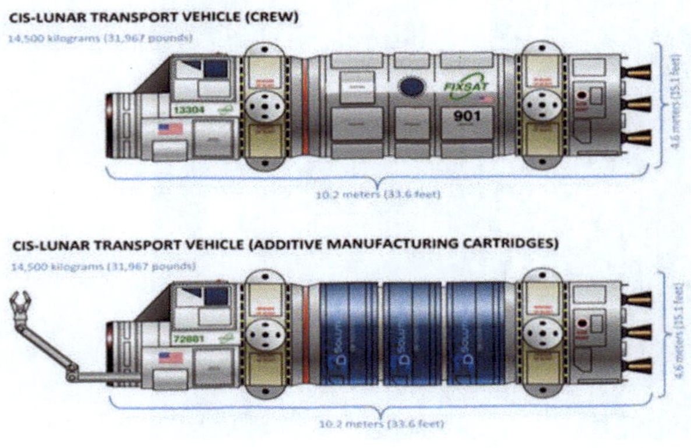

Fig 5. A Notional LEO-GEO Tug, Operated by Space Tourism Company (Credit: Phil Smith)

On arrival at the "Gateway Earth" complex, the tugs will rendezvous and dock, and the supplies will be transferred to their storage locations using the Gateway's manipulating arm. Human cargoes will, of course, enter the complex via one of the docking ports. The tugs operate in both directions, and will also carry materials and people back to Earth via transfer in LEO at the ISS node, before taking on their next load of people and supplies for returning back to GEO.

BUSINESS CASE FRAMEWORK

For this proposal to succeed, a combination of government and private industry resources, working together – towards the same aims but not necessarily, or even likely, in unison – will be needed. This is a non-traditional way of working in a number of respects, but offers great benefits if it may be achieved. It will require working agreements between, eg, NASA, other space agencies, and the members of The Commercial Spaceflight Federation.

In the case of the ***governmental*** work, this may be achieved by a judicial phasing of a series of tasks within budget projections over a sufficiently long timeframe to be politically supportable. The tasks needing to be achieved and funded by governments include (1) basic research and design development of LEO fuel depots and fuel transfers for a range of fuels, (2) further in-orbit testing of 3-D manufacturing capabilities, (3) Earth-based demos of ISRU capabilities for a range of end products, and (4) various planning and design activities – especially those related to interface specifications for power cabling, fluid transfers, access ports, and the structural elements needed for the governmental part of the "Gateway Earth" station – including (5) the costs of launching the elements into GEO, etc. The Agency planners therefore need to begin to sketch out the likely costs of these activities, and determine how they might be phased and shared between separate international agencies. It has been pointed out by Grey (Ref 27) that the difficulties of funding long-term expensive governmental space activities can be handled by accepting that the *total* costs are unknown, but that the costs can nevertheless be covered by budgeting a regular fixed *proportion* of the total budget, with the duration of the activities being the independent variable. The proportion itself would be determined by what to Congress seems acceptable. This allocation agreement will be made more palatable by a simultaneous recognition that commercial revenues are assisting in the overall funding plan, as described below.

For the ***commercial*** part of the architecture, it will be necessary to allow sufficient time for the new GEO-based commercial businesses to emerge. And a pre-requisite for that to happen is for a highly rigorous market assessment of likely demand to be performed. In particular, this is an extension of our knowledge of existing space tourism markets to a new destination. Previous results of financial evaluations of this "Gateway Earth" – based infrastructure are provided in Ref 3, where it was determined that the costs of interplanetary travel would conservatively be at least a third cheaper by using this approach. The Appendices to this paper contain the material that further addresses the market demand question. First of all there are analogs which provide some guidance on relative demand levels for different tourist attractions at increasing heights and increasing price levels. Then data is provided that is relevant to the now well-established requirement for getting from LEO to GEO in delivering communications satellites into their GEO slots. And finally an initial data set is provided that pulls together what we know about the various commercial business opportunities supporting the seven steps of the proposed "Gateway Earth" infrastructure. This work is clearly very preliminary and is based solely on desk research, and some elements remain of necessity completely unknown at this point. New statistically valid demand-based primary market research will be needed to further explore the validity of the notion that a combined governmental/commercial complex at "Gateway Earth" is viable, and therefore this need is noted in the next section. For a complete business case assessment, of course, we shall also need cost data – and this is also noted at appropriate places in the Appendices to assist future economics software modelers. As a first step, some of the most critical revenue elements have been scoped. And therefore we have been able to obtain some insight just by using the desk research described in the Appendices. What we find is that there is probably a peak market for about 150 tourists a year at "Gateway Earth", and they would generate about $4.5B revenue, assuming an optimum sized tug and hotel, both supporting 6 occupants at a time, and a ticket price of $30M to GEO.

NEXT STEPS

There are a number of identifiable next steps. As has been stated earlier, the *time frame* for instituting this "Gateway Earth" – based infrastructure must be fluid, because of historically low governmental space budgets and the need to allow time for the commercial businesses to emerge and develop their revenue streams. Furthermore, it should be acknowledged that the primary need for this capability is premised on a *very long term view* of planetary and solar development that does not require an urgent response. That having been admitted, there are nevertheless many nearer-term reasons for building the capability for regular low-cost interplanetary travel, including the beginnings of using the resources of space (such as the mineral content of asteroids) as part of Earth's economy, and the continued extension of public access to space through the development of a geostationary space tourism market segment. And there is the need to begin to develop methods to protect Earth from future catastrophic asteroidal or cometary impacts, with imperfectly known urgencies. Therefore, at the very least, we should endeavor to maintain the momentum by conducting a range of low cost, low risk, preparatory activities related to the establishing of the "Gateway Earth" infrastructure at the edge of Earth's gravity well. Listed below, in no particularly relevant order, are a number of tasks needing to be funded, performed, documented, videod and authenticated in order to convince Congress and the public that the proposed architecture is achievable. Some of them require relatively small amounts of funding, and might even be fundable via crowd-sourcing and crowd-funding approaches. Some would be the responsibility of the commercial sector, some would be in the remit of governmental space planners; some would benefit from joint activities. However it is done, we make a plea for transparency and sharing:

- Commercial operators conduct statistically relevant and valid market research into demand amongst wealthy individuals for space tourism in GEO to confirm or challenge the estimated $4B market opportunity, with results placed in the public domain (estimated $150K)
- Commercial and/or governmental space planners conduct full economic model analysis of the "Gateway Earth" infrastructure proposal, using software to compare the costs of doing space exploration via "traditional" methods, compared with doing so using the proposed infrastructure to confirm or challenge that the approach produces major cost savings, with results placed in the public domain
- Joint governmental/commercial operations continuing to develop and space-rate 3-D manufacturing facilities, leading to the extension of capabilities, the development, orbital trials and demonstrations of industrial-scale equipment
- Earth-based demos of ISRU capabilities for creating water, oxygen, rocket fuels, building bricks, solar cells, etc, from soil samples (governmental and commercial interests – eg markets for water on Earth)
- Joint governmental/commercial work to establish the terms of a Heads of Agreement (including interface specs) that could be used to regulate the joint international, governmental and commercial interests involved in setting up and operating the "Gateway Earth" infrastructure
- Further preliminary design work for the tugs and station modules, including the layout and *modus operandi* for the space hotel modules to be installed in GEO (governmental and commercial operators)
- Governmental and/or commercial providers conduct essential design work for the LEO-based propellant depots, and the demonstration in orbit of effective and safe fuel transfers
- Further work, including market research and business case analysis, is needed to explore the feasibility of a commercial geostationary communications satellite repair and refueling service, with some capabilities co-located at the "Gateway Earth" complex (commercial operators)

It would seem that crowd-sourcing and crowd-funding approaches might yield results for eg modeling the economic assessment of using a "Gateway Earth" - based manufacturing approach compared to one that launches interplanetary vehicles from the Earth, or for conducting the new statistically valid primary market research, and the author will be pursuing this approach in 2016 via an Incubator website currently being constructed for the purpose.

CONCLUSIONS

We have demonstrated some features of the "Gateway Earth" station proposed to be installed in geostationary orbit, and of the commercially designed, built and operated GEO/LEO tugs needed as a logistical supply channel, and have provided some preliminary business case analysis (suggesting a $2B - $4B market), then identified further steps required to refine the concept.

The main idea behind the "Gateway Earth" infrastructure is to use the combined resources of government and private industry to fund future interplanetary space travel and exploration via the shared use of a facility near the edge of Earth's gravity well. By so doing, the costs of interplanetary travel will be significantly reduced in comparison with more 'traditional" ways of proceeding. By its very nature, this will require a very different approach to managing future space developments and exploration, and it is an open question whether there exists enough of the combined technical and negotiation skills to enable the approach to succeed. In any event, there is an absolute *a priori* requirement for the nascent space tourism businesses (both in LEO and in suborbital experiences) to succeed in order for the revenue stream to emerge which could make the venture possible. There is another key requirement, and that is for the continued existence in LEO of a base station – which is currently considered to be the ISS, or any subsequent replacement beyond 2020 or 2024. The preliminary business case framework presented in this paper indicates that there is enough of a viable opportunity to consider the idea further (and indeed the author intends to progress this work in 2016 via an Incubator website approach currently under development). There is, moreover, short-term work enough for both governmental and commercial players to undertake, if a choice is made to pursue and refine the approach. Is $4B enough of a potential market to encourage the space tourism firms to consider supporting the concept? If so, will government agencies begin to do their part by making minor deviations to their current plans to make the "Gateway Earth" approach possible? Can both parties find ways to jointly move forward informally and explore these concepts?

Many authors over the years, and even centuries, have pointed out why the human race needs to open up interplanetary space to economic development in order to counteract depletion of resources on Earth and to provide a back-up plan for life itself in the very long term. The problem was initially one of technology. That was resolved by the Apollo program during the 1960's. The problem then became one of funding and political will, which to a certain degree were interrelated. The joint governmental/commercial "Gateway Earth" infrastructure, described in this paper, represents an attempt to address the funding issue, and incidentally provides a new focus for national governmental endeavors related to space development. Building the "Gateway Earth" complex in GEO, if this could be achieved as a joint governmental/private commercial venture, would provide a new platform for the economic continued exploration of space during the 21st century, and would thereby be a fitting contribution from the first generation of space explorers to their grandchildren and descendants.

ACKNOWLEDGEMENTS

Thanks are due to Sa'id Mosteshar for his guidance on the legal and regulatory aspects of placing "Gateway Earth" in the geostationary orbit, and also to Apollo astronauts Edgar Mitchell and Harrison Schmitt, who have both been at the putative "Gateway Earth" location in GEO, and who provided responses to queries about that site's likely acceptance by future space tourists as a destination for a space hotel. Thanks also to Phil Smith for his artwork which so effectively and dramatically captures the essence of the "Gateway Earth" concept.

REFERENCES

1. Webber, D. October 2012. "Space Tourism – Essential Step in Human Settlement of Space". IAC Naples, Italy
2. Webber, D. May 2013. "An Architecture for Survival". ISDC 2013, San Diego, California
3. Webber, D. Jan 2014. "Bridgehead – Interplanetary Travel Becomes Routine". IAA Space Exploration Conference, Washington, DC. IAA-WAS0420
4. Webber, D. March 2013. "Leaving the Cradle". *Space News*
5. Webber, D. April 2013. "Getting to the Edge". *Spaceflight Vol 55*
6. Webber, D. April 2014. "Everest, the Camps and the Sherpas". *The Space Review*
7. Webber, D. December 2012. "Inserting the "s"-word: a Modest Proposal". *The Space Review*
8. Webber, D. May 2015. "Two Small Steps for Humankind". *The Space Review*
9. Webber, D. December 2015. "Seven Steps to Space Settlement". *Spaceflight*
10. Webber, D. 2010. "The Wright Stuff – The Century of Effort Behind Your Ticket to Space". Apogee Books
11. Beard, S, Starzyk, J, Webber, D et al. October 2002. "Space Tourism Market Study". Futron/Zogby
12. Webber, D and Reifert, J. 2006. "The Adventurers' Survey", Spaceport Associates
13. Kothari, AP and Webber, D. Sept 2010. "Potential Demand for Orbital Space Tourism Opportunities Made Available via Reusable Rocket and Hypersonic Architectures", AIAA Space 2010, Anaheim, California
14. Tauri Group. 2012. "Suborbital Reusable Vehicles: a 10 Year Forecast of Market Demand". FAA and Space Florida.
15. Werner, D. March 2015. "Reimagining Satellite Construction". *Aerospace America*
16. Stetson, D. June 2015. "A Perfect Launch and a Dramatic Finish for *Lightsail*'s Test Flight". *The Planetary Report*
17. McConnell, B. May 26 2015, "A Stagecoach to the Stars". *The Space Review*
18. Bienhoff, D. May 2010. "Propellant Depots and a Reusable Cislunar Transportation Architecture". ISDC 2010
19. Crane, R. June 2014. "NASA Commissions $17.8 M Space Home Tests", CNN
20. Svitak, A. July 2015. "In Return". *Aviation Week and Space Technology.*
21. Morring, F. March 2015, "Extensible". *Aviation Week and Space Technology*
22. Mihalic, T and Gartner W.C., Editors. 2013 "Tourism and Developments – Issues and Challenges". Nova Books – Chapter 7
23. Money, S. 2014. "Here be Dragons". Apogee Books
24. Morring, F. 2014. "Orbital MRO" *Aviation Week and Space Technology*
25. Klotz, I. 2014. "NASA Planning for Mission to Mine Water on the Moon", *Space News*
26. Le Goff, T and Moreau, A. 2011. "Astrium Suborbital Spaceplane Project". IAA Conference on Private Human Access to Space, Arcachon, France.
27. Grey, J. August 2015. "Setting Arbitrary Cost, Schedule, Will Never Get People to Mars", *Space News*

APPENDICES

Primary market research is needed to assess the likely demand and revenue potential for a new space tourism destination in GEO.

Until this research is funded and conducted, we can, however, make some progress by pursuing desk research using market analogs, as developed below. We must of course not assume too high a level of confidence in the resulting findings – and this point is discussed in the course of the analysis.

- *I* *MARKET ANALOGS*

How attractive will a hotel in GEO be for space tourists? Given an equivalence of other factors, the main difference compared with a LEO hotel will be the view.

View from a LEO space tourism hotel (left) and from a GEO space tourism hotel (right) credit: NASA

Would tourists be willing to pay a sufficient price increment over and above the already high price to LEO in order to reach the GEO hotel? Though clearly an imperfect analogy, in the absence of more relevant primary market research, the following terrestrial tourist attractions provide some insight into price elasticity of demand with increasing altitude. Basic data is collected and presented with some analysis for both the CN Tower and the Tour Eiffel experiences.

1) CN Tower

The CN Tower in Toronto, Canada, is 553m high, and receives 2 million visitors annually. Tourists are offered an opportunity to see the view of downtown Toronto from each of two observation platforms – the main platform is at 346m, and the Skypod is even higher at 447 m altitude. Whilst all tourists are taken up to the 346m level, there is an additional charge to the tourists who wish to go even higher, to the Skypod level at 447m. Distant horizon views of up to 160km are possible from the Skypod. Views are dramatic from each of the observation levels. What are tourists willing to pay for the *additional* ride up to the Skypod level, and how do the views differ? Below is the CN Tower and the tourist entry ticket.

CN Tower showing main observation deck and higher Skypod deck, and entry ticket (credit: author)

Below is a comparison of the views obtained by tourists, looking at the same section of downtown Toronto, from each of the two observation deck levels.

Contrasting views from (left) Main deck and (right) Skypod at CN Tower (credit: author)

Are the two views significantly different? Certainly not so significantly different as are the views from LEO and GEO as shown above. At any rate we can note that the Skypod is **29% higher** than the Main deck, and that the tourists who opt to go up the incremental stage from the Main Deck pay **41% more** for the experience (ie a $12 increase over the $29 basic charge). In the space tourism context, GEO is **100 times higher** than LEO, and of course, importantly, offers a whole-Earth hemisphere view rather than just a slice with a slightly curved horizon.

2) Tour Eiffel

Tour Eiffel showing the three levels for visitor access. (Credit: French Tourism Agency)

We gain a similar insight from viewing the Eiffel Tower's tourist statistics. The tower, in Paris, France, is 324 m high, and has over 6 million visitors per year, the most visited paid monument in the world. Tourists again pay more as they ascend higher in the structure. The first floor is at 57 m, the second floor is at 115m (**2 times higher** than the first floor) and the third floor is at 276m (**1.4 times higher** than the second floor). Initial prices in 1889 were 2 francs for the first level, 3 for the second and 5 for the top (ie the second floor cost **2 times** the price for the first floor, and the top/third floor cost **1.5 times** the price to the second floor). More recent pricing (2015) has simplified the price structure, and now the first ticket goes all the way to the second floor at 9 Euros, and all the way to the top for 15 Euros (ie it now costs **1.6 times** the price to the second floor to reach the top).

We may summarize the market analogs data with the following charts, which indicate that tourists will certainly pay for a more "enhanced" experience at greater heights:

Summary Data on Tourist Pricing at CN Tower and Tour Eiffel				
TOURIST DEST'N	VISITORS/YR	PRICE 1st LEVEL	PRICE TOP LEVEL	DELTAS
CN TOWER	2 million	$29	$41	41% price increase for 29% height increase
TOUR EIFFEL	6 million	9 Euros	15 Euros	1.6X price increase for 1.4X height increase

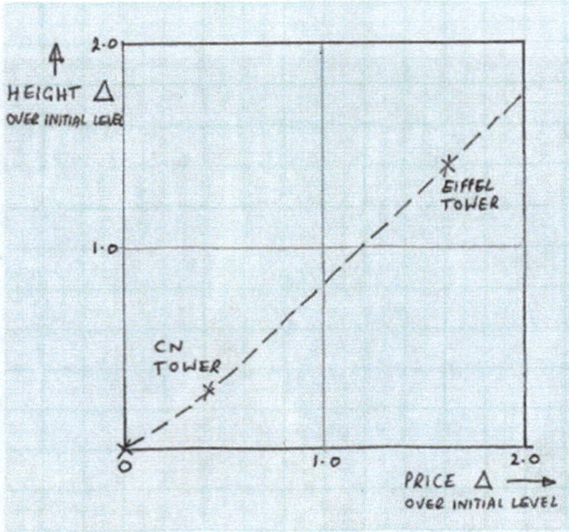

Price elasticity chart for increasing elevations at tourist destinations

- ***COMSAT DATA***

Some useful insights are also provided by analyzing, in energy terms, the relative costs of inserting satellite communications spacecraft cargoes into both LEO and GEO orbital destinations. Because of the use of the geostationary orbit for commercial communications and broadcasting satellites, a great deal of experience has been built up over the decades in getting between LEO and GEO. And this experience can provide some perspective into the relative distances and difficulties in getting to GEO, right at the edge of Earth's gravity well, compared with simply getting to LEO. Sometimes the pricing data is hard to come by, because of commercial confidentiality, but a good surrogate is the relative payload masses. Note that GTO refers to geostationary transfer orbit, not GEO itself. Spacecraft require to carry part of their mass for circularization burns on reaching the apogee of the GTO. Therefore the ratios developed in the table below would be even higher if this were taken into account. The data source for the following table was the FAA's Annual Compendium of Commercial Space Transportation (Feb 2013 version).

RATIO of MASSES to LEO and GEO ORBITS				
Orbit	Falcon 9	Ariane 5 ECA	Proton M	Atlas V
LEO	13,150kg	21,000kg	23,000kg	18,510kg
GTO	4,850kg	9,500kg	6,920kg	8,900kg
RATIO	2.7X	2.2X	3.3X	2.0X

Thus, we can see that it is almost **three times as hard**, in energy terms, to get to GEO (near the edge of Earth's gravity well) than it is to simply get into LEO (still relatively near the bottom of the well).

- ***ADVENTURERS' SURVEY COMPARISON DATA (ref 12)***

This survey was conducted in 2006 amongst approximately 1,000 visitors to the website of the adventure travel firm Incredible Adventures. Amongst the questions asked was an assessment of what respondents considered to be "fair" prices for a range of possible space adventures. The question about a visit to GEO was unfortunately not foreseen at that time as being of interest, so that specific question was not asked. However, in addition to the LEO orbit experience, a question was asked about a circum-lunar tourist flight, and so it is possible to draw some conclusions from comparisons of the respective responses, as tabulated below:

COMPARISON OF LEO AND CIRCUM-LUNAR "FAIR PRICES"				
SOURCE	PRICE	PROPORTION AGREEING THAT PRICE IS 'FAIR"		MULTIPLIER
		LEO	CIRCUM-LUNAR	
Adventurers' Survey (Ref 12)	$20M	8%	18%	**2.2X**
	$5M	14%	n/k	n/k
	$1M	30%	30%	1X

Thus, we can see that it was considered by twice as many respondents to be fair at a given price to go into lunar orbit than to go to LEO, or alternatively, approximately the same proportion thought $5M to be fair for LEO and $20M fair for circum-lunar (ie **4X more**).

- ***OTHER ANALOG SOURCE***

One other source of circumstantial evidence exists in the public domain, for pricing of tourist flights above LEO. The space tourism firm Space Adventures announced a price of $150M for a circumlunar ticket (for each of two tourists) for an Apollo 8 type mission proposed using Soyuz technology, at a time when LEO ticket prices were about $50M.

So that we can commence our demand assessment, we now attempt to pull together, in the following table, our findings on a reasonable assumption for pricing of a space tourism trip to the "Gateway Earth" station in GEO:

CONSOLIDATED PRICING DATA FOR SPACE TOURISM TO GEO		
SOURCE	MULTIPLIER	COMMENTS
ANALOGS	>2	from this paper
COMSAT COST BASE	~3	see above
ADVENTURERS' SURVEY	2.2 – 4.0	applies to Moon, not GEO
SPACE ADVENTURES' PRICE QUOTE	~3	applies to Moon, not GEO ($150 M quoted cf $50M LEO)

We shall therefore use, conservatively, for the purposes of developing this preliminary business case framework, an assumed ticket price for space tourists spending 2 weeks at the "Gateway Earth" space hotel in GEO, of **three times the LEO prices in the same time period.**

- *II PRELIMINARY MARKET ASSESSMENT AND COMMERCIAL ASPECTS (and Sources)*

In considering the non-governmental business opportunities that will emerge if the "Gateway Earth" architecture is introduced, we begin to see a framework of commercial funding sources which can augment governmental budgets at each of the seven stages of development. This therefore becomes the new funding mechanism for 21st century interplanetary exploration; a blending of governmental funds with revenues generated from commercial operations. The degree to which commercial funds can boost the governmental budgets will vary between the different segments of the architecture, and indeed through time for any given sector. In some cases we cannot know *a priori* whether a commercial operation will develop. Work may, eg, be initially carried out under governmental funding, and then transfer to commercial operations further down the road. So, the indications recorded here are necessarily broad-brush, and designed mainly to suggest the constituent elements. Much more work will be required to provide a better basis for the assumptions developed here. However, that being said, we can make some progress if we set out, and consider in turn, the separate phases of the operation. In some areas we have pretty good data; in others we have very little, and some new primary market research is needed. Where appropriate, we can judiciously use some of the data from the market analogs developed above. This paper's focus, it should be remembered, however, is mainly on the interrelated steps 3, 4 and 5 which in practice are developed simultaneously. Note that, as suggested by J Grey (Ref 27), it is not necessary for a business case to close at each step of the way; governmental expenditure may be provided by assuming simply a fixed percentage of the agencies' annual budget, with a duration of "as long as it takes" to get the job done. The commercial revenues developed in this section therefore merely help out the overall funding so that the "fixed percentage" figure can be set at an acceptably low figure for congressional oversight purposes.

Step 1: EARTH to LEO (Musk data Ref 20, 23)

Revenues and costs for this step come from commercial launch experience, with projected cost and price reductions due to gradual introduction of reusability. Examples of reusability which could be used are the SpaceX Falcon 9 (entire first stage) and Airbus Adeline (engine only).

Step 2: AT LEO NODE (ISS operations – orbital space tourists Ref 10, 19)

The costs of the operation of the ISS are available from NASA budget data. The cost of the Bigelow inflatable habitat being attached to ISS for more internal volume is $17.8M.
Revenues will come from conducting research at the ISS, and from orbital space tourists. Ticket prices for tourists in LEO vary from $20M to $60M for two weeks as orbiting space tourist.

Space Tourist Anousha Ansari (credit: Space Adventures)

Step 3: AT "GATEWAY EARTH" in GEO

This is the focal point of the whole commercial space exploration architecture. Cost information will need to take account of two different kinds of modules. Bigelow "Sundancer" modules may be assumed as the costing basis for the *commercial* space hotel part of the complex (Ref 19). For the *governmental* part of the complex, cost data can be assumed using a subset of the ISS costing information (source NASA).

Bigelow "Sundancer" Module Prototypes (credit: Bigelow)

For this paper, the focus has been more on the revenue opportunities than the overall business case closure. In exploring the business cases for the *commercial* elements of "Gateway Earth", the revenue elements generated at the "Gateway Earth" complex are likely to come from two distinct businesses, namely space tourism and satellite repair and servicing. For the current paper, the focus has been on the former revenue source.

SPACE TOURISM (Extending public access to space at GEO – Ref 11, 12, 13, 14, 22, 26)

The following revenue analysis explores how many tourists are likely to want to go to a GEO space hotel at the assumed price. The hotel is assumed equipped with high quality and magnifying optical viewing ports. Note that Apollo astronaut H. Schmitt reported to the author that views of Earth from around GEO, a region through which he passed during Apollo 17, were fine using a monocular. It is assumed, therefore, that the "Gateway Earth" space tourist hotel will be equipped with suitable optical aids for its occupants.

The assumed starting point for the revenue analysis is the Futron/Zogby Survey (Ref 11). This was a statistically valid survey of potential space tourism demand amongst millionaires, conducted in 2002. The forecasting methodology took into account the numbers of millionaires available, and, despite the answers to the questions asked, only included in the calculation those for whom the ticket price would represent no more than 1.5% of their net worth. This means that eg at a price of \$1M per ticket, we only assume that the demand can come from a population with an individual net worth in excess of \$66M. The F/Z forecasts, moreover, only used those respondents who responded "Definitely likely" to the demand question at the stated price level. For context, we should note that the Futron/Zogby survey only considered sub-orbital space tourism and LEO orbital demand, however. At the time of the survey only a few space tourists (Akiyama, Sharman and Tito) had gone into space, all to LEO using Soyuz. Since then, LEO has become the regular destination orbit for space tourists. We now proceed to work with this base data, and our assessment of likely GEO prices, to infer demand for space tourism at the "Gateway Earth" complex in GEO.

"GATEWAY EARTH" PASSENGER DEMAND ASSESSMENT AND TUG TRIPS			
LEO PRICE ASSUMPTION	GEO PRICE ASSUMPTION (3 X LEO price)	PEAK ORBITAL PASSENGER DEMAND (F/Z)	LEO/GEO TUG LOADS at 6 PASSENGERS
\$20M	\$60M	60	10
\$5M	\$15M	600	100
\$1M	\$3M	9,000	1,500

We have determined, therefore, that at a price of \$60M for 2 weeks in GEO there would be a peak demand for around 60 passengers. If we assume that the tug carries 6 passengers, then that implies 10 tug rides between LEO and GEO carrying passengers to the "Gateway Earth" hotel. If economies of scale make it possible to reduce orbital tourism prices into LEO to only \$5M, then that would equate to a $15M ticket to "Gateway Earth" and 600 passengers to GEO in a peak year, or 100 tug journeys. The business case model, however, assumes that the GEO tourists will stay two weeks at "Gateway Earth", and that the hotel there can only support 6 tourists (plus attendant company astronaut). It is furthermore assumed that the tugs can only carry 6 tourists at a time. With these constraints, what would be the best pricing zone for the operation? The following chart makes it clear that for 25 tug trips a year to the GEO hotel, carrying a total of 150 GEO space tourists, the business case closes at a price level of about $10M for LEO and $30M for GEO:

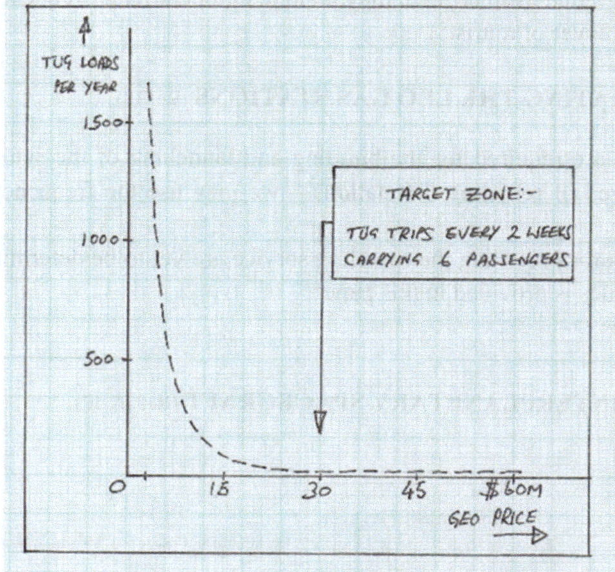

Price Elasticity of Demand for tickets to the "Gateway Earth" Space Hotel

At this target price of $30M for GEO, the "Gateway Earth" complex, and its supply chain, would generate **$4.5B/ year in revenues**.

What is the level of confidence in these results? The original Futron/Zogby study primary market research (Ref 11) involved interviews with 450 millionaires, so that the findings at that time were expressed as being representative at a confidence level of +/- 4.7% of millionaires in general. Much has changed since 2002, some positive and some negative factors, adding to the uncertainty. Furthermore we have here extended the methodology by inserting additional steps with clearly stated but unverified assumptions. Nevertheless, it seems we are unlikely to be wrong in our assessment by as much as an order of magnitude – it might be reasonable to assume a factor of about two, though (ie the range in uncertainty in the $4B figure would probably be between $2B and $8B).

(GEO COMSAT SERVICING (Ref 24 – DARPA/Phoenix experiments)
At present, we do not have any useful data, either on the cost or revenue side, to provide for this potential commercial opportunity.)

Step 4: OPERATING THE TUGS (Ref 21)

For cost data with regard to refueling of the tugs in LEO (Ref 18, 21)
For revenue assumptions we need to separately assess each direction of the tug operation:

LEO TO GEO (service for transport of crews, tourists and cargoes)
For revenue estimates we may assume the above GEO tourist forecasts

GEO TO LEO (service for removal of trash)
For revenue estimates we may assume an operational frequency of once per two weeks period (at changeover of tourists).

Step 5: OPERATING THE LEO GAS STATIONS (Ref 18)

Costs are required for the building and launching of the stations into LEO orbit. Costs are also required of refueling the stations – we may use the frequency based on the tug trips calculated above.
Prices for providing the refueling service are yet to be determined, so no assessment of associated revenues is provided in this paper.

Step 6: THE INTERPLANETARY SPACECRAFT (Ref 16, 17)

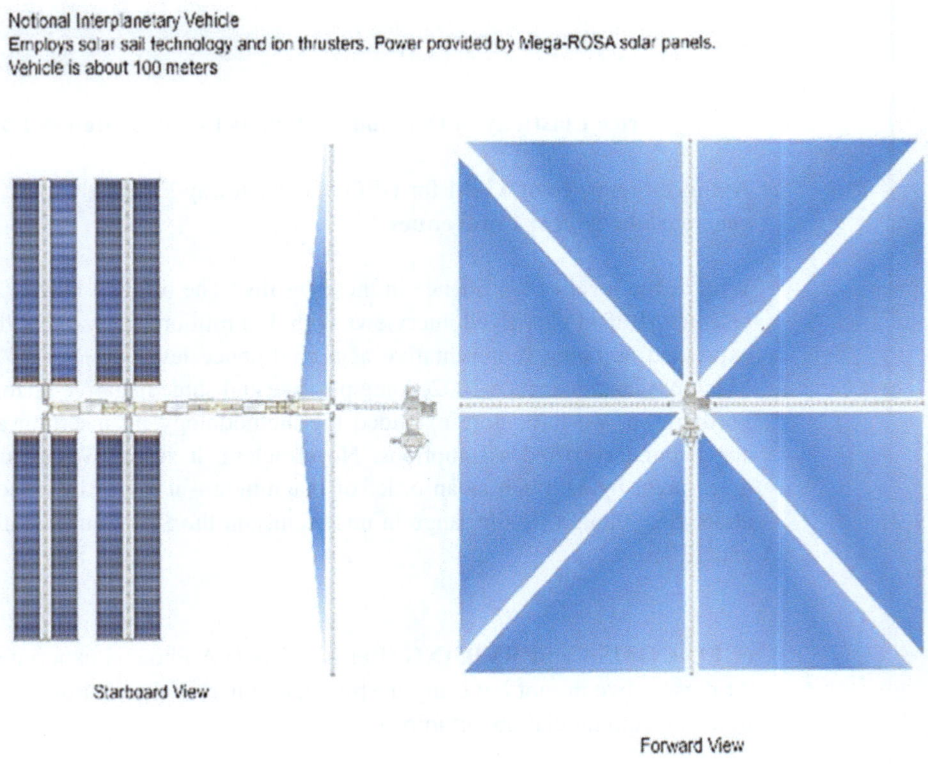

Notional Interplanetary Spacecraft (Credit: Phil Smith)

Costs are required for building the craft (raw materials to GEO, 3-D manufacture, astronaut assembly)

No revenues are developed in this paper. They would ultimately depend on the prices charged for providing a service for crews and for (eventually) tourists.

Step 7: THE PLANETARY LANDER AND ISRU CAPABILITIES (Ref 25 – MOXIE/RESOLVE)

No useful cost or revenue data is available at this stage, except for the possible use of Bigelow surface infrastructure habitats (with associated published pricing).

(Note: Asteroid and planetary resource revenues have been excluded from the analysis at this stage)

- *III* *COMBINED SUMMARY RESULTS of COMMERCIAL REVENUE OPPORTUNITIES*

- Previous Results (Ref 3):-

  (Note: this analysis could be refined by software modeling, maybe using crowd-sourced software) The costs of interplanetary travel would be at least a third cheaper by using this approach, than by using the traditional methods.

- Combined new results from the current analysis:-

  - A price to GEO of about $30M for two weeks is supported by market data
  - Logistically, 6 passengers occupy the space hotel at a time, and 6 passengers will fit in a LEO-GEO tug
  - Tugs arrive and depart every two weeks, ie 25 tug trips/year each way
  - **150 GEO space tourists at peak, generating $4.5B/year in revenues ($2B to $8B range)**
  - GEO Comsat servicing, or asteroid resource extraction, revenues not included

New superlight class of Launch Vehicles from Yuzhnoye

Olexander Degtyarev
General Designer – General Director, Yuzhnoye SDO
3, Krivorozhskaya Str, Dnepropetrovsk, Ukraine, 49008, tel.: +380 56 770 04 47
space@yuzhnoye.com

Mykola Lytvyn
Head of Space Launch System and Vehicle Design Department, Yuzhnoye SDO
3, Krivorozhskaya Str, Dnepropetrovsk, Ukraine, 49008, tel.: +380 56 770 04 47
space@yuzhnoye.com

Oleh Ventskovsky
Head of European Representative Office, Yuzhnoye SDO
3, Krivorozhskaya Str, Dnepropetrovsk, Ukraine, 49008, tel.: +380 56 770 04 47
e-mail: space@yuzhnoye.com

ABSTRACT

Study of configurations of superlight class of ILV which allows to launch micro- and nanosatellites into required orbits and to essentially reduce costs of a single mission (as compared to injection by more heavier ILV) was made by Yuzhnoye SDO within research works.
The study was based on using prospective technologies, particularly structures made of carbon-filled plastic and onboard avionics on the basis of the up-to-date microelectronic technologies. In line with modern requirements regarding development of ecologically friendly rocketry, liquid oxygen and kerosene are used as ILV propellants.
The report includes different options and peculiarities of superlight ILVs currently under development at Yuzhnoye.

INITIAL PROVISIONS

Initial purpose of SRC development based on superlight class ILV is a goal of designing a competitive launch vehicle with a minimum cost of single launch cost and competitive launch cost.

The market of space launch service has needed such vehicle since its forming. However the level of task complexity and technology limitation applied by launch system designers bring us to a conclusion that current goal is still in an engineering process

Periodically, systems with similar or close goal have been developed. However their applying was stop when increased cost makes their level of self-cost is not uncompetitive (for example Scout LV in USA, Mu-III, MU-V LV in Japan and etc.)

It should be taken into account that the space vehicle launch cost depends on works structure and designer management thus can be significant different. It is obviously that applying of economic oriented approach directed to minimize launch cost can provide development and inexpensive space launch system of light class.

At present, there is not any space launch system with a cost of single launch less than 15-20 million USD at worldwide launch market. Though a lot of light and superlight class spacecrafts are designed that oriented to a passing and cluster launch at heavy class ILV. At the same time it is existed a range of tasks that needed independent launch into specific orbits but they cannot be realized with a present high price to a single launches.

S. Hatton (ed.), *Proceedings of the 13th Reinventing Space Conference*,
https://doi.org/10.1007/978-3-319-32817-1_16

Therefore -

1) It is necessary to create a space launch system based on superlight class ILV for taking of this market niche. Space launch system should provide such payload injection into required orbits that allows to assure a recoupment of system operation and its competitive at the launch service market.

2) The main target of this system is spacecraft injection (single and cluster) into the circular and sun-synchronous orbits with the altitude not less than 500km.

3) ILV should use ecologically friendly and inexpensive propellant components.

4) Applied technology and material in space launch system and launch vehicle should be acceptable for using. It should be used already developed, applied or being developed components for reducing SRC and ILV development cost and also increasing of their reliability (Of course if it's not bring to worsening indices of SRC competitiveness and economic efficiency).

SPIKE ILV SUPERLIGHT CLASS

As main propellant components of superlight class ILV is selected liquid oxygen and kerosene (if it is reasonable it can be selected other hydrocarbon fuel). This propellant pair has ecological safety and low-cost conditions. It has been already developed in technologies of enterprises. Besides nontoxic condition of propellant simplifies and reduces the price of SLS and ILV operation.

The task of launch cost minimization is a quite complex. In the first place the cost is determined with a main target. In the second place cost depends on ILV scale. The main target of light and heavy vehicle is fairly similar. It is an injection of required payload mass into near-Earth orbit with acceptable level of reliability and accuracy. The task of injection into sun-synchronous orbit is complicated additionally with problem of upper stage engine re-ignition. Besides the general characteristic velocity must be about 9800m/s-9900m/s. The ILV upper stage with a payload should bring velocity to this range.

The index of final mass ratio by ILV stages should be pretty high for getting such velocity range. Furthermore than less engine specific impulse and stage number than higher final mass ratio index must be. Besides, if happens propellant mass decrease it will have a tendency to worsening (influence of scale factor).

Provision of required index of final mass ratio by stages is crucial and complex task when superlight class designing ILV. It is especially important for an upper stage because it carries control system (CS) and measurement system (MS) main units mass. These systems provide a fulfillment of flight scheme in a part of ILV control and transfer data to a tracking station about ILV systems functioning during a flight.

The last years progress in microelectronics, communications system and design of compact orientation devise particularly in gyroscopes and accelerometer based on laser and fiber-optic hardware and also micro electromechanical scheme (MEMS) in aggregate with miniature satellite navigation equipment allow to design control system with a minimum mass (such as a few kilograms). Devices based on MEMS are widely adopted at mass equipment. It provides a cost reduction up to affordable value due to a considerable expansion of their production (one or a few thousands USD for *industrial* category). The same happens with microprocessors that can be used as base for designing flight control computer (as a core of Control System) and telemetry main block of measurement system. It can be reduced the level of development cost for CS and MS of superlight class ILV due to applying of popularity designs with *industrial* digital components. For example, they are successfully applying at small spacecrafts. They also can be used at vehicles with regular launch (design of basic on-board software kit with typical flight task that can cover a lot of possible missions).

During an examination of necessary ILV stages parameters and specifics of their operations was made a conclusion about reasonability of ILV design under three-stage scheme.

Applying of three-stage scheme allows to reduce requirement of final mass ratio index by ILV stages. It provides a reduction of technical risk as well as increasing of payload mass (sufficient mass to repay the appearance of additional stage) in comparison with two-stage scheme. At the same time, the phrase "appearance of additional stage" doesn't accurate to describe a complication of three stage ILV in comparison with two stage vehicle.

The number of propulsion systems and control system at ILV is not changed in comparison with two-stage ILV based on ecologically friendly propellant components. It is necessary additional propulsion system (low- thrust propulsion system, LTPS) for carrying out an apogee maneuver (according to injection into SSO). It can be started in weightlessness state and provide orientation at passive path flight.

It has been conducted comparative ballistic calculations to determine an assessment of ILV stage number influence on required final mass ratio index. At the same time, ILV launch mass, engine thrust by stages, aerodynamics and payload mass were constant. Relative final mass ratio values by stages (μfinal =mfinal/mlaunch) have been chosen as maximum as possible for getting an assigned task that is –payload injection into SSO with 500km altitude. Calculation results are given at Table 1.

Table 1. Influence of stage number on required final mass ratio index

Relative final mass of stages μ	First stage	Second stage	Third stage
Three-stage ILV	0.09	0.111	0.276*
Two stage ILV	0.079	0.105*	-
*Note: Index of final mass ratio (μ) is determined according to influence of scale factor and estimation of an on-board avionics (CS+MS) in assembly with ILV upper stage			

Calculations show that ILV with three-stage configuration has a significant advantage in required final mass ratio index than ILV with two-stage configuration. It means that with similar index of final mass ratio by stages three-stage configuration will have more a payload mass. Increment of payload capacity will repay increasing launch cost provoked by appearance of additional stage. Besides a two-stage configuration can be applied for the missions where two engine ignition of upper stage or maximum payload is not needed.

Characteristics of materials that applied in a structure are important factor. It allows bringing accessible values of stage mass to required mass (based on calculated values of final mass ratio index).

One of the basis conceptions of superlight ILV design is using a structure (including propellant tanks) out of polymer composite such as carbon fibre-reinforced plastic.

At the same time, using of polymer composite structure (incl. carbon fibre-reinforced plastic) brings variety of problems that are required descisions and tests (verification) in the future. One of the main problems is structure operability during a contact with liquid oxygen.

Complexity of tanks structure design out of polymer composite for cryogenic components (particularly liquid oxygen) is in the following:
– sufficient elasticity (plasticity) of material for structure cracking elimination under acting loads and possible shocks;
– sufficient integrity and strength for material exfoliation elimination during cryogenic component loading and draining;
– sufficient chemical inertia during a contact with liquid oxygen.

It is considered ILV configurations in a part of superlight class ILV conception research tanks and dry bays are manufactured out of polymer composite (carbon fibre-reinforced plastic) with 1.0m in diameter. It uses liquid oxygen and kerosene as propellant components. Main engines by stages are designed under an open scheme. It is simpler in production. It is going to use at the second stage main engine the same as at the first stage but with a different nozzle.

The key point of new ILV design is an engine selection. The selection accent is applying of inexpensive variants. It can be engine with opened scheme or pressure-feed system, hybrid engine or other alternatives.

The preliminary assessments show that more inexpensive engine brings specific impulse to significant reduction. As a result of it ILV characteristics are worsened (payload mass or required launch mass). In addition engine mass is not reduced appreciable as well as its costs at serial production. The gain in cost is tangible rather at small lots than at serial production.

The actual engine advantage with simpler scheme is a cost reduction (incl. time) on its design and test as well as low required to production. In the final analysis it is going to reduce a payload mass. ILV dimensions (engines thrust and their sizes accordingly) should be increased for compensation of reuction. Also main structure of final mass ratio index should be increased however it has a limitation and it needs an increment of development cost and serial production cost. In addition, increment of final mass ratio index can be carried out for ILV with engine under a closed scheme operation.

At the same time, Yuzhnoye's specialists have experience in the first stage engine design especially with an engine under a closed scheme based on liquid oxygen and kerosene propellant components. Applying of others even more simple scheme will need research, development and experimental works thus it increases time and development cost.

It is considered variety of engine design approaches based on pressure feed system and turbopump assembly (turbopump is considered with opened and closed scheme).

It is considered applying the same engines at the first and second stage of superlight ILV according to chosen conception with time and cost minimization (at the first stage can be used an cluster with 7-9 engines). In accordance with liquid-propellant rocket engine preliminary analysis follows that it is required engine thrust about 2 tone-forces (tf). Therefore design of new engine under closed scheme operation for superlight class ILV is not reasonable (because of high-level development cost and sufficiently heavy mass).

Using of the same engines provides a significant reduction of development time and cost also engine dimensions. Therefore a reduction of testing cost is expected. Because of it is necessary less size of ground equipment for testing and production. Similar decision with the same engine (without a nozzle) at the first and second stage has been applied at several projects of superlight launch vehicles as well as Falcon 9 middle class ILV.

Besides significant number manufacture of identical engines provides costs reduction per one engine production due to using of serial production. Also it is not necessary to design an independent roll control system (as compared with one engine). However the general reliability of engine cluster is brought down. It appears a possibility to continue a flight even with one engine failure with help of other engines out of cluster. In case of underexploitation of ILV load-carrying ability it has ability to payload injection into required or low-altitude orbit.

Therefore it is chosen LPRE of new design for the first and second stage. It is going to use seven engines at the first stage and one engine at the second stage. These engines have different nozzles. Engine at the first stage has a shortened nozzle than at the second stage. It is also adapted to ignition at the sea level.

Engine with pressure feed system is much more acceptable for applying at ILV third stage. It has a minimum number of elements, less cost and higher reliability.

ILV main characteristics are given at Table 2.

Table 2. The main characteristics of Spike ILV superlight class

Characteristics	First stage	Second stage	Third stage
ILV launch mass without payload, kg	9988	2132	366
Propellant	LOX+ kerosene	LOX+ kerosene	NTO+ UDMH
Engine thrust, sea level / vac, kgf	11900/ 13227	- /2110	- /250
Usable propellant mass, kg	7150	1500	255
Stage diameter, m	1.0	1.0	0.95
Payload fairing diameter, m		1.0	

Spike ILV is shown at Figure 1.

Figure 1. Spike ILV

The values of injected payload mass into standard orbits by Spike ILV are given at Table 3.

Table 3. The values of injected payload mass into standard orbits by Spike ILV

Orbits	LEO Hcir=200 km, i=51°	SSO Hcir=500 km, i=97,4°
Payload mass, kg	150	100

VECTOR ILV SUPERLIGHT CLASS

As a result of launch service market analysis with SC mass distribution for the last 5 years and a future prognosis follow that estimated SC number at mass range 50-150kg is about 6 units per year with a tendency to increment (for different mass range up to 60-80%). It supposes that SC average number at current mass range can increase in 3-7 times in next 4 years. However proposition grows as well as demand grows (with launch possibility).

If it is supposed maintenance of present price level then current ILV will able to cover up to 20-25% market. Launch cost is expected to be 2-3 million USD (20 thousand per a payload kilo with possibility to add passing payload up to maximum load-carrying value). It can be 4-10 launches per year or general annual income 10-30 million USD. It is even without costs on a launch services.

ILV payback at payload mass range (up to 100kg into SSO with 500km) is quite risky because of costs level of all necessary work for ILV and space rocket complex designing. Therefore it is more reasonable a gradual achievement of required technology level at the same time to design other vehicles that have a less level of development cost and higher level of income.

ILV with small SC should be designed with more competitive payload mass range such as 350-500kg on the assumption of ratio assessment between required costs level and small SC prices on launch.

Vector ILV is proposed as a solution for this problem. It is based on more traditional technical solutions.

ILV first and second stage have a monoblock diagram with a monocoque cylindrical tanks. They have smooth shells with elliptical (torus-spherical) domes. Loadbearing shell structure can be reinforced with rings or others load-carrying elements (it's depend on acting loads). Stage structure is made out of aluminum alloys with a possibility to apply polymer composite at dry bays structure (it's depend on characteristics and production cost).

First stage engine, RD835-1 is placed at the aft-bay and installed on frame that is fastened with a lower ring of fuel tank. Engine provides stage control under the pitch at yaw channels due to a rocking in gimbal mount. Gas-jet nozzles provide ILV control under a roll channel. They are operated on hot gas out of engine gas generator. Pneumohydraulic supply system includes a gas bottle pressurization system with cold helium that is drowned in oxidizer.

Second stage structure with RD809K is similar with first stage under main technical decisions. It has the same tanks diameter and domes dimensions. Engine is also installed in gimbal mount on frame that is fastened with a lower ring of fuel tank. It provides ILV control under a pitch at yaw channels. Control under a roll channel is realized by the gas-jet nozzles with hot gas.

Payload fairing and interstage are made out of carbon fibre-reinforced plastic. They have diameter equal to ILV diameter. Payload fairing with increased dimensions can be designed in the future perspective.

Vector ILV main characteristics are given at Table 4

Table 4. Vector superlight class ILV main characteristics

Characteristics	First stage	Second stage	Third stage
ILV launch mass without payload, kg	35900	8770	650
Propellant	LOX+ kerosene	LOX+ kerosene	NTO+ UDMH
Engine thrust, sea level / vac, kgf	43200/ 47600	- /10000	- /450
Usable propellant mass, kg	24700	6760	460
Stage diameter, m	1.7	1.7	1.4
Payload fairing diameter, m	1.7 (2.0)		

Vector ILV is shown at Figure 2.

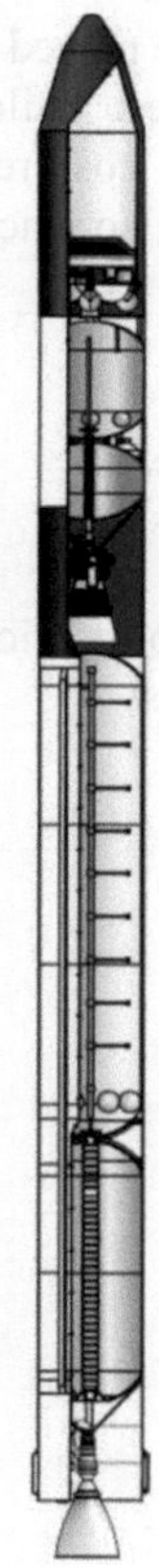

Figure 2. Vector ILV

The values of injected payload mass into standard orbits by Vector ILV is given at Table 5.

Table 5.The values of payload mass injected into standard orbits by Vector ILV

Orbits	LEO Hcir=200 км, i=51°	SSO Hcir=500 км, i=97.4°
Payload mass, kg	710	500

GROUND COMPLEX

ILV ground complex includes:
- processing complex;
- launch complex;
- command post;
- means of propellant and ILV components transportation;
- telecommunication support system.

Processing complex includes:
-SC and PLU processing complex;
-LV and ILV processing complex;
-fueling-neutralization station.
All processing complexes are placed in one building, ILV assembly, integration and test building. It is done for building cost and operating cost reducing. Building of transport storage (garage) is also placed there.

Launch complex include:
-launch pad with a trigger;
-propellant temporary storage;
-control and communication station;
-power supply system;
-lighting system, lightning protection system, television observation system and others/

Launch complex diagram with main components is shown at Figure 3.

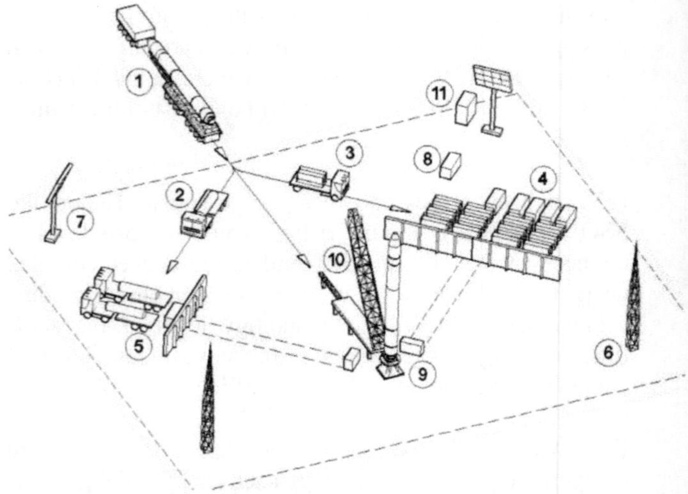

Figure 3. The main components and buildings at launch complex

1 – Transportation of assembled ILV with mobile thermostating system to launch complex
2 – Transportation and mounting of fuel filing system
3 – Transportation and mounting of gas supply system and oxidizer and nitrogen filing system
4 – Gas and oxidizer filing system
5 – Fuel filing system
6 –Lightning protection system
7 – Lighting system
8 – Control and communication station
9 – Launch pad
10 – Transporter/Erector
11 – Power supply system

Reinventing Space Conference
BIS-RS-2015-60

Small satellite launch vehicle from a balloon platform

Kieran Hayward
Cranfield University

José Mariano López-Urdiales
Founder, zero2infinity

13th Reinventing Space Conference
9-12 November 2015
Oxford, UK

S. Hatton (ed.), *Proceedings of the 13th Reinventing Space Conference*,
https://doi.org/10.1007/978-3-319-32817-1_17

Small Satellite Launch Vehicle from a Balloon Platform

Kieran Hayward[1] and Jose Mariano Lopez Urdiales[2]

[1] Cranfield University, Cranfield, UK
[2] zero2infinity, Barcelona, Spain

Abstract

In the last decade there has been growing use of smaller satellites (0-100kg) to conduct Earth observation and science missions and this industry is growing. 2014 saw a small satellite launch increase of 72% compared with 2013. Companies such as Planet Labs are starting to launch large numbers of small satellites. However, to date the use of small satellites has been restricted due to the limited launch availability to this class of satellite. Due to their small size these satellites are normally launched as secondary payloads on larger launch vehicles (such as Falcon 9 or Ariane 5). This restriction has severely limited the launch dates available to these small satellites and also limits their orbit selection.

Due to the restrictions mentioned above and building on its experience in high-altitude ballooning zero2infinity has begun designing a new small satellite launch vehicle, called bloostar. This three-stage vehicle is designed to put a 75kg payload into a 600km sun synchronous orbit. It is launched from a high altitude helium balloon at 20km, at this altitude the atmospheric density is low enough (<7% of sea level density) that aerodynamic drag is negligible. Traditional launch vehicles, which are launched from sea level have to pass through the densest part of the atmosphere, incurring large amounts of aerodynamic drag. By launching above the denser part of the atmosphere bloostar avoids these drag losses resulting in a significant ΔV saving. This reduction in atmospheric drag removes the need for an aerodynamic fairing around the payload, as the payload no longer needs to fit into an aerodynamic fairing, the constraints regarding payload volume are much reduced compared to existing launch vehicles. Additionally, the acoustic and shock environments are more benign which reduces the minimum thicknesses of components and overall satellite structural weight. This will allow light-weight, high volume satellites, such as small Earth observation telescopes which need larger diameter mirrors, to be launched.

Launching at an altitude where aerodynamic drag is negligible also leads to a launch vehicle that is no longer required to be slender but instead is a series of concentric tori. This novel shape has a number of advantages, including allowing all engines to fire at the same time reducing inert mass of the first and second stages. The control system has to be adapted to the new geometry and mass distribution of the rocket. How much thrust vectoring and how much differential throttling of 13 engines is needed to optimize the trajectory even in engine out situations is the focus of some ongoing research.

Introduction

Nano- (1-10kg) and micro-satellites (10-100kg) have proven their capabilities to perform increasingly complex missions effectively, affordably and responsively. Multiple factors have contributed to enhance their performance such as miniturisation of electronics and enhanced precision in small mechanical systems. There is an increasing interest in missions performed by nano/microsatellites with a whole new industry value chain emerging around them. For example, 2014 saw a small satellite launch increase of 72% compared with 2013 (Crisp, Smith, & Hollingsworth, 2014). Companies such as Planet Labs (Marshall & Boshuizen, 2013) are starting to launch large numbers of small satellites. The use of these smaller satellites has been restricted by the limited launch availability for this class of satellite. To date these satellites have been required to be launched as secondary (or tertiary) payloads on large launch vehicles such as Europe's Ariane 5 or SpaceX's Falcon 9. Despite many competing efforts to develop a launch vehicle solely for the purpose of launching small satellites (Niederstrasser, Frick, 2015) none has yet been fully developed. Those launchers which were being developed (for example Falcon 1) have been cancelled as they were viewed by their manufacturers as non-profitable (or less profitable than launching larger satellite payloads). Being restricted to secondary payload spaces has severely limited the launch opportunities available to small satellites and also limits the selection of orbit location and launch date to existing planned launches.

After identifying that the true revolution in space capabilities would come not just from microsatellites but from the combination of microlaunchers and microsatellites (via responsive access and constellations) (Palerm, Barrera, Salas, 2013) zero2infinity, a Barcelona based start-up company, has begun development of a new launch vehicle specifically for this market, bloostar. It is a three stage launch vehicle designed to put a 75kg payload into a 600km sun-synchronous orbit. It will be dropped from a helium balloon at an altitude of 20km. At this altitude the atmospheric density is low enough (<7% of sea level density) that aerodynamic drag can be neglected. This reduction in aerodynamic drag results in a significant ΔV saving (Beerer, 2014) and also means that instead of the long slender shape of ground launched vehicles bloostar is a series of concentric tori, see Figure 1. This shape has a number of advantages over a slender body: this shape is easier to hoist by balloon; and it allows all rocket engines on every stage to be ignited in parallel reducing inert mass in the lower stages of the vehicle.

Figure 1: bloostar

Advantages of Stratospheric Launch

As mentioned in the introduction bloostar will be launched from a balloon at an altitude of 20km. Based on Astos simulations it is expected that the launch vehicle will need approximately 9.28km/s of ΔV to reach a 600km polar orbit, compared to a ΔV of 10.12km/s for a ground launch (calculated from Astos simulations and Vega launcher data). While this is may seem only a small difference (8%) using the Tsiolkowsky equation it can be shown by assuming a specific impulse of 3,250m/s (typical of a methane fueled rocket) that the initial mass of bloostar at ground launch would be approximately 30% larger than if launched at altitude.

The benefits of launching at altitude are further shown by considering the graph in Figure 2 which shows that by launching at 20km drag losses can be as much as 22.5 times less. In addition to the reduced ΔV and atmospheric drag, launching at altitude also results in reduced payload vibration, which reduces the impact of launch on sensitive scientific instruments, and also results in reduced heat transfer, shown in Figure 3.

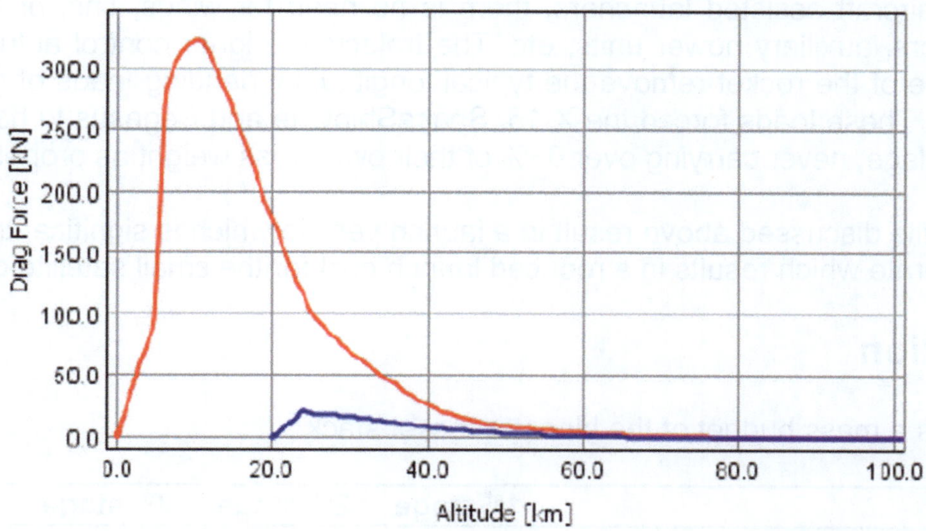

Figure 2: Drag vs. altitude for a generic launcher ignited from ground (red) and 20km (blue)

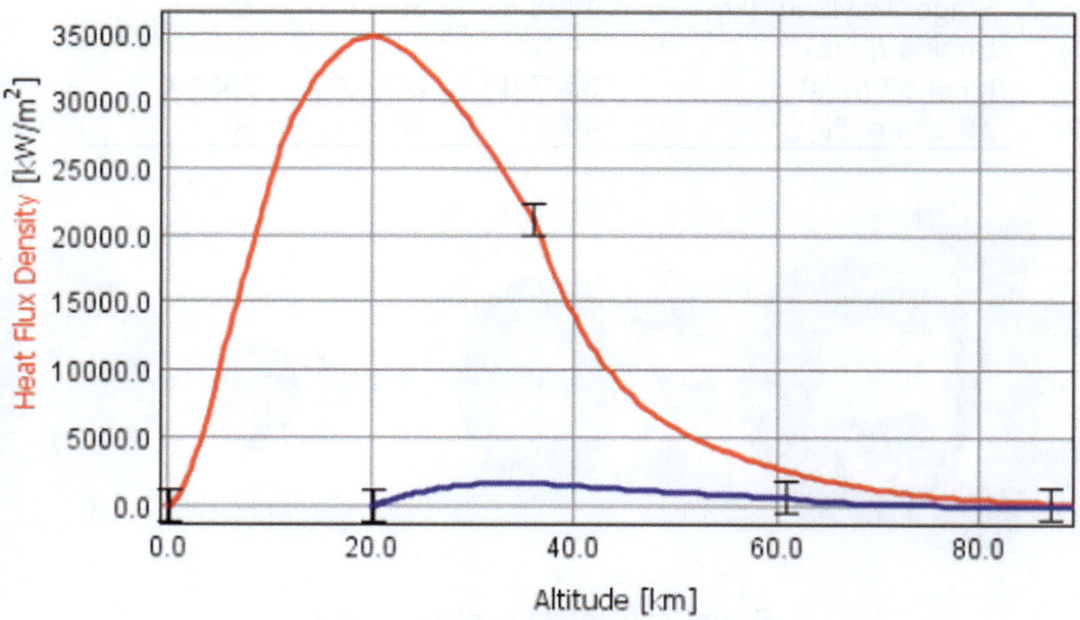

Figure 3: Heat flux vs. time for generic launcher ignited from ground (red) and 20km (blue)

The only existing air launched orbital rocket, Orbital's Pegasus experiences a maximum dynamic pressure (maxq) of 67kPa. With bloostar the maxq is only 5kPa. This has implications on the structural weight of the launcher.

Compared to aircraft assisted launchers, there is no need for wings, fins, or heavy control system actuators, auxiliary power units, etc. The trajectory, higher control authority, and the structural shape of the rocket remove the typical longitudinal bending loads of air launch pull up maneuvers. These loads forced the X-15, SpaceShipOne and Pegasus to have a heavier, reinforced fuselage, never carrying over 63% of their own gross weight as propellant.

All of the benefits discussed above result in a launch vehicle which is significantly simpler and cheaper to operate which results in a reduced launch cost for the small satellite community.

Configuration

The following is a mass budget of the bloostar rocket stack.

	1st stage	2nd stage	3rd stage
Structural mass (kg)	552.7	118.8	103.7
Fairing (kg)	25	0	0
Propellant mass (kg)	3284.6	622.6	218.7
Total stage mass (kg)	3862.3	741.4	322.4
Stage "Payload" (kg)	1138.8	397.4	75
Engine I_{sp} (s)	342	342	342
Ideal ΔV (m/s)	3587.8	2654.6	2681.4
ΔV share (%)	40.2	29.8	30

Figure 4: bloostar exploded view

Concept of operation

bloostar will be launched from a ship reducing the risk of launch delays. Several launch windows can be mapped over the surface of the ocean and the one with least chances of weather delays can be selected. Additionally, the ship can move at the speed of the wind and thus compensate ground winds. A near zero wind column is generated in which to inflate and release the balloon from the deck. The ship itself does not need any significant adaptation for the operation. Any ship with a sufficiently large flat area to fit the ISO containers in which all the system is packed could be rented to perform the flight. The payload is mounted near the balloon inflation area. The effective launch area should be around 50x17 meters.

Initial launches will be conducted to the south-west of the Canary Islands due to the calm seas and low wind speeds generated by the constant weather patterns and the geographic characteristics of the islands. This location is also excellent to choose a desired orbit since most azimuths are available.

The first phase of the flight is a balloon ascent to Near Space (20 km) taking approximately 90 minutes. The second phase starts once the rocket has been dropped from the balloon. Due to the reduced ballast on the balloon it stabilizes itself at higher altitude where it acts as a telemetry relay for the remainder of the flight. The first stage lasts around 110 seconds and takes the microsatellite launcher up to 80 km at an inertial speed of 2.3 km/s, during this stage the engines are producing thrust with a total vacuum impulse of 104 kN. The second stage raises the vehicle to 400 km in 230 seconds, at the end of this stage the vehicle is flying at 4.4 km/s. The last stage performs several upper stage firings to optimally orbit the payload. The first burn lasts for 340 seconds and allows the payload to reach 600 km altitude while still slightly below the target orbital speed. Then, the upper stage coasts for 250 seconds before performing a circularization burn and then releasing the satellite. Lastly, the upper stage de-orbited in order to minimize the amount of space debris left by the mission. During the ascent, the maximum axial acceleration reaches approximately 7 g's. The sequence of the bloostar flight that has been described above is shown in Figure 5, distances are not scaled.

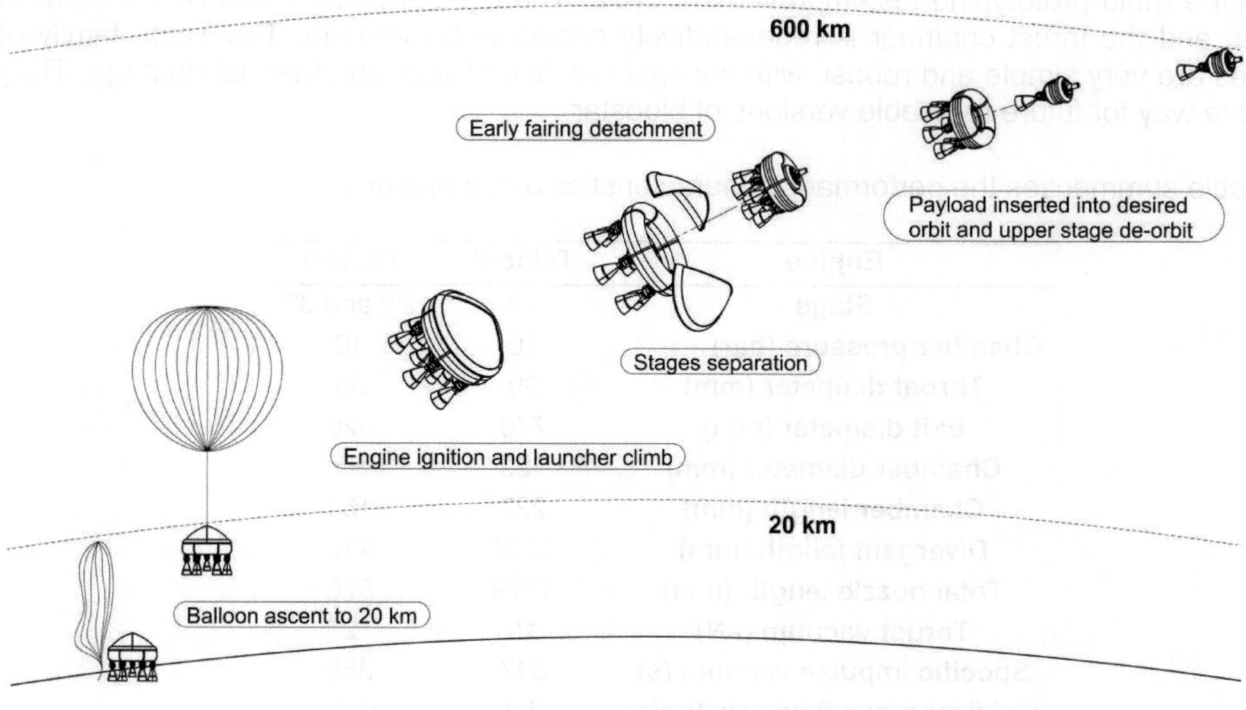

Figure 5: bloostar ascent phases

Rocket Engine Design

In order to maximize the cost saving associated with the launch vehicle simplicity is a key driver for the rocket design. Launch from the stratosphere has already been shown to significantly reduce the launch vehicle mass, this in turn reduces the required thrust of the rocket engines. Furthermore, as the engines are initially ignited at altitude they operate at close to vacuum conditions throughout flight and therefore thrust losses are minimized.

The propellants used are liquid oxygen and liquid methane. This bi-propellant combination provides a good mix of performance, simplicity and is a green propellant. Being liquid at similar temperatures allows for less insulation and the use of a common bulkhead.

The specific impulse depends on the ratio between thrust chamber pressure and the external pressure. In order to increase this number traditional rockets launched from sea level either use heavy, expensive and prone to failure turbo pumps or require very thick and heavy tanks in order to pressure feed the engine. In bloostar's case the tanking can be lightweight and the engine simple since the pressure ratio is kept high not by increasing the chamber pressure but by the effect that altitude has in lowering the external pressure.

There are 6 engines in the first stage with a thrust of 15 kN each, another 6 engines in the second stage with 2 kN of thrust, and a core engine in the upper stage with also 2 kN of thrust.

Due to the engine size they can be effectively produced using 3D printing techniques which eases the rapid prototyping-test-improvement cycle. The injector plate is a classical co-axial system and the thrust chamber is regeneratively cooled with methane. The Teide family of engines are very simple and robust, with minimal creation of soot and thermal damage. They pave the way for future re-usable versions of bloostar.

This table summarizes the performance characteristics of the engines:

Engine	Teide-2	Teide-1
Stage	1st	2nd and 3rd
Chamber pressure (bar)	10	10
Throat diameter (mm)	99	36
Exit diameter (mm)	770	320
Chamber diameter (mm)	222	95
Chamber length (mm)	227	161
Divergent length (mm)	1051	415
Total nozzle length (mm)	1279	576
Thrust vacuum (kN)	15	2
Specific impulse vacuum (s)	347	355
Oxidizer mass flow rate (kg/s)	3.4	0.44
Fuel mass flow rate (kg/s)	1.03	0.133

Contraction area ratio	5	7
Expansion area ratio	60	80
Contraction angle (deg)	30	30

Figure 6: Test of the Teide-0.1 rocket burning propane and oxygen

Propellant Tanks

All tanks carry a certain amount of external MLI to limit the boil-off, which has been estimated to be 5kg of liquid methane and 15kg of liquid oxygen for a 2-hour cruise to altitude. These small amounts don't justify carrying an extra tank on the balloon gondola to top off the tanks on the rocket.

The first and second stages are toroidal carbon fiber filament wound tanks. They are kept at pressure and provide structural rigidity to the rocket. They are made in one piece by filament winding, and they are equipped with an internal liner to avoid micro-cracking in the composite walls. T1000G fibers provide the right strength to weight ratio for this application. Internal baffles have been added to prevent low frequency sloshing modes. Several axial reinforcements made by hand lay-up will be placed at the attachment points between the stages, tanks and fairing.

For the third stage, where every kg saved in dry mass is a kg of payload, an even more optimized solution has been adopted: Ultra High Performance Vessels (UHPV). These are produced by Thin Line Aerospace in Canada and have been previously tested in flight with NASA and zer2infinity. They are a light and flexible cryogenic tank that stores the liquid methane and oxygen in separate tanks, with a common bulkhead, through the use of a multilayer leak-proof isotensoid. Propellant is fed to the rocket engines using a pressure fed system with pressure in the tanks maintained by helium gas. As well as providing pressure for the feed system the helium pressurant gas also helps to maintain the structural rigidity of the tanks as propellant is used. Using UHPVs results in a significant mass saving while also reducing costs.

Figure 7: Flexible multilayer tank tested at 27km altitude.

Fairing

The purpose of the fairing is to keep the payload protected during the balloon ascent and to prevent damage of the sensitive parts of the payload because of the aero heating from the first phase of the rocket propelled flight.

Traditional fairings also need to withstand significant structural and acoustic loads, this is not the case for bloostar.

A flexible, retractable fairing, with rigid ribs and a multilayer canvas, mostly made of betacloth (Teflon covered fiberglass), has been designed.

Figure 8: Opening of the fairing.

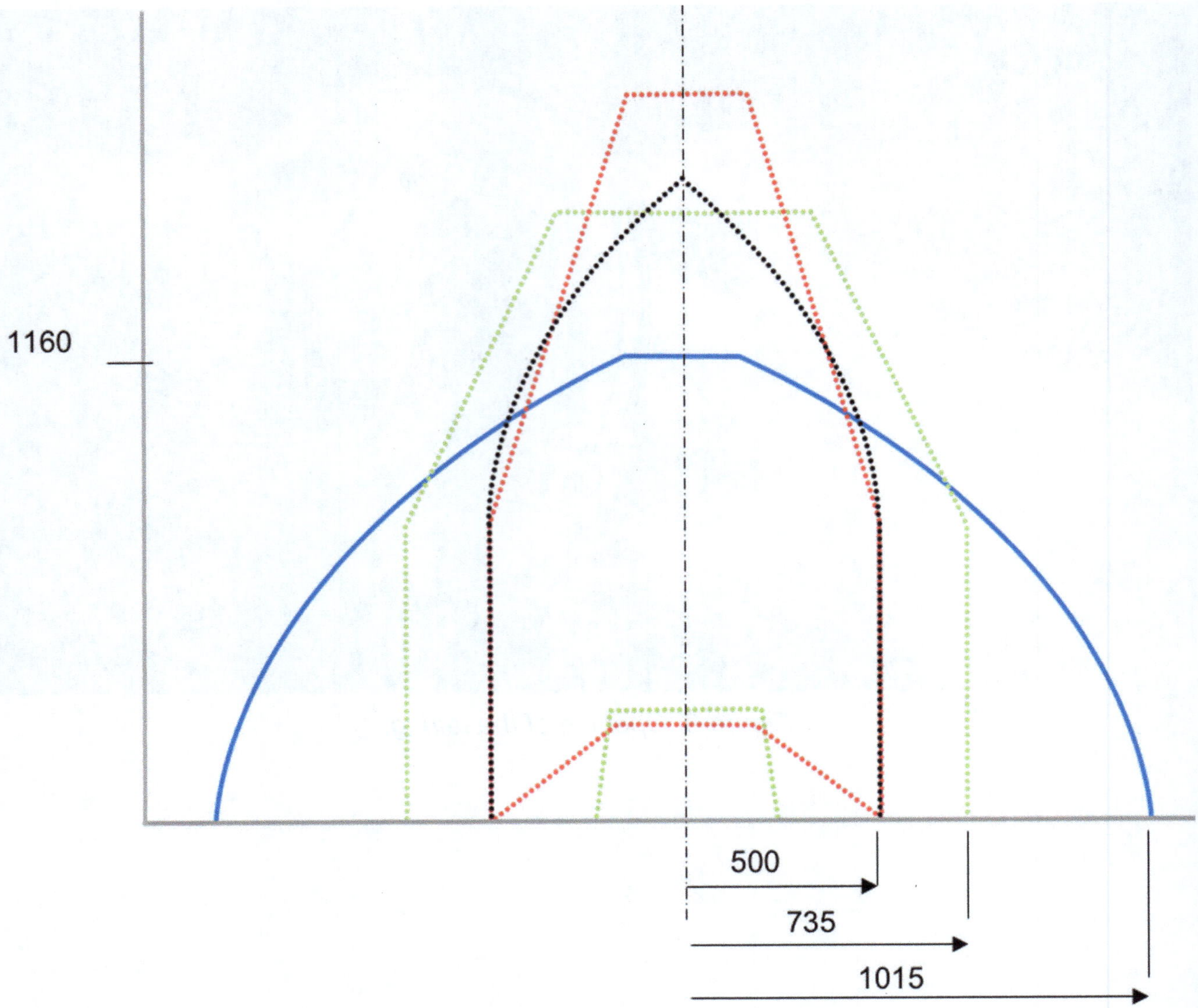

Figure 9: Usable space under the fairing for several microsatellite launchers (bloostar in blue).

The volume (2.4m^3) available under the fairing is unique amongst the proposed microsatellite launchers. Near Space launch avoids the heavy acoustic loads that occur at launch from the ground and beyond transonic phases of atmospheric flight. bloostar is, without comparison, the quietest ride to Space. This in turn allows satellite designers to fully take advantage of the extra volume to introduce elements with large surface and therefore high performance relative to weight and cost. The following are some of the benefits provided:

- Higher on-board computing power, thanks to larger radiators that can dissipate heat efficiently.
- Higher resolution imagery, thanks to larger imaging components (mirrors, lenses and arrays).
- Higher communications throughput, thanks to antennae with larger surface and higher gain.

- Higher on-board power, thanks to larger solar array surface.

Control System Design

The novel architecture of bloostar results in a launch vehicle shape for which a control system has not previously been designed. In order to research possible control methods a MSc research project has been conducted (Hayward, 2015). The main objective of the project was the preliminary design of the control system for this launch vehicle. Historically launch vehicles are controlled through Thrust Vector Control (TVC), deflecting the thrust vector in some manner to produce a torque about the vehicle Centre of Mass. It is possible that a similar system could be designed for bloostar. The novel shape of bloostar also presents a new method of control which has typically only been used on small Unmanned Air Vehicles such as quad-rotors. This method uses differential throttling. As the radius of bloostar is significantly larger than its height (especially for the first stage) it is possible to use differential throttling of the rocket engines to produce the required torques in a similar fashion to that employed by quad-rotors. Investigation of the application and suitability of these two control methods formed the bulk of the research conducted in the project. In order to test the control systems a model of bloostar was produced in MatLab/Simulink and a series of simulations conducted testing the control systems in various flight phases.

The project found that if the throttle response rate of the engines was fast enough the differential throttling controller performed better than a TVC system. A brief description of the preferred control system is outlined below.

First and Second Stage Control

Each of the first two stages is equipped with 6 equally spaced engines. This layout makes it possible to easily implement differential throttling to control the attitude of bloostar. The differential throttling control system is based on that found in quad-rotors, whereby in order to produce a torque the thrust of engines either side of the required axis is changed.

Analysis of the developed control system has shown that the effectiveness of the controller is highly dependent on the rate at which the thrust of each engine can be changed. As part of the research conducted for the initial control a number of throttling methods were researched, as the depth of throttling required is low simple throttling methods can be used. The method selected for bloostar is simple propellant flow regulation. As the propellant is pressure fed into the engines this can be easily achieved using fast variable-amplitude valves (Stone, 1995). The combination of this propellant feed system and regulation allows the thrust level to be adjusted quickly enough to maintain good control of the vehicle.

Third Stage Control

As the third stage of bloostar is only equipped with one engine nozzle it must be controlled using TVC. While there a number of methods of producing the thrust vector deflections (including hot gas injection) due to its relative simplicity and because it avoided the production of shockwaves within the rocket nozzle a gimbal system is used (Sutton, 2001).

Traditional launch vehicles are long and slender and therefore the rocket nozzles are generally a long distance away from the vehicle Centre of Mass, thus a small deflection of the thrust vector will produce a sufficiently large torque. However, because of the shape of bloostar it will be necessary to use larger TVC deflections to produce the same size torques, as TVC deflections are generally limited to small values this will make control of bloostar with TVC more challenging.

Research on the effect of the mass of the TVC system showed that due to the low dry mass of the vehicle TVC actuator mass could easily become a large component of the overall mass. It is therefore necessary to use actuators with as low mass as possible. For this reason Electro Mechanical Actuators are proposed as the actuator types.

Control Immediately Following Launch

bloostar will be launched from a balloon. In order to ensure it does not impact the balloon shortly after launch it is dropped at an angle so that it can fly around the balloon before commencing its climb. The flight phase immediately following launch is complex, it is not possible to ignite the rocket engines while the vehicle is still hanging below the balloon. Therefore, it is necessary to try and ignite all the engines simultaneously in order to avoid the vehicle entering into a spin. However, it is unlikely that all 13 engines can be ignited simultaneously, therefore testing has been conducted to determine whether the control system is capable of controlling the vehicle while all the engines fire and throttle up. In order to simulate this a delay was introduced between the time the vehicle was dropped from the balloon and the ignition of each engine individually. The results of this test are shown in Figure 10. This initial test shows that bloostar is still controlled despite the non-simultaneous engine start, however launch remains a complex area for the control system and therefore remains an area of significant further work.

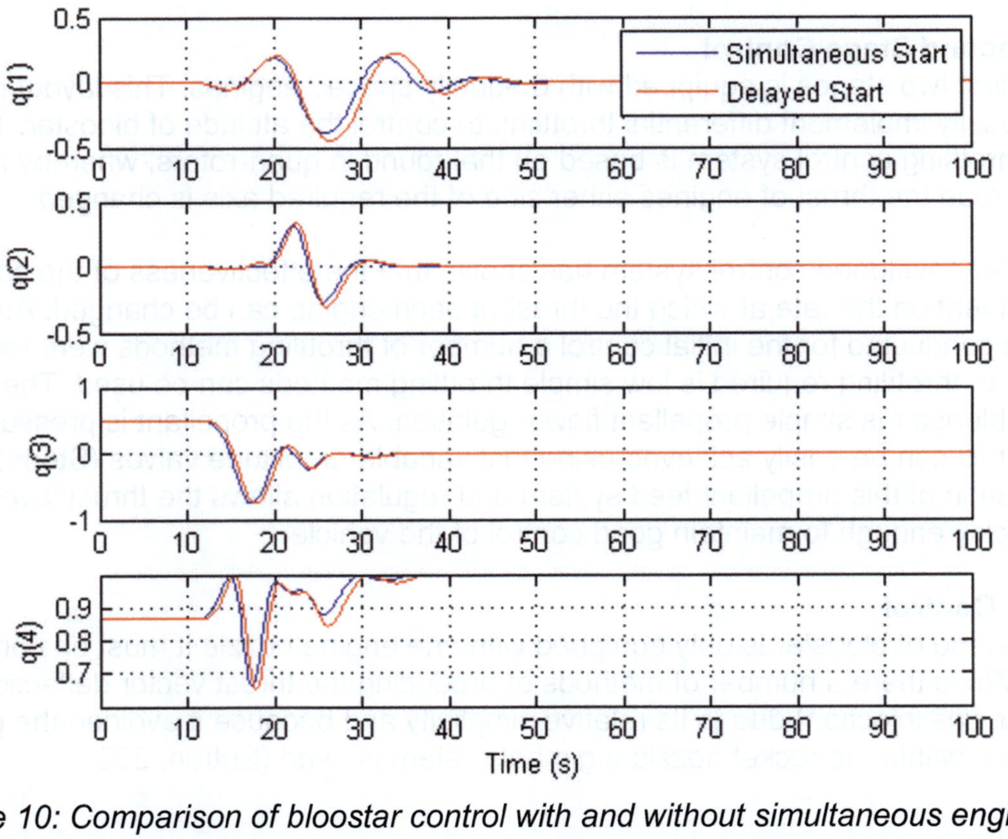

Figure 10: Comparison of bloostar control with and without simultaneous engine start

Conclusion

For the last 30 years we've been launching payloads to orbit either from the ground or from an aircraft. This is about to change, since a new way is coming. By decoupling the problems of acceleration and getting above most of the atmosphere bloostar can service the nascent micro and nanosatellite markets with significant advantages. The cost of a single launch is €4M, discounts for bulk orders are possible. The performance that can be achieved from our base in the Canary Islands is the following:

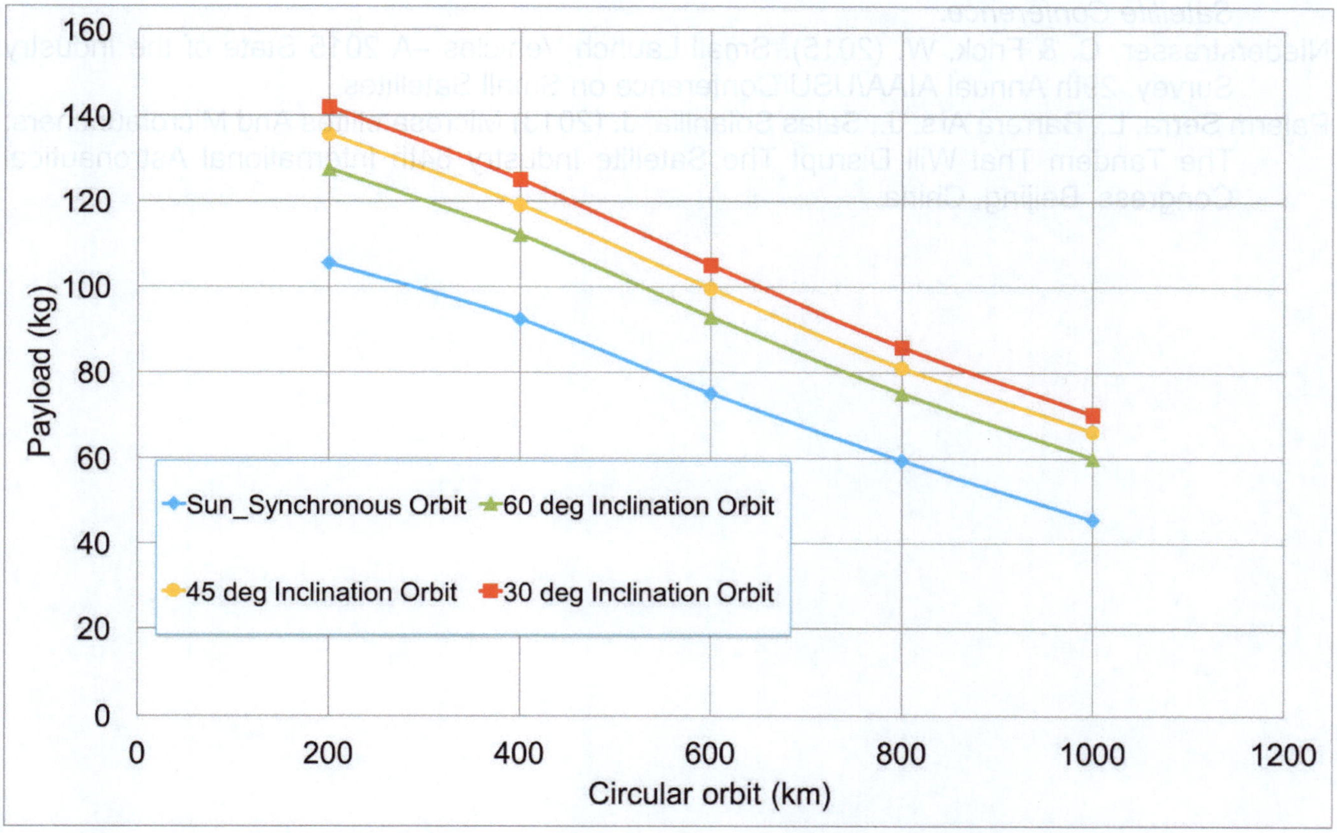

Figure 11: Payload delivered to circular orbits from the Canary Islands

References

Beerer, I. (2014). Modeling dispersions in initial conditions for air launched rockets and their effects on vehicle performance. *Masters Thesis*. Massachusetts Institute of Technology.

Crisp, N., Smith, K., & Hollingsworth, P. (2014). Small Satellite Launch to LEO: A Review of Current and Future Launch Systems. *Transactions of the Japan Society for Aeronautical and Space Sciences, Aerospace Technology Japan, 12*, 1-9.

Hayward, K. (2015). Control System Design for a New Balloon Launched Small Satellite Launch Vehicle. *MSc Thesis*. Space Research Centre, Cranfield University, UK.

Marshall, W., & Boshuizen, C. (2013). Planet Labs remote sensing satellite system. *Small Satellite Conference.*

Niederstrasser, C. & Frick, W. (2015). Small Launch Vehicles –A 2015 State of the Industry Survey. 29th Annual AIAA/USU Conference on Small Satellites

Palerm Serra, L., Barrera Ars, J., Salas Solanilla, J. (2013) Microsatellites And Microlaunchers: The Tandem That Will Disrupt The Satellite Industry 64th International Astronautical Congress, Beijing, China.

13th Reinventing Space Conference, Oxford, UK
Presented 10 November 2015

Reinventing constellations: the effectiveness of rideshare approaches for constellation deployment

Joseph W. Gangestad
The Aerospace Corporation, 2310 E. El Segundo Blvd., El Segundo, CA, 90245,
joseph.gangestad@aero.org

James R. Wilson, Kristin L. Gates, and John V. Langer
The Aerospace Corporation, 2310 E. El Segundo Blvd., El Segundo, CA

ABSTRACT

CubeSats have achieved growing credibility among government and commercial stakeholders as a valid architecture for future space systems, and many groups have proposed fielding CubeSat constellations for applications ranging from space-weather monitoring to space-based surveillance. Due to their small size and mass, a large number of CubeSats can be lofted to orbit to yield resilient constellations with short revisit times and global coverage. However, the challenge of reaching the application-specific orbits necessary for some proposals is often neglected, and although several small-satellite launchers are in development, rideshare is likely to remain the most reliable access to space for CubeSats for the foreseeable future.

This paper evaluates the near-term feasibility and performance of several multi-satellite CubeSat mission concepts by constraining constellation designs to those that could be assembled using the current and prospective rideshare manifests. For several promising CubeSat constellation applications, we assess how different combinations of these rideshare opportunities yield better or worse performance and also how many CubeSats in such rideshare-initiated constellations are necessary to achieve the mission goals. We also identify launches in upcoming years that, if outfitted with rideshare capability, would have the most positive impact on enabling these applications. With this characterization of the trade space for future constellations, acquisition agencies can design more streamlined space architectures that take judicious advantage of limited rideshare openings, and rideshare intermediaries will have some guidance on where the CubeSat community could most benefit from the implementation of rideshare capability.

Introduction

The class of very small satellites called CubeSats is part of a revolution in space. Enabled primarily by the establishment of a standardized container which is being incorporated in an ever-growing number of launch vehicles, the CubeSat approach has made rideshare access to space a commodity. As a result, this tiny form factor has enabled dramatic steps forward in the advancement of space technologies, rapid prototyping, and proof-of-concept endeavors. This paper focuses on the next step: can these "new space" techniques extend to enable missions of national or commercial significance?

One of the key differences between a technology demonstration and a space system mission is that the latter is typically characterized by the constellation it requires. If a space system's requirements compel a specific constellation, and if this constellation in turn requires a dedicated launch or launch regimen, then the cost and schedule benefits of the ridesharing CubeSat paradigm are challenged. On the other hand, populating a constellation merely by taking whatever rideshare opportunities are at hand would produce a constellation whose geometric properties are unknown beforehand. Figure 1 illustrates this difference in two cases that

S. Hatton (ed.), *Proceedings of the 13th Reinventing Space Conference*,
https://doi.org/10.1007/978-3-319-32817-1_18

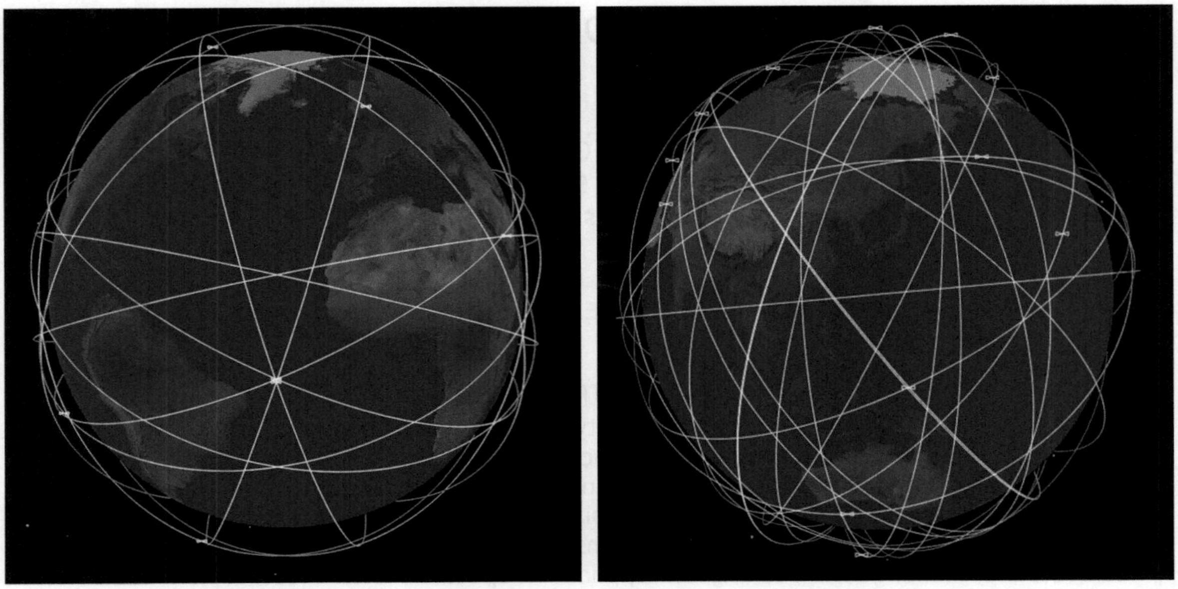

Figure 1: The orbits of a planned, dedicated launch constellation on the left, show typical symmetry. The rideshare-initiated constellation on the right appears random and may be more challenging to analyze.

will be discussed later in this paper. Most earth-orbiting space missions require a significant level of certainty in the constellation geometry, typically to ensure visibility for observation, connectivity for communications, or both.

This paper develops an approach to determining the potential mission utility of a rideshare-initiated constellation. It starts by exploiting a fundamental characteristic of the otherwise random constellation: at any given time, the manifest of rideshare opportunities is fairly well known for the next two years. Next, this paper examines appropriate statistics for assessing the resulting constellations. These two items are combined via a robust framework to identify and characterize the mission utility of a rideshare-based approach.

Rideshare Manifest

The analysis in this study requires a reference manifest: a schedule of rideshare opportunities against which a constellation based mission might be assessed. In developing such a manifest, it is essential to reflect the realities facing a mission planner at a given point in time.

One of the challenges of planning a rideshare constellation is knowing what rideshare opportunities are available. At the time a mission planner might look at opportunities, the upcoming 12 months will be fairly well known, and the subsequent 12 months somewhat less so. Beyond that, rideshare opportunities become harder to predict, with perhaps one per year that have high certainty.

The reference manifest used for the analysis in this paper reflects this level of uncertainty. The manifest, outlined in Figure 2, is set in early 2015. It comprises a total of 24 launch

Date	Apogee	Perigee	Inclination
Apr 2015	423	418	51.6
May 2015	830	670	98.4
May 2015	700	350	55
Jun 2015	423	418	51.6
Jun 2015	500	500	45
Jun 2015	390	390	63
Jul 2015	720	720	86
Aug 2015	720	500	65
Sep 2015	423	418	51.6
Nov 2015	550	550	31
Nov 2015	423	418	51.6
Dec 2015	560	560	97.6
Dec 2015	423	418	51.6
Dec 2015	836	836	98.8
Feb 2016	421	418	51.6
Mar 2016	720	720	24.5
May 2016	520	520	24
May 2016	421	418	51.6
Jun 2016	620	620	98.9
Jun 2016	680	680	98.1
Jun 2016	600	500	63.4
Dec 2016	836	836	98.8
Nov 2017	550	450	94
Jun 2018	700	350	55

Figure 2: A manifest based on early predictions of available rides for 2015-2016, plus one additional ride in each of 2017 and 2018.

opportunities, with 14 in the first year, 8 in the next, and just one launch in each of the two out years. Reflecting the known cubesat-carrying opportunities, the manifest includes a significant number of sun-synchronous orbits and International Space Station resupply missions. The constellation shown on the right side of Figure 1 is a visualization of this manifest, propagated to 2019. Details as to how this manifest was selected, how initial conditions were determined, and how orbit propagation is modeled can be found in [1].

For the remainder of this paper, the manifest will be held fixed. In an operational environment, it would be prudent to review the manifest at regular intervals, updating analyses and plans accordingly.

Performance Metrics

When developing a mission-focused constellation, the mission planner has to address a number of metrics. These include cost and schedule, which is often associated with the number of satellites or launches, and geometric concerns such as visibility to various points on the earth or to space-based objects. This paper assumes a mission where a the constellation must

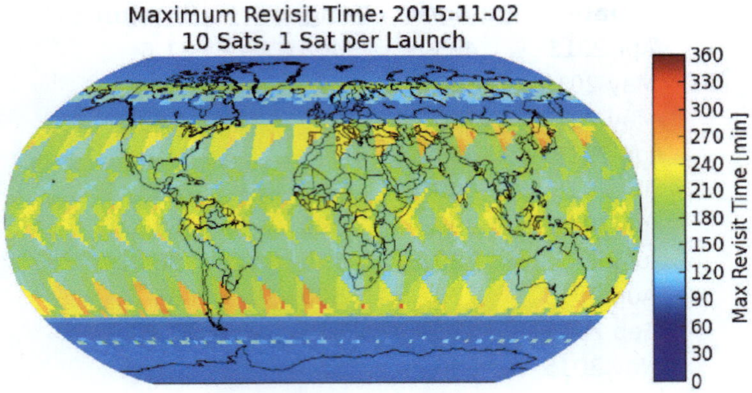

Figure 3: Maximum revisit time for the 7-day period starting 2 Nov 2015 with the reference rideshare manifest, assuming 1 satellite per rideshare.

provide a high degree of availability to every point on the globe. Specific metrics of interest in such missions are

- **Revisit Time**: the time interval between two periods of visibility to a ground point from one or more satellites. During a propagation, many such intervals of varying duration accumulate, and revisit time can be expressed as a statistic of these intervals. For example, "maximum revisit" would be the longest period of time over a propagation that a ground point does not see a satellite.
- **Daily Visibility**: the total amount of time each day that one or more satellites is visible to a ground point. When considering propagations over many days, this metric could be expressed as an average, minimum, maximum, or other statistic.

For example, Figure 3 shows the maximum revisit time for the 7-day period starting 2 Nov 2015. Each point on the Earth is color-coded according to the maximum revisit time: regions with short revisit are in blue, and regions with longer revisit are in orange and red. The constellation used to generate this example employed the first 10 launches in the reference manifest, assuming one successful satellite deployment per rideshare opportunity.

Figure 4 shows this same statistic, but for the entire four-year interval of the manifest. In one case, the plot assumes one satellite for every rideshare opportunity. In the other, the assumption is two satellites per rideshare. Each of the plots shows, on the left vertical axis, the worst case revisit for the constellation at that point. On the right vertical axis is shown the total number of satellites in the constellation at that time.

Note that the random aspects of the rideshare deployment lead to intervals of more and less favorable orbit phasing, which in turn leads to irregular performance over time. In some cases, even as the constellation adds more satellites, the global performance worsens. Even with two satellites successfully deployed on each of the 24 rideshare opportunities, the global max revisit remains on the order of two hours.

Figure 5, on the other hand, looks at the same constellation with an alternate metric. Rather than insisting on the extreme performance associated with a global maximum, these plots reflect the 95th percentile revisit time. This may be an acceptable mission requirement for a

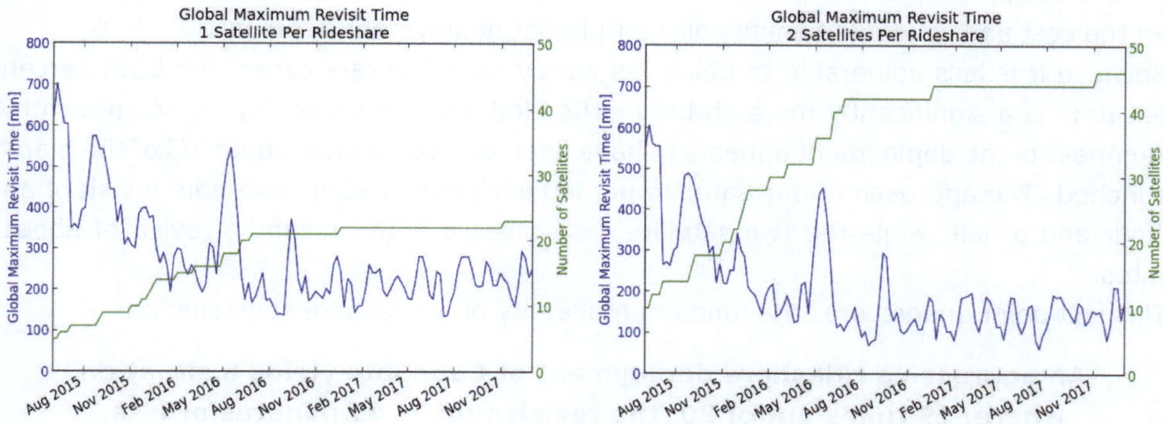

Figure 4: Global maximum revisit time (in minutes) for the representative rideshare manifest. The left plot assumes the successful deployment of one satellite per launch, the right plot assumes two satellites per launch.

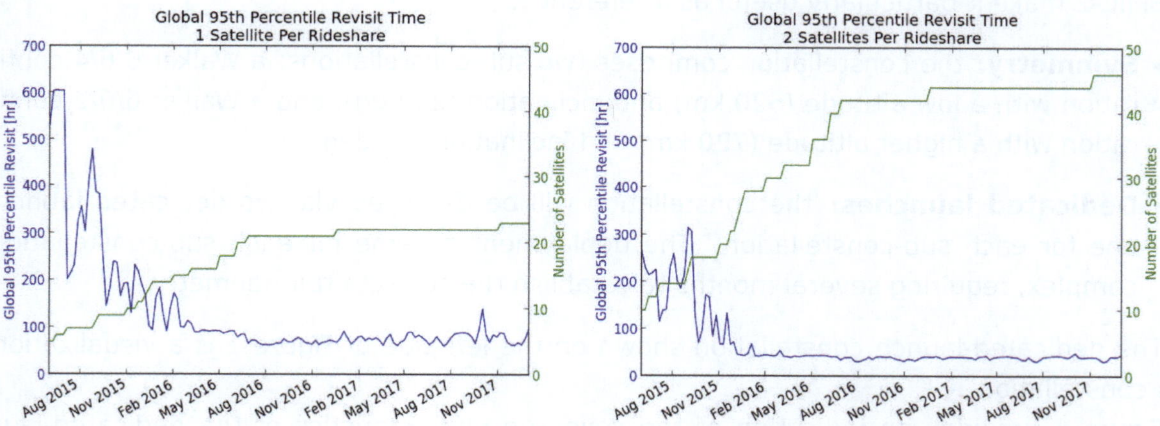

Figure 5: Global 95th percentile revisit time (in minutes) for the representative rideshare manifest. The left plot assumes the successful deployment of one satellite per launch, the right plot assumes two satellites per launch.

CubeSat class system, where prudent concessions in performance are considered reasonable given the cost and schedule benefits of the CubeSat deployment approach.

Because it is less vulnerable to being led astray by a few rare cases, the 95th percentile revisit time is a significantly more stable metric. Indeed, the variability associated with the randomness of the deployment appears to have been overcome after about 2/3 of the manifest is launched. The approach of one satellite per launch yields a 95th percentile revisit of about an hour and a half, while the two satellite case gives a 95th percentile revisit of about 45 minutes.

This last point underscores the fundamental utility of a rideshare constellation:

An aggressive rideshare deployment of CubeSats yields a situation where, 19 times out of 20, the revisit time is 45 minutes or less.

Rideshare versus Dedicated Launch

Deeper insight into the utility of a rideshare-initiated constellation may be found by comparing to a traditional dedicated-launch constellation. This paper considers a constellation modeled on the upcoming COSMIC-2 system.

The primary mission of COSMIC-2 is radio occultation of Global Navigation Space System signals, rather than earth coverage. But the resulting requirements for global availability are similar in character to those pursued in the previous sections of this paper.

See [4] for details about the COSMIC-2 constellation design. However, two key elements of COSMIC-2 make it particularly useful as a reference:

- **Symmetry:** the constellation comprises two sub-constellations: a Walker 6/6/4 configuration with a low altitude (520 km) and inclination (24 deg); and a Walker 6/6/2 configuration with a higher altitude (720 km) and inclination (72 deg).
- **Dedicated launches:** the constellation will be deployed via two dedicated launches, one for each sub-constellation. The deployment scheme for each sub-constellation is complex, requiring several months to establish the correct orbit geometry.

The dedicated-launch constellation shown on the left side of Figure 1 is a visualization of this constellation.

Figure 6 provides an indication of the coverage characteristics of the dedicated-launch constellation in terms of the two revisit type performance metrics. For assessing mission utility, there is little difference whether the revisit statistic is viewed as a maximum or a 95th percentile. This is because the symmetric structure does not allow for the regions of anomalous visibility seen in the rideshare-initiated case. The dedicated launch constellation has revisits on the order of 90 minutes in the mid-latitudes, and revisits closer to 45 minutes at the poles and equator.

Figure 7 shows the revisit statistic versus latitude. Each plot compares the dedicated-launch constellation with the one- and two-satellite per launch rideshare-initiated constellations. The left plot uses the maximum revisit time, and the right plot uses the 95th percentile revisit.

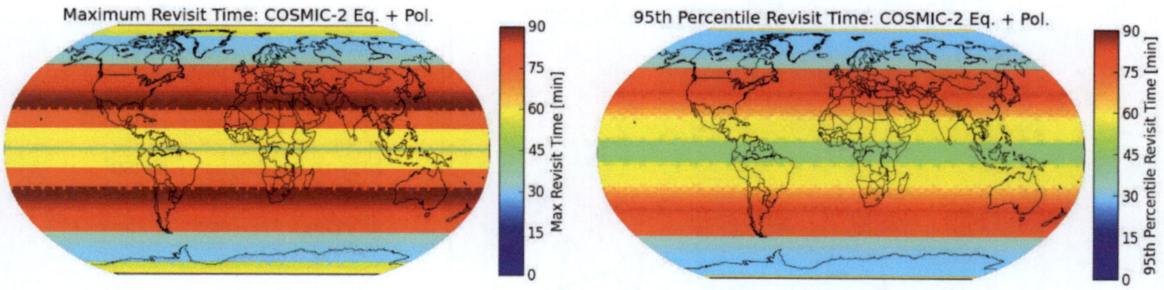

Figure 6: The left contour map shows maximum revisit time (in minutes) for the dual Walker reference constellation, the right contour maps shows 95th percentile revisit time for the same constellation.

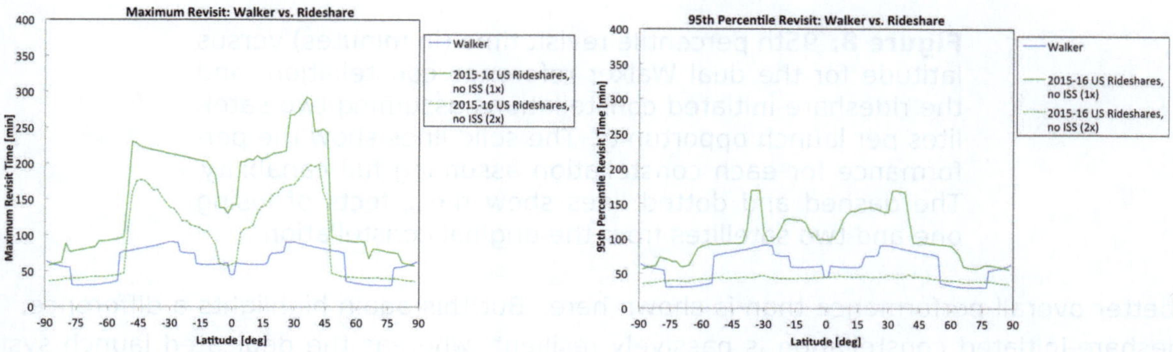

Figure 7: A comparison of the performance using the maximum revisit metric on the left and the 95th percentile metric on the right. In each case, symmetric reference constellation is compared against rideshare-initiated constellations assuming one- and two-satellites per launch.

Note that, with two-satellites per rideshare, the rideshare-initiated constellation actually outperforms the dedicated launch constellation.

Resilience

In addition to improved performance, the redundancy in the rideshare-initiated constellation provides a measure of resilience against the loss of a spacecraft. The dedicated launch constellation assessed here provided comparable utility with a comparatively smaller number of satellites in large part because of the constellation's symmetry. This same characteristic works against the constellation performance when one or two satellites are lost. Figure 8 shows the effect of losing one or two satellites from each constellation. The loss of a single satellite from the dedicated-launch constellation shows in immediate degradation in performance. The rideshare-initiated constellation, on the other hand, shows virtually no loss.

This example shows the effect immediately following the loss of one or two satellites. In reality, a dedicated launch constellation would re-phase in this situation, perhaps resulting in

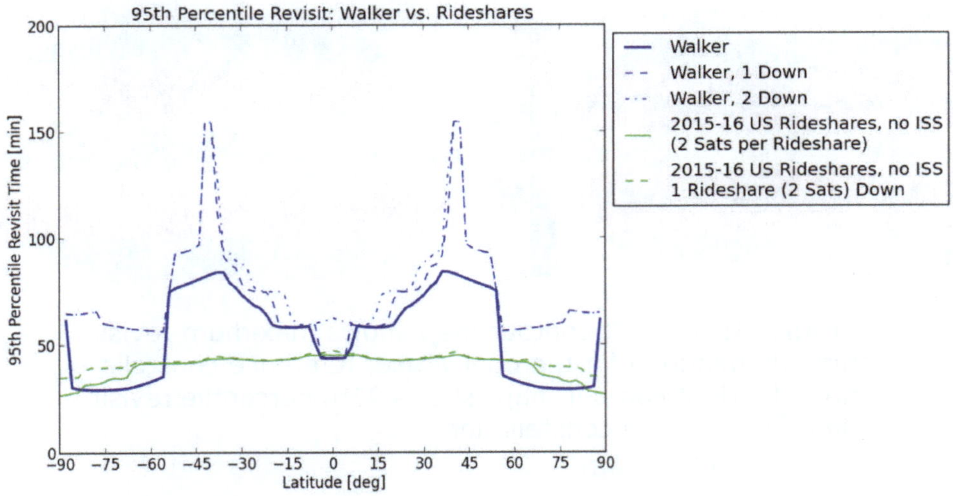

Figure 8: 95th percentile revisit time (in minutes) versus latitude for the dual Walker reference constellation, and the rideshare-initiated constellations assuming two satellites per launch opportunity. The solid lines show the performance for each constellation assuming full capability. The dashed and dotted lines show the effects of losing one and two satellites from the original constellation.

a better overall performance than is shown here. But this again highlights a difference. The rideshare-initiated constellation is passively resilient, whereas the dedicated launch system requires planning and maintenance.

Combining the results on performance versus latitude and resilience yields the following:

An aggressive rideshare deployment of CubeSats yields a situation where, 19 times out of 20, the constellation will outperform a comparable dedicated-launch constellation both in coverage and in resilience.

Strategies for Rideshare Constellations

The intent of the discussion in the previous sections was to illustrate, in terms of mission utility, qualitative aspects of rideshare-initiated constellations: their utility is seen in cases where 95th percentile ("19 times out of 20") type requirements are acceptable; their performance, given a large enough manifest, can be comparable to some dedicated-launch type constellations; they exhibit an inherent resiliency due to the variability resulting from many rideshare opportunities.

These qualitative perspectives of necessity overlook key constraints on mission design. It may not be practical to negotiate rideshare access with every single opportunity. Moreover, even with CubeSat based systems cost and schedule issues come into play. This section looks at strategies for determining the most effective use of rideshare opportunities to achieve constellation mission requirements.

The Aerospace Corporation has developed a strategic analysis framework based on a multiple-

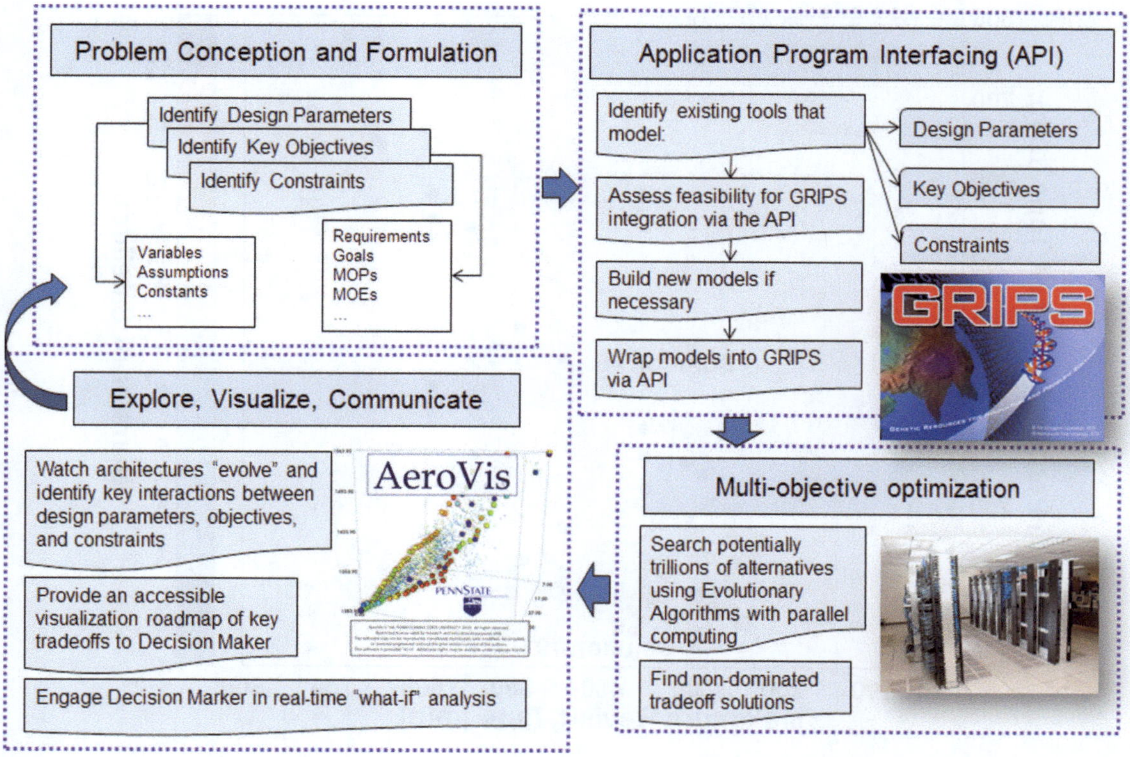

Figure 9: The GRIPS decision support process.

objective optimization approach and a suite of associated visualization tools. The framework, as well as the underlying optimization algorithm, are both referred to as GRIPS – an acronym for Genetic Resources for Innovation and Problem Solving. Specific details of the optimization algorithm and the underlying approach for GRIPS are discussed in [7], [8], and [9].

Figure 9 provides an overview of the GRIPS decision support process. GRIPS uses an evolutionary algorithm to find non-dominated solutions to a problem subject to several different, often conflicting, objectives. In the context of this paper, the goal is to find a series of rideshare selections (the solutions) that provide the best performance on several different coverage metrics (the objectives), such as maximum revisit time and average revisit time. The "non-dominated" solutions identified by GRIPS are equally optimal across all objectives simultaneously. That is, each solution cannot be changed to improve one metric without suffering on one or more other metrics. For example, if one non-dominated solution (a series of rideshare selections) provides the lowest maximum revisit time but a high average revisit time, another solution with a lower average revisit time must necessarily have higher maximum revisit time.

To better understand the value of the opportunities in the reference launch manifest, GRIPS was configured to select constellations based on subsets of that manifest. Specifically, GRIPS was constrained to constellations of no more than 12 satellites, using the reference manifest, that are optimal in terms of the following:

- low Global maximum revisit time
- low 95th percentile revisit time
- low mean average revisit time

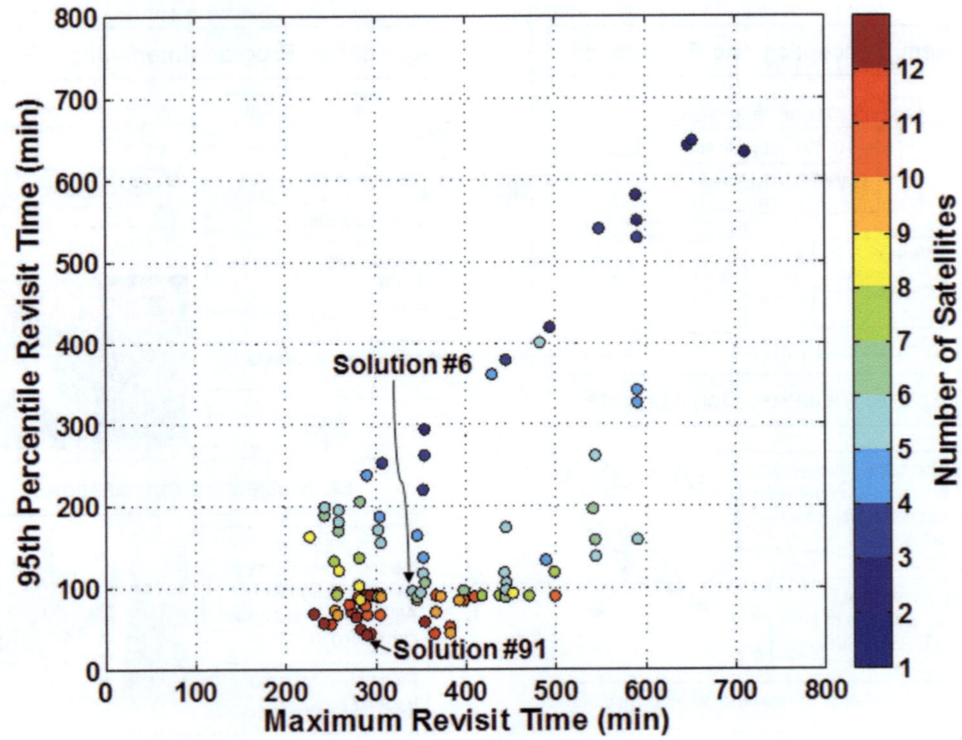

Figure 10: A visualization of the 96 non-dominant rideshare constellations identified by GRIPS. The axes show maximum and 95 percentile revisit times, and the colors reflect the number of satellites in the rideshare-initiated constellation. Solutions #6 and #91 are identified.

- high "minimum total visibility"
- low number of satellites

In this case, out of many millions of possible permutations, GRIPS converged on 96 solutions. Figure 10 shows these solutions, visualized against three attributes: the 95th percentile revisit time (vertical axis), the global maximum revisit time (horizontal axis), and the number of satellites (color).

This visualization identifies a number of characteristics of this manifest. In particular, there are no solutions with a maximum revisit time of less than about four hours, and solutions with a 95th percentile revisit under 90 minutes will require at least nine satellites.

For better insight into the GRIPS result, two solutions – #6 and #91 are examined more closely. Solution #6 comprises five satellites over five launches, and produces a 95th percentile revisit time of about 90 minutes. Solution #91 represents 12 satellites in six launches, and provides 95th percentile revisit times of 45 minutes – about half that of #6. Figure 11 shows this performance on a contour plot. In addition to the 95th percentile revisit, GRIPS also looks to optimize the average revisit. The average revisit performance for these two GRIPS solutions is shown in Figure 12. For both of these revisit statistics, in is interesting to note that the GRIPS solutions provide coverage that is largely uniform coverage across latitudes.

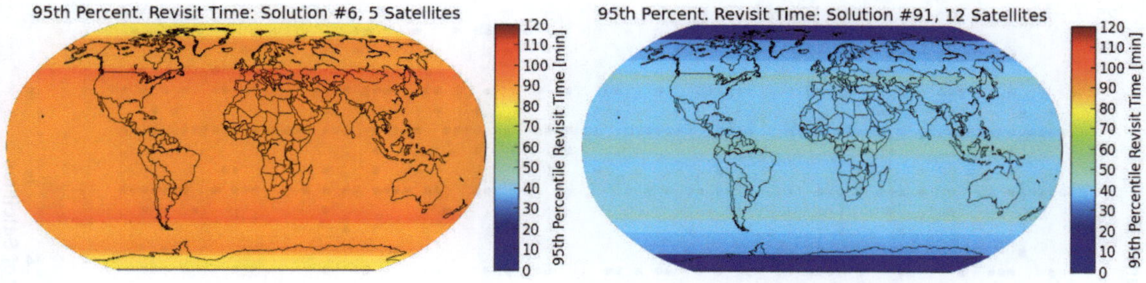

Figure 11: The 95th percentile revisit time for two rideshare-initiated constellations found by GRIPS.

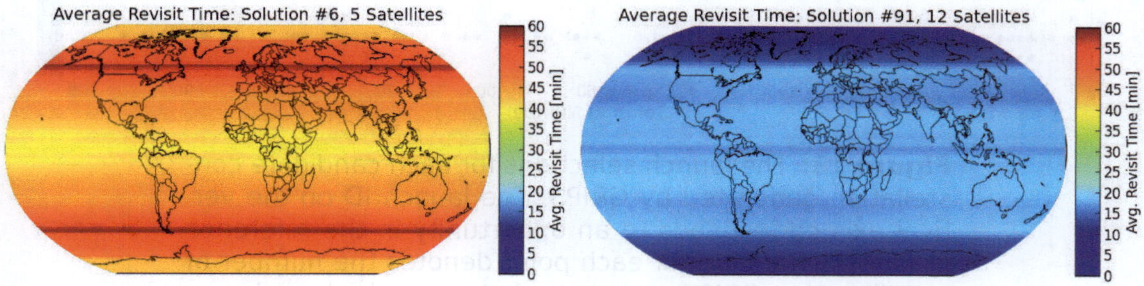

Figure 12: The average revisit time for two rideshare-initiated constellations found by GRIPS.

More details on the specific launch schedules and other attributes of these solutions are discussed in [1]. The 95th percentile revisit performance of solution #91, however, compels the following observation:

With 12 satellites on six rideshare launches, each point on the earth will see a revisit, 19 times out of 20, of 45 minutes or less.

Figure 13 shows the GRIPS rideshare selections for all 96 solutions. The influence of specific rideshare opportunities is illustrated visually by the density of solutions associated with six particular launches: 2, 8, 12, 16, and 19. Deeper review of the orbit attributes associated with these launches (see [1]) shows that these represent four sun-synchronous orbits and two low-inclination orbits. This suggests that, as a mission developer proceeds into deeper analysis, these six opportunities are worth specific attention.

Conclusion

Opportunities for access to space are increasing dramatically, particularly as standardized deployment schemes become more widespread. However, this access comes with a fundamental restriction – as a secondary payload, a rideshare based mission must accept (or decline) an orbit determined by the primary launch mission. While this constraint may be acceptable for single-launch technology demonstrations, it presents a challenge to the design of a mission-oriented constellation.

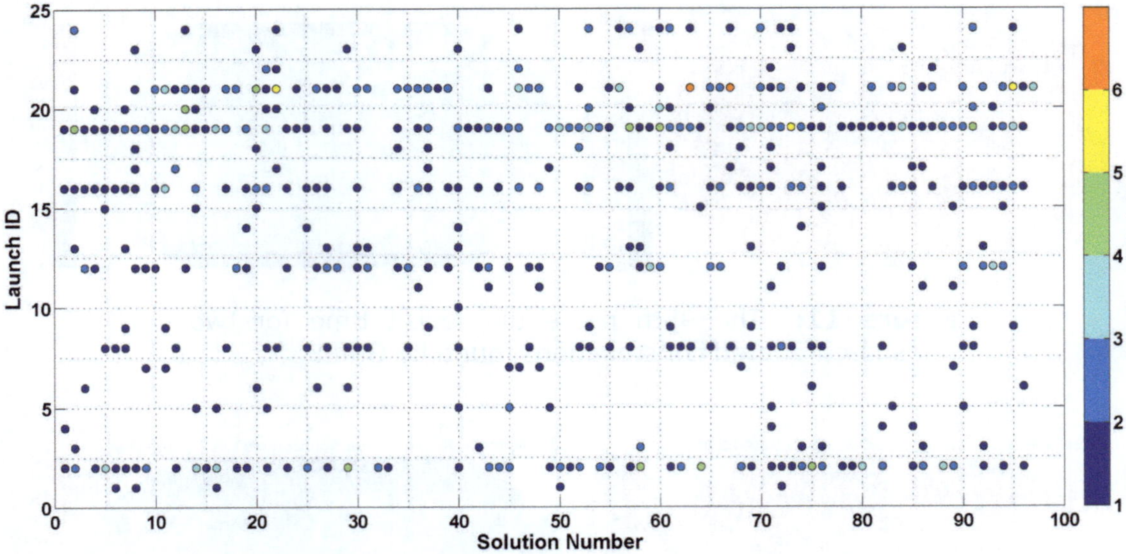

Figure 13: The launch selections for each candidate constellation generated by GRIPS. The launch ID on the vertical axis corresponds to an opportunity in the reference manifest. The color of each point denotes the number of satellites that GRIPS assigns to that particular launch.

This paper shows that rideshare-initiated constellations can provide significant mission utility if the requirements can tolerate occasional exceptions. For example, when revisit statistics are framed as 95th percentile – or "19 times out of 20" – a rideshare-initiated constellation can be competitive in performance to a dedicated-launch, symmetric constellation. Moreover, rideshare-initiated constellations can be inherently resilient.

Additionally, the power of evolutionary algorithms makes it possible to explore the large tradespace of rideshare selections and combinations to identify scenarios that achieve the best performance with the fewest number of satellites. Specifically:

The GRIPS evolutionary algorithm found a 12-satellite, 6-launch rideshare-initiated constellation that matched the performance of launching two satellites on each of the 24 rideshare launches in the reference manifest.

Every space system mission has unique aspects, and many objectives to optimize. A rideshare-initiated constellation may or may not be the best approach. The examples discussed in this paper, however, indicate that a rideshare-based approach may be worth considering when assessing options in space system mission design.

References

[1] Gangestad, J.W., Wilson, J.R., Gates, K.L., Langer, J.V., "Rideshare-Initiated Constellations: Future Cubesat Architectures with the Current Launch Manifest," in *31st Space Symposium, Technical Track*, Colorado Springs, CO, 2015.

[2] Buchen, E., "Small Satellite Market Observations," in *AIAA/USU Conference on Small Satellites*, Logan, UT, 2015, SSC15-VII-7.

[3] Nugent, R., Munakata, R., Chin, A., Coelho, R., and Puig-Suari, J., "The CubeSat: The Picosatellite Standard for Research and Education," in *AIAA SPACE 2008 Conference and Exposition*, San Diego, CA, 2008, AIAA Paper 2008-7734.

[4] Juang, J.-C., Tsai, Y.-F., and Chu, C.-H., "On Constellation Design of Multi-GNSS Radio Occultation Mission," *Acta Astronautica*, Vol. 82, Issue 1, 2013, pp. 88–94.

[5] Silva Curiel, A., Lambert, M., Liddle, D., Sweeting, M., Chu, C.-H., Fong, C.-J., and Chang, G.-S., "Introduction to FORMOSAT-7/COSMIC-2 Mission," in *AIAA/USU Conference on Small Satellites*, Logan, UT, 2013, SSC13-VI-1.

[6] Gangestad, J.W., Hardy, B.S., and Hinkley, D.A., "Operations, Orbit Determination, and Formation Control of the AeroCube-4 CubeSats," in *AIAA/USU Conference on Small Satellites*, Logan, UT, 2013, SSC13-X-4.

[7] Ferringer, M., and Spencer, D., "Satellite Constellation Design Tradeoffs Using Multiple-Objective Evolutionary Computation," *Journal of Spacecraft and Rockets*, Vol. 43, No. 6, 2006, pp. 1404–1411.

[8] Ferringer, M., Clifton, R., and Thompson, T., "Efficient and Accurate Evolutionary Multi-Objective Optimization Paradigms for Satellite Constellation Design," *Journal of Spacecraft and Rockets*, Vol. 44, No. 3, 2007, pp. 682–691.

[9] Ferringer, M., DiPrinzio, M., Thompson, T., Hanifen, K., Whittecar, W., and Reed, P., "A Framework for the Discovery of Passive-Control, Minimum Energy Satellite Constellations," in *AIAA SPACE 2014 Conference and Exposition*, San Diego, CA, 2014, AIAA Paper 2014-4158.

MicroLaunch: The Electric Rocket

A. Bacon – Thales Alenia Space UK Ltd
andrew.bacon@thalesaleniaspace.com

13th Reinventing Space Conference 9-13th November 2015, Oxford, UK

ABSTRACT

MicroLaunch is a potentially game changing development for access to space for nano and pico satellites (1 to 50kg) which today are being used in ever greater numbers by commercial providers (e.g. Skybox Imaging, Planet Labs), universities, domestic space agencies and military research organisations.

These satellites are poorly served by today's launch marcet, generally having to 'piggy back' as secondary payloads on large launchers where commercial prices can reach up to €50,000/kg, the number of launch opportunities is unpredictable, the range of possible orbits are limited and the schedule is set by the primary payload (for example; problems with the primary payload delayed UKube-1's launch by two years). Dedicated launches can be purchased to overcome some of these issues, but at a greatly increased total cost.

MicroLaunch represents novel developments in the field of microwave beamed launchers; a type of orbital launch system which generates thrust from wireless energy transmitted to it from a specialised ground station. By removing the need to store energy in the propellant, inert fuels and thruster fuel efficiencies several times greater than chemical rockets can be achieved; gains which enable a single-stage-to-orbit system that can be reused hundreds of times.

The proposed design could offer dedicated launches for payloads up to 1,000kg, very high launch rates (five per day) and significantly lower cost access to polar orbits (targeting <€3,000/kg). All elements of the MicroLaunch system are highly suited to development, build and launch in the UK and could provide a significant boost to the UK's fast growing small satellite capabilities.

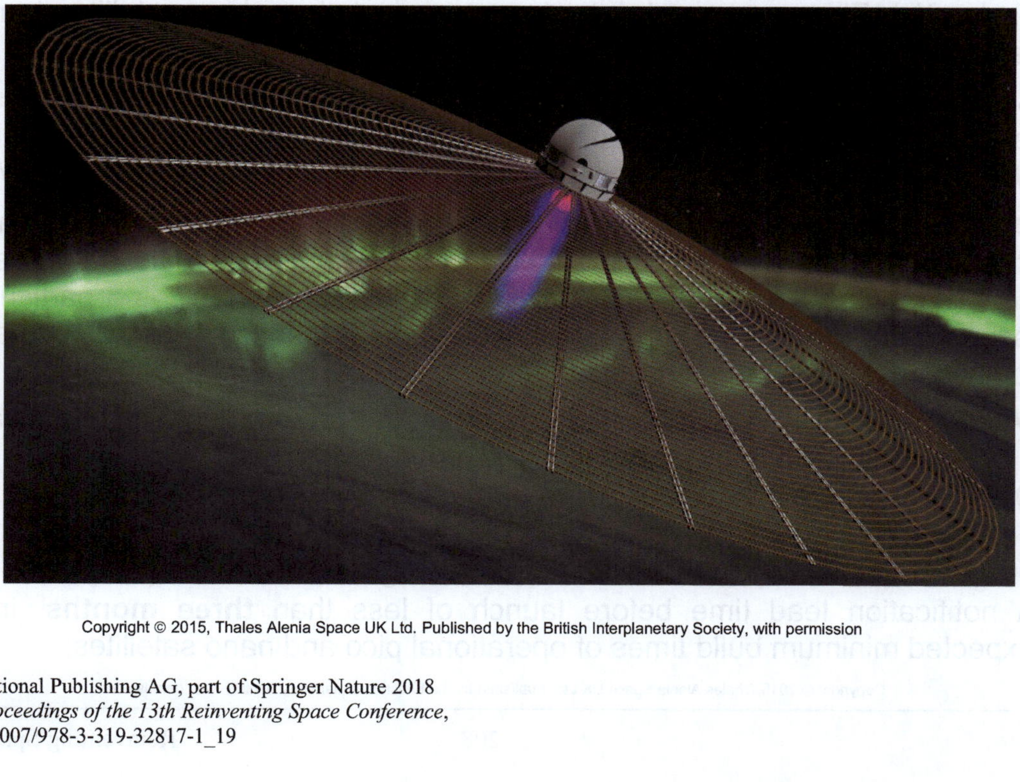

S. Hatton (ed.), *Proceedings of the 13th Reinventing Space Conference*,
https://doi.org/10.1007/978-3-319-32817-1_19

1. THE SMALL SATELLITE LAUNCH PROBLEM

Pico, nano and micro class satellites (1 to 200kg) are being used in ever greater numbers by commercial providers (e.g. Skybox Imaging, Planet Labs), universities, domestic space agencies and military research. These classes of satellites have the greatest potential for growth in Earth Observation (EO) and telecommunications applications as they are highly suited for rapidly establishing high temporal and moderate spatial resolution constellations. However, they are poorly served by today's launcher marcet for the following key reasons:

- **High relative prices:** For large satellites (>1,000kg) launch prices to Low Earth Orbit (LEO) are typically in the range of €10,000/kg to €20,000/kg and over €30,000/kg for launch into Geostationary Earth Orbit (GEO)[i]. However, launch prices do not scale down well for smaller payloads due to the large overheads of handling, qualification and certification. Commercial launch prices per kilogram for pico satellites (1 to 10kg) can be double that of small satellites (200 to 1,000kg) for launch into LEO and over four times more for GEO. This non-proportional scaling of launch prices to payload mass makes it difficult to justify the fractionation of capabilities into several smaller satellites rather than a single larger satellite, despite the advantages for coverage and redundancy.
- **Low availability:** The smallest payload capacity of any currently available launcher is 433kg (Pegasus XL) and so pico to micro class satellites are generally flown as secondary payloads alongside large primary payloads. This severely limits the number of launch opportunities available for small payloads as it is rarely financially viable to fund a launch with less than the maximum payload capacity. Whilst the number of launchers offering places for secondary payloads has significantly increased over the past decade, the total number of launches per year has not and so there is still a limit to the availability of places.
- **Lead time:**Launch providers generally require notification of at least 18 months [ii] for acceptance of secondary payloads due to the need to ensure compatibility with primary payload(s). This creates a barrier to responsively launching satellites to meet a specific temporal need such as an emerging conflict or natural disaster.
- **Schedule risk:** When launching as a secondary payload there are schedule risks as the launch can be delayed by issues with the primary payload(s), as was the case of the UK's TechDemoSat and UKube-1 which had their launch delayed by two years.
- **Orbit restrictions:** The target orbit for launches is, in most cases, primarily dictated by the needs of the primary payload which means that secondary payloads often have to deal with sub-optimal orbits for their particular missions. This is a particularly significant issue for the rapid launching of multi-plane constellations for global coverage as it is rare to find subsequent launches into different planes at the same altitude.

Thales Alenia Space UK (TAS-UK) have analysed the requirements for the ideal launch small satellite system based upon the expected requirements post 2025:

- Dedicated launch of satellites **up to 200kg** into Low Earth Orbits (LEO) of 150 to 1500km altitude;
- Frequent launch opportunities on the order of greater than **300 per year**;
- A notification lead time before launch of less than **three months**, in line with the expected minimum build times of operational pico and nano satellites;

- Access to inclinations from 80 to 120° for near polar (ideal for telecommunications constellations) and sun synchronous orbits (ideal for Earth observation missions);
- A cost to the customer of less than **<€5,000 per kilogram** of payload into a 600km altitude 90° inclination orbit.

Whilst the development of several small satellite launch vehicles (Virgin Galactic's LauncherOne and Rocket Labs' Electron being notable examples) are currently underway none of these will come close to meeting the cost and frequency of launch required by the small satellite market in 2025. Current costs are projected to be up to €50,000/kg and launch frequencies no greater than 52 a year.

TAS-UK's MicroLaunch concept exploits novel developments in the field of microwave beamed electric propulsion, a potentially game changing type of orbital launch system for access to space, that can meet and potentially exceed all of these requirements.

2. THE MICROLAUNCH SOLUTION

All current satellite launchers store their energy inside the propellant which is released via exothermic chemical reactions. The need for the propellant to act as an energy store and reaction mass places fundamental limitations on the minimum mass (>20,000kg) and specific impulse (<500s IsP) for launchers leading to an average payload to total take-off mass ratio of <1%.

Beamed propulsion concepts can overcome these limitations by transmitting the required energy from the ground to the launcher whilst it is in-flight which, as they only require on-board propellant for reaction mass, enable thruster efficiencies of >1000s IsP and payload fractions of at least 3%.

To achieve a thrust to weight ratio high enough to enable launch from the Earth's surface, beamed propulsion for a small satellite launch vehicle requires over a Giga Watt (GW) of continuous power delivered from the ground to the launcher whilst it is under acceleration and consists of three primary elements;

1. Creation of a beam of wireless energy by a fixed ground station that tracks the launch vehicle;
2. The capture and conversion of the incident wireless energy on the vehicle;
3. The conversion of the captured energy into useful thrust.

The two primary forms of wireless energy that have previously been considered for beamed propulsion systems are lasers and microwaves. High power lasers currently have a very low input power to light conversion efficiency (typically <20%), require mechanical steering, have a low duty cycle, and encounter significant atmospheric refraction problems. High power microwave generators (klystrons, gyrotrons and now solid state) can achieve conversion efficiencies exceeding 70%, can be electronically steered via phased array techniques and are not strongly affected by clouds or atmospheric refraction. Therefore microwave based systems present the best choice for a beamed propulsion launch system and 10s of MW scale microwave generators already exist in long range radar systems.

TAS-UK's MicroLaunch concept takes advantage of advancements in the fields of microwave generation, rectification and electric propulsion to propose a reusable Single Stage To Orbit (SSTO) launcher capable of serving the needs of the pico, nano and micro class satellite market in the post 2025 timeframe.

2.1 Vehicle Design

The MicroLaunch ground to orbit vehicle consists of a single stage made up of the key elements shown in Figure 2-1 below. The function of each is described in the following sections.

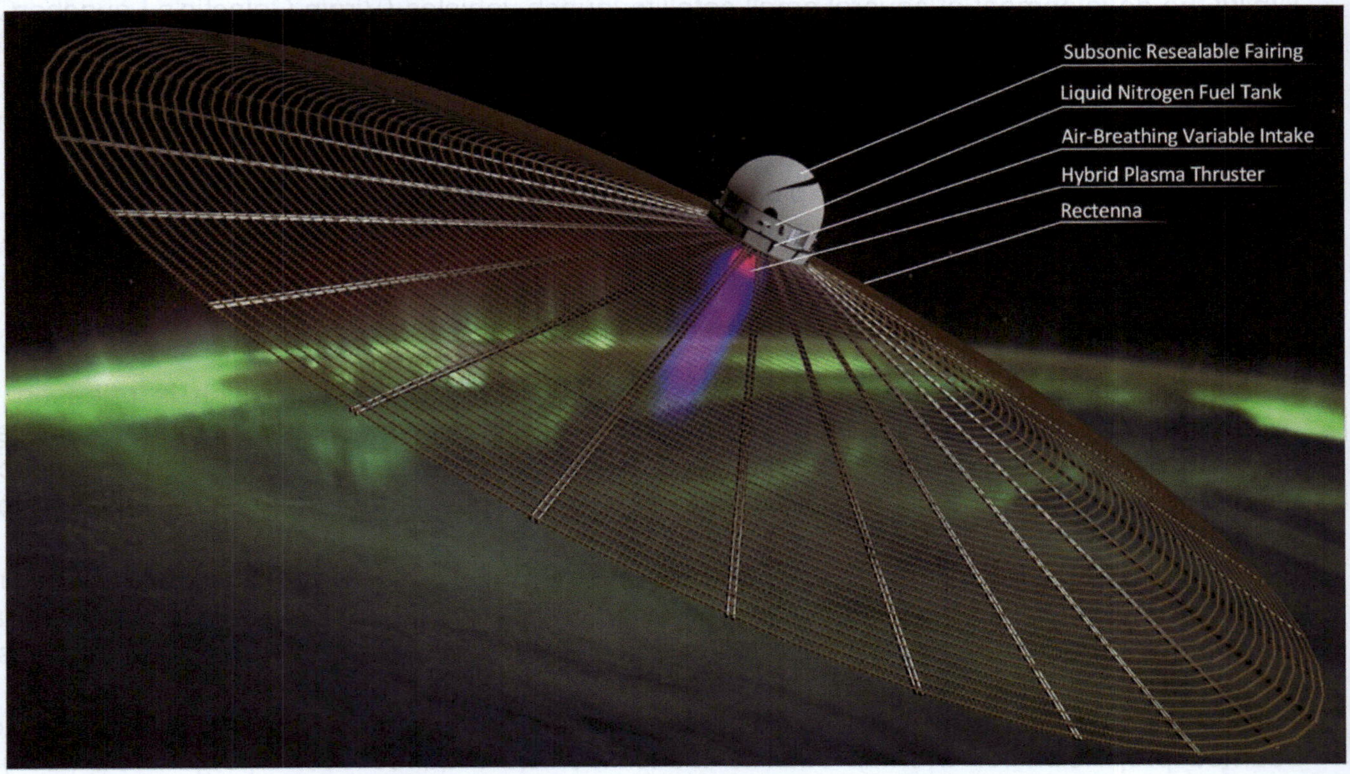

Figure 2-1 Key elements of the MicroLaunch Vehicle

2.1.1 Rectenna

The rectifying-antenna (rectenna) is used to convert the very high power density incident microwave beam into DC electronic power to supply the vehicle's electric propulsion system. The rectenna is made up of hundreds of thousands of independent wavelength-scale elements, each capable of outputting over a kilowatt of power, which are made up of an RF antenna, full wave rectifier and filters. The DC output power of the independent elements is then combined and conducted towards the centre of the vehicle via the rectenna's webbed structure.

The operational frequency of the rectenna has been selected, via extensive trade-off, to be in in the 5.8GHz Industrial, Scientific and Medical (ISM) band which has fewer restrictions on its use than conventional spectrum. As this form of wireless power transfer does not require modulated signals, the operating bandwidth of the rectenna is very small; only dictated by the Doppler shift at orbital velocities which is on the order of 20kHz. This significantly reduces the complexity of the rectenna elements, as well as the design of the ground based transmission antenna.

In order to provide power for the required thrust to weight ratio, rectenna is required to withstand and efficiently convert very high microwave power densities on the order of 1MW/m^2, have a total conversion efficiency of 72% or better, operate in a vacuum with radiative cooling and have a surface density of <2kg/m^2.

To achieve this will require high power rectifier diode components that can operate efficiently at two orders of magnitude higher frequency than is currently possible with devices available today. Towards this end TAS-UK has partnered with the University of Warwick who are developing next generation wide band gap devices that have the potential to meet the requirements of the MicroLaunch rectenna.

The rectenna also acts as a drag sail that greatly increases the altitude of peak deceleration when compared to a re-entering US space shuttle. Deceleration at higher altitude reduces the peak temperature of exposed surfaces which allows non-ablative thermal protection to be used, greatly reducing maintenance costs and enhancing lifetime. This enables the vehicle to be fully reusable with a potential turnaround time comparable to a high performance jet fighter.

2.1.2 Hybrid Plasma Thruster

The vehicle uses a single hybrid arcjet / MagnetoPlasmaDynamic (MPD) electric thruster which provides all the required thrust for the vehicle from sea level to orbit whist achieving a specific impulse of up to 1600s, which is over four times more fuel efficient than conventional launcher engines. When in atmosphere, the thruster is in air-breathing arcjet mode where the thrust mechanism is primarily electro-thermal; a high current arc heats a neutral gas to tens of thousands of degrees kelvin which creates thrust from the subsequent pressure increase.

Figure 2-2 Hybrid Plasma Thruster and Thrust Vectoring Assembly

When the vehicle has left the significant portion of the atmosphere (above 30km altitude) it switches to use of internally stored nitrogen fuel and operates in MPD mode where thrust is generated by a mixture of electro-thermal and electromagnetic processes. This raises the specific impulse and is the point where the peak power level is required and maximum thrust is achieved. The thruster stays in this mode until orbital velocity is reached and it can be shut down.

Whilst arcjets in air and MPD thrusters in vacuum have both been tested previously, the MicroLaunch thruster will combine the characteristics of both into a single cathode/anode set whilst operating at a significantly higher thrust (100s of kN) than has previously been demonstrated. The majority of MPD thrusters developed for space applications are designed to operate for 10s of thousands of hours before failure, which strongly drives the design of the key cathode and anode parts. In order to meet cost targets, the MicroLaunch thruster requires a mean time before failure of only 100 hours as it can be inspected and worn parts replaced after landing. This allows the thruster to operate at higher thrust levels than currently available systems.

2.1.3 Liquid Nitrogen Fuel Tank

The nitrogen required to supply the thruster when in MPD mode is stored in a liquid storage tank that forms the core of the vehicle body. Liquid nitrogen (LN2) was chosen as it offers a high storage density and moderate specific impulse at a temperature much higher (77K) than that required by more commonly used liquid hydrogen (20K). LN2 presents no explosion hazard and so is considerably safer to handle than traditional rocket propellants (e.g. kerosene, oxygen and hydrogen) which reduces costs and handling time.

2.1.4 *Air-Breathing Variable Intake*

Due to the large drag area of the rectenna it is important for the vehicle to maintain sub-sonic velocities when below 30km altitude which causes it to have much higher gravity losses than traditional launch vehicles (which try and exit the atmosphere as fast as possible). These gravity losses would cancel out the benefits in relative propellant fraction from the increased specific impulse. When in atmosphere and operating in arc-jet mode, the vehicle eliminates the need for on-board propellant by extracting ambient air using a variable area ram intake placed around the circumference of the fuel tank. The variable area is required to maintain a constant propellant mass flow rate and a high pressure ratio compressor is used to allow the arcjet to operate over wide range of air flow rates and pressures that will be encountered during the ascent. When the thruster is in MPD mode the intake is sealed and the system switches over to internal propellant.

2.1.5 *Subsonic Re-sealable Fairing*

The fairing is used to protect the payload(s) from both the aerodynamic forces encountered at lower altitudes and the high intensity microwaves incident on the vehicle. Unlike traditional launcher fairings, the shape is optimised for sub-sonic velocities as the vehicle will not break the sound barrier until it is above 30km altitude. Additionally, whist launchers generally eject the fairing above this altitude to save mass, the higher specific impulse of the MicroLaunch concept eliminates this need and allows it to be re-sealable in a similar fashion to the Space Shuttle. This also gives the vehicle the ability to re-enter with cargo (e.g. microgravity and life science experiments) for return to Earth.
The blunt sub-sonic fairing shape is suited to the deployment of multiple nano-satellites, shown in Figure 2-3.

Figure 2-3 Deployed Fairing and Variable Intake

2.1.6 *Other Sub-Systems*

In addition to those described above the vehicle has other sub-systems, including:

- The propellant management sub-system that controls the flow of propellant to the thruster and includes a gas expander to convert the stored nitrogen from its liquid to gaseous form;
- Flight control avionics that manage the electrical systems including the vectoring of the thruster. It should be noted that these avionics will have more in common with those found on a satellite in terms of radiation hardening as each vehicle will spend a combined lifetime of over a year in orbit;
- Solar power generator to allow long duration stays in orbit which is a requirement for some mission types;
- S-Band communications to maintain contact with ground at all stages during the ascent and descent. In order to prevent damage to sensitive receivers from the microwave beam, multi stage reflecting analog filters will be used for protection.

2.2 Transmission Antenna and Spaceport Design

Unlike traditional launchers, the majority of the cost and complexity for MicroLaunch is in the build and operation of the spaceport. The key part of the spaceport is the transmission antenna which is made up of a large phased array of identical elements that autonomously track the vehicle in flight to transmit the required GW of microwave power from the ground to the rectenna via a highly directional beam.

The diameter of this antenna determines the maximum distance that power can be sent to the rectenna without significant losses due to divergence of the beam to a size greater than the diameter of the rectenna. An extensive trade-off has shown that a 700m diameter transmission antenna is optimal to meet the vehicle maximum range requirements when at 5.8GHz.

A 700m diameter antenna is considerably larger than anything in operation today (Arecibo is 300m), but would be only 1/3rd of the active area of the planned Square Kilometre Array (SKA) which is due to be completed in 2030.

This antenna will require on the order of 40 million individual elements, each rated to radiate up to 100W narrowband RF, and so building and installing these elements will be the dominant setup cost of the whole system. However, these costs can be controlled by making sure the vast majority of these elements are identical and are suited to mass production techniques.

Additionally, a traditional phased array approach where all the elements are externally phase controlled would have a prohibitively large processing requirement to manage 40 million elements simultaneously to accurately point the beam. This can be solved by using self-phasing elements that autonomously adjust the phase depending upon the arriving phase of a beacon signal placed on the launch vehicle. This allows all processing to be done internally and so the only external input is the DC power which does not need to vary between elements.

Figure 2-4 Potential layout of a MicroLaunch spaceport, suitable for deployment in the UK and thousands of launches per year

The transmission antenna and surrounding spaceport require direct access to the national power grid in order to support the up to 7GW peak input power requirements of the antenna. As the transmission elements are expected to dissipate up to 50% of this energy locally as heat, a nearby source of coolant such as a lake or coastline would be ideal.

The choice of spaceport location will also be driven by the need to access 80 to 120° orbital inclinations and not have an excessive proportion of rainy days which would lead to depletion of beam energy. The north-eastern coastline of the UK is well suited to the siting of such a spaceport, and up to 7GW power is within the capability of the UK's national grid to provide at non-peak usage times.

2.3 Configurations

Hundreds of different configurations of the MicroLaunch system have been tested via monte-carlo analysis to determine optimal baselines for different payload sizes. Baseline designs for three payload classes; 200kg, 400kg and 1,000kg have been selected and are summarised in Figure 2-5 below:

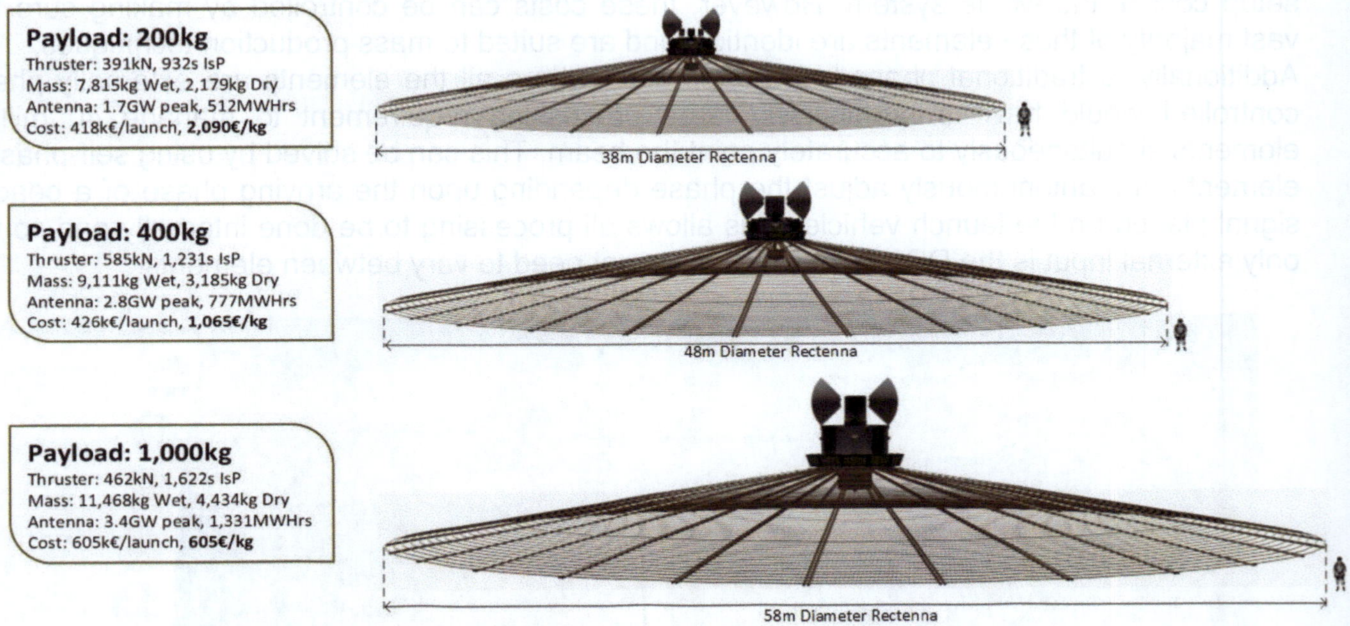

Figure 2-5 Baseline configurations for MicroLaunch Vehicles using a 700m diameter transmission antenna

These three baselines are all compatible with the spaceport described in section 2.2 and could all be developed and operated to offer a range of capabilities to a customer. Payload configurations smaller than 200kg have been found to be less economical due to the increasing cost per kilogram, although they may be used for early test flights. Payloads larger than 1,000kg require more peak power than it is feasible for the current UK power grid to provide, although the future use of energy storage systems may change this.

The estimated cost for launch of a 200kg payload, including recovery of the vehicle, is €418k which is a cost per kilogram of €2,090. This is **two orders of magnitude lower cost** than the small satellite launchers currently in development. The 1,000kg configuration reduces the cost

per kilogram by a further factor of 3.45, which is a figure comparable only to that projected for a fully productionised Falcon 9R.
It should be noted that these costs include those for fuel, electricity, operations, payload handling, launch vehicle build and maintenance as well as the R&D and build cost of the spaceport assuming a 10 year return on investment and an average of five launches per day by 2035 (1825 launches/year). These costs will likely reduce further as the investment is paid off.

3. OPERATIONS, SAFETY AND ENVIRONMENTAL IMPACT

3.1 Flight Profile

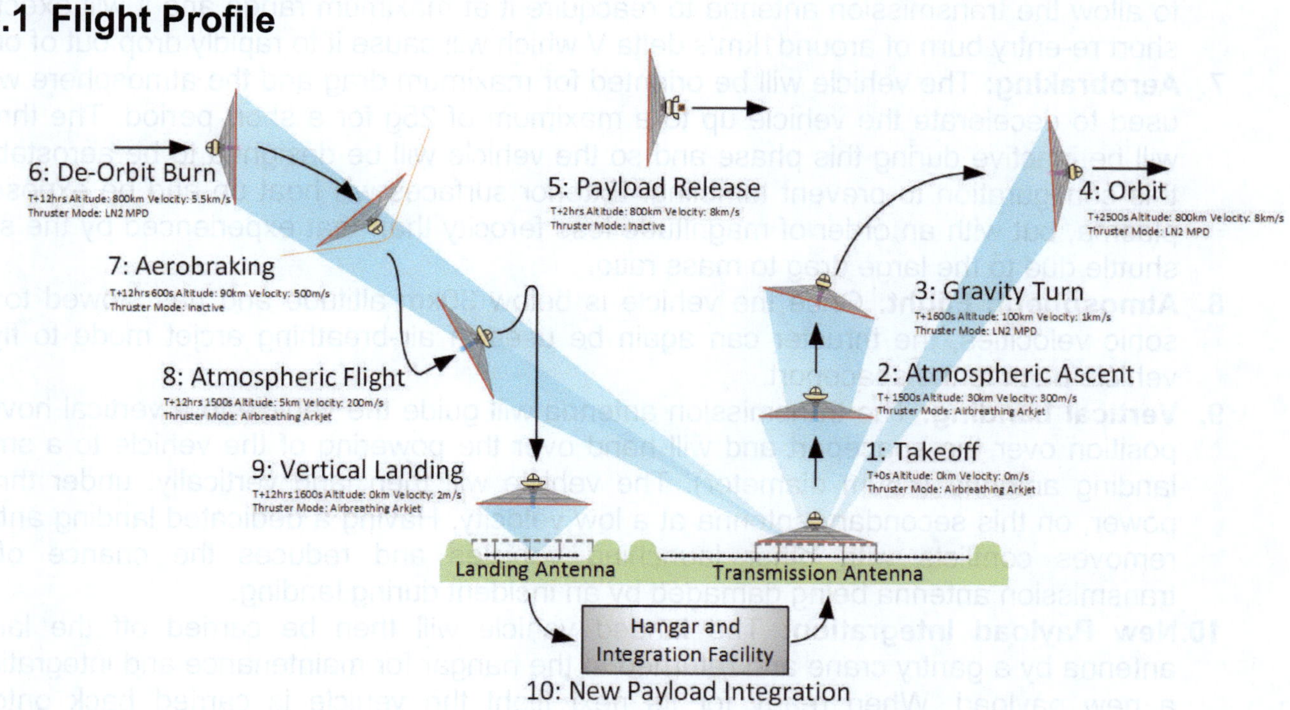

Figure 3-1 Typical MicroLaunch Flight profile

A typical flight profile of a MicroLaunch vehicle can be seen in Figure 3-1 above, which consists of the following major steps:

1. **Takeoff:** A fully loaded vehicle launches in arcjet mode from the centre of the transmission antenna. Special high power elements directly underneath the launch pad are used to push the vehicle up to ≈2km altitude where all the transmit antenna elements on the outer edge can focus on the vehicle beacon.
2. **Atmospheric Ascent:** The vehicle climbs to 30km altitude whilst maintaining a sub-sonic velocity and using only propellant acquired from the ambient air.
3. **Gravity Turn:** Above 30km altitude the vehicle closes the air intake, switches to internal fuel and the thruster changes to MPD mode. The vehicle slowly pitches over at a rate of ≈1°/s and the thrust increases to maximum.
4. **Orbit:** As propellant is expended, the velocity and acceleration climb to a peak of ≈10g and 7,900m/s. When the correct orbital velocity and altitude are reached the transmission antenna shuts down and the vehicle ceases to accelerate. For most missions the

transmission antenna will be visible one orbit later and this window of opportunity can be used to circularize the orbit or correct any injection errors. After this the phasing beacon on the launcher will be shut down to prevent interference with other launches.

5. **Payload Release:** When the orbit has been verified the fairing can be opened and the payload(s) activated. The vehicle is able to stay in orbit for up to a week, allowing each payload to be tested to ensure that it has survived the launch before release. If it has not the fairing can be re-sealed and it can be returned to the Earth for repair.
6. **De-Orbit Burn:** Once the fairing has been re-sealed the vehicle must wait for a favorable return approach which could take up to 12 hours. The launcher will reactivate its beacon to allow the transmission antenna to reacquire it at maximum range and it will execute a short re-entry burn of around1km/s delta V which will cause it to rapidly drop out of orbit.
7. **Aerobraking:** The vehicle will be oriented for maximum drag and the atmosphere will be used to decelerate the vehicle up to a maximum of 25g for a short period. The thruster will be inactive during this phase and so the vehicle will be designed to be aerostable in this configuration to prevent tumbling. Exterior surfaces will heat up and be exposed to plasma, but with an order of magnitude less ferocity than that experienced by the space shuttle due to the large drag to mass ratio.
8. **Atmospheric Flight:** Once the vehicle is below 30km altitude and has slowed to sub-sonic velocities, the thruster can again be used in air-breathing arcjet mode to fly the vehicle back to the spaceport.
9. **Vertical Landing:** The transmission antenna will guide the vehicle to a vertical hovering position over the spaceport and will hand over the powering of the vehicle to a smaller landing antenna (<50m diameter). The vehicle will then land vertically, under thruster power, on this secondary antenna at a low velocity. Having a dedicated landing antenna removes conflicts with other launching vehicles and reduces the chance of the transmission antenna being damaged by an incident during landing.
10. **New Payload Integration:** The landed vehicle will then be carried off the landing antenna by a gantry crane and returned to the hangar for maintenance and integration of a new payload. When ready for its next flight the vehicle is carried back onto the transmission antenna by another gantry crane.

This complete cycle is expected to take, on average, 5.5 days for each vehicle. At any one time several vehicles can be in orbit awaiting their next phase.

3.2 Atmospheric interactions

As the microwave beam is operating at a very high power density its level of interaction with the atmosphere is of great importance. The 5.8GHz operational frequency was selected as it provides the optimal compromise between rectenna capabilities, antenna sizes and amount of absorption by the atmosphere. 5.8GHz is in the C-Band range of frequencies which is commonly used for robust satellite to ground communications due to its high transmissivity through the atmosphere and lower rain attenuation when compared to X and K bands.

For this reasons 5.8GHz is the most commonly investigated frequency for use in beaming down GW of power from space to the ground in Space Based Solar Power (SBSP) concepts.

For very high power density beams, atmospheric breakdown becomes a concern whereby a threshold is reached that ionisies the neutral atmosphere, greatly increasing the absorption and reflection of beam energy. This breakdown threshold varies depending upon operating frequency and altitude; the minimum for 5.8GHz occurs at 50km altitude where the threshold has been empirically determined[iii] to be ≈2.3MW/m^2. This is what drives the limitation of incident power density on the rectenna which in turn limits the minimum payload threshold to >150kg. It may be possible to "throttle back" the power density when the launcher is passing through this key point, allowing for a higher power density to be used at later stages of the flight (for example the threshold at 100km has risen to 6MW/m^2). However the practicality of doing this has not yet been fully investigated and it may not be required for most missions.

It should be noted that microwave beams are not greatly affected by atmospheric refraction in the way that lasers are.

3.3 Efficiency

Beamed propulsion launch systems are commonly described in terms of "wallplug efficiency" which describes the ratio of input power to the ground station to the power of the useful thrust. For MicroLaunch the efficiency is dominated by the losses in the transmission antenna but beyond 200km vehicle range the antenna cannot focus the full power of the beam onto the rectenna, leading to divergence losses. Figure 3-2 below shows that the estimated wallplug efficiency of the 400kg configuration at peak thrust is 10.8%:

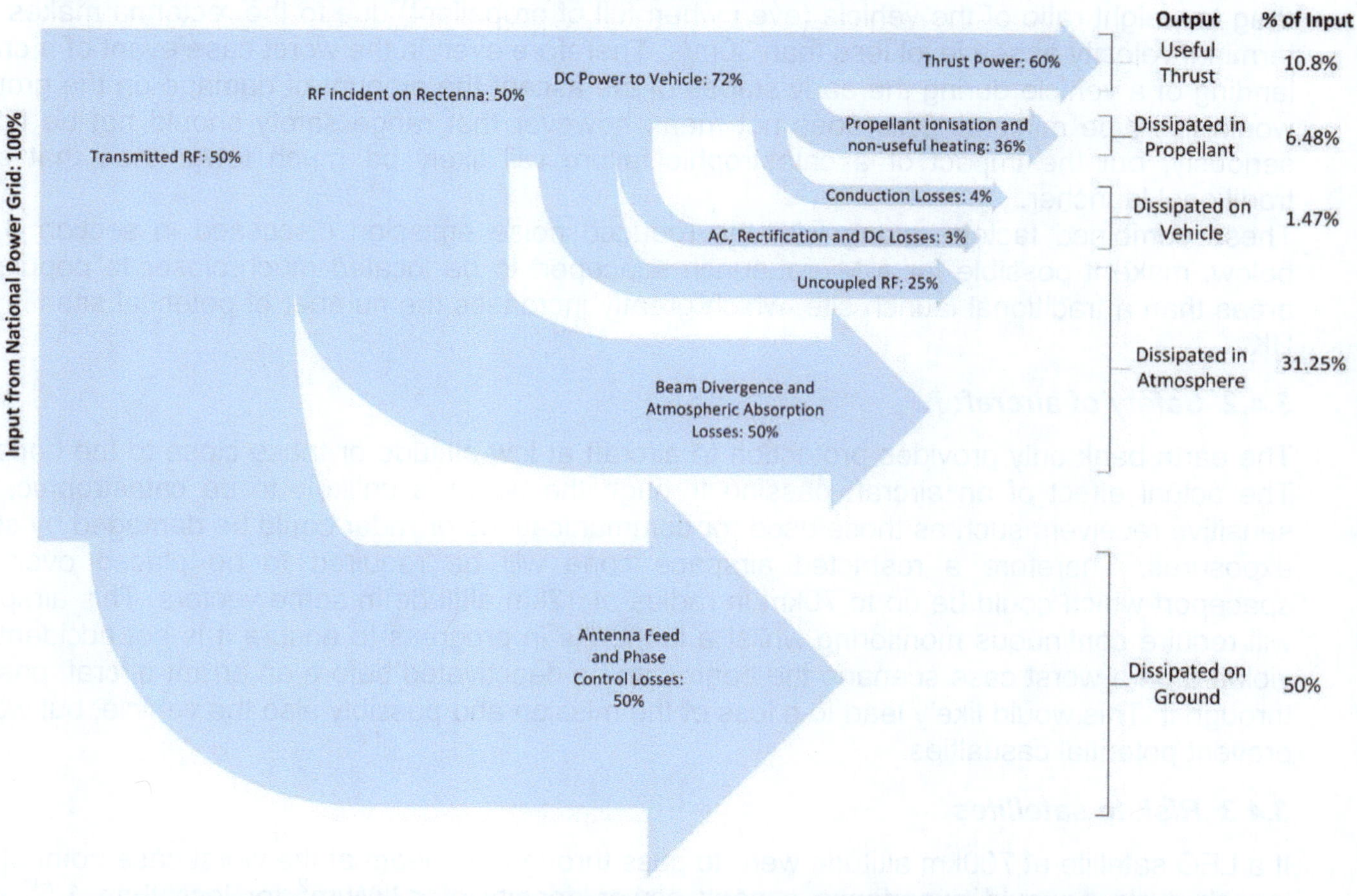

Figure 3-2 Wallplug to thrust efficiencies and losses at maximum thrust

This efficiency may be most straightforwardly improved by either improving the input power to RF conversion efficiency of the transmission antenna or by recovering some of its waste heat.

3.4 Safety considerations

As with all satellite launch systems, there are several considerations to be made for the safety of people and property that could be affected by a launch.

3.4.1 Safety of personnel and local residents

The presence of a such a high power microwave source presents a danger to local residents and spaceport personnel. However the danger to people on the ground can be effectively mitigated by the use of earth bank walls that form a ring around the transmission antennae that block them from line of sight. These walls are capable of preventing any passage of microwaves through them and as at GHz frequencies refraction in air is negligible there should be no impact at ground level out to the horizon.

There is however potential for non-isotropic scattering of microwaves off the vehicle during the first seconds of flight. Whilst this is not expected to be of a significant level it does require further study.

The advantage of MicroLaunch over traditional launch systems is that the use of non-explosive propellant greatly reduces risk of injury to site personnel from mishandling. In addition, the large drag to weight ratio of the vehicle (even when full of propellant) due to the rectenna makes the terminal velocity at sea level less than 30m/s. Therefore even in the worst case event of a crash landing of a vehicle during the early stages of the ascent the amount of damage on the ground would likely be minimal. This does not mean however that range safety should not be taken seriously, but the impact of a catastrophic failure will likely be much lower than that of a traditional launcher.

These combined factors, along with the reduced noise emission discussed in section 3.5.1 below, make it possible for a MicroLaunch spaceport to be located much closer to populated areas than a traditional launch site, which greatly increases the number of potential sites in the UK.

3.4.2 Safety of aircraft

The earth bank only provides protection to aircraft at low altitude or those close to the horizon. The actual effect of an aircraft passing through the beam is unlikely to be catastrophic, but sensitive receivers such as those used for communications or radar could be damaged by short exposures. Therefore a restricted airspace cone will be required to be placed over the spaceport which could be up to 70km in radius at 12km altitude in some vectors. This airspace will require continuous monitoring whilst a launch is in progress to ensure it is not accidentally violated. In a worst case scenario the beam can be deactivated before an errant aircraft passes through it. This would likely lead to a loss of the mission and possibly also the vehicle, but would prevent potential casualties.

3.4.3 Risk to satellites

If a LEO satellite at 700km altitude were to pass through the beam at the worst case point of the launch cycle it would experience a peak power density of <2kW/m^2 for less than 1/5th of a second. This is the energy flux equivalent of ≈1.5 suns and so is unlikely to damage the satellite by excess heating, however it may affect sensitive receivers; particularly those operating in C-Band. This could well lead to a permanent loss of satellite functionality and so will have to be avoided. An exclusion radius around all active satellites should be considered whereby a launch will be timed to ensure that there is a high confidence that the beam does not pass within the exclusion zone of any satellite. The size of this exclusion zone will vary depending upon the altitude and type of satellite (e.g. a C-Band telecommunications satellite may require a larger zone) but a 5km radius should be sufficient to reduce the flux to a safe level for most satellites at 700km altitude. If a satellite in GEO is accidentally caught in the beam it would experience a peak flux of <100W/m^2.

The risk to receivers on future satellites can be eliminated altogether by use of a narrowband notch filter on their inputs. Whilst this adds slightly to the mass of future satellites, the benefits of the reduced cost of launch are expected to greatly outweigh this.

It should be noted that due to the self-phasing requirement of the transmission antenna, it is not possible to focus or steer the beam without the presence of a vehicle beacon and so it cannot be deliberately aimed at satellites.

3.5 Environmental Considerations

Along with the safety of local people and property, modern launch systems should also account for their effect on wildlife, pollution and climate change.

3.5.1 Local Wildlife

The most obvious concern of the operation to MicroLaunch is to birds that could fly through the beam. It is currently unknown if a bird would voluntarily fly into a potentially fatal microwave beam but it is possible that they could be caught in a sweeping beam without time to escape. As the vast majority of birds fly at altitudes under 150m[iv], a 850m radius exclusion zone using techniques already developed by airports, to avoid bird strikes on engines, can be used. These include physical and audible systems that mimic predators to scare away animals. Some migratory birds can fly considerably higher than 150m and so during migration periods optical confirmation that no flocks are nearby will likely be required before the antenna is activated.

However, MicroLaunch vehicles have an advantage over traditional launchers in regard to noise emission as they are expected to be significantly quieter in operation due to the lower nozzle pressure and lack of turbopumps. This, coupled with the bird scaring devices, may make the spaceport ideally suited as a sanctuary to ground based wildlife.

3.5.2 CO_2 Emissions

It can be difficult to estimate the carbon dioxide emissions associated with a traditional launch system that uses a disposable vehicle, as large portion will be associated with the energy required to build the vehicle rather than that just given off in its exhaust (which can be near zero for hydrogen/oxygen based rockets). Both the emissions associated with the exhaust and the build of the vehicle will be low for MicroLaunch as it can be reused at least 100 times and doesn't use a carbon based propellant.

Therefore the per launch CO_2 emissions will be dominated by the transmission antenna input power which has been estimated for the three baseline configurations given in Figure 2-5. The calculation of emissions is shown in Table 3-1 below:

Energy Source	CO_{2eq}/MWh[v] [kg]	CO_2 Emission for Payload Configuration (Energy Required) [tn CO2]		
		200kg (512MWh)	400kg (777MWh)	1,000kg (1,331MWh)
Biomass	18	9.2	14.0	24.0
Solar (PV)	46	23.6	35.7	61.2
Hydro	4	2.0	3.1	5.3
Ocean	8	4.1	6.2	10.6
Wind	12	6.1	9.3	16.0
Nuclear	16	8.2	12.4	21.3
Natural Gas	469	240.1	364.4	624.2
Oil	840	430.1	652.7	1118.0
Coal	1001	512.5	777.8	1332.3
UK Average	515	263.6	400.2	685.5

Table 3-1 Estimated CO2 emissions per launch for various energy sources

The table shows that the amount of CO_2 emitted per launch can vary by two orders of magnitude depending on the primary energy source. As of 2010 the UK's national grid is 47% natural gas, 28% coal, 16% Nuclear and 7% Renewable and 1% oil[vi] and so UK average is ≈515kg of CO_2 per MWh. Assuming a five flights a day of the 400kg configuration this would give total emissions for a spaceport of 730ktn of CO_2 per year. This is roughly equivalent to a medium sized airport such as Bristol[vii].

Despite the reusability and use of clean propellant, this is likely to be a higher emissions rate than for a similar number of flights using traditional rocket technology. Therefore, the use of low emissions sources is highly preferable and the spaceport's likely location next to a northern coast may suit the deployment of dedicated ocean, wind or hydro power to reduce these emissions.

Additionally, as ≈50% of the input power is wasted as heat in the transmission antenna, it may be possible to recover some of this energy by extracting it from a water based coolant. AC energy could be extracted from hot water by using a steam turbine and the resulting moderate temperature water could be used for heating local homes.

3.5.3 NO_x emissions

As MicroLaunch uses high temperature air as a propellant during the first phases of the flight there is potential for it to generate significant quantities of nitrogen oxides in its exhaust which is a common pollutant in cities that can affect the health of residents. Further study is required to determine the actual rate of production, but in the absolute worst case (100% of the exhaust is converted to nitrogen oxides) each flight could generate up to 14 tonnes per flight (25.6ktn/year). This can be compared to Heathrow airport which generated an estimated 5.8kt in 2008[viii]. It is not known how much of this NOx will reach ground level and the majority will be emitted over the North sea, but methods for reducing this rate of emission will be investigated.

4. DEVELOPMENT PLAN

There are three key technologies required to enable the MicroLaunch concept; the rectenna, the thruster and the transmission antenna. All of these technologies are well established, but have not been demonstrated together or at the power levels required by MicroLaunch. Therefore a development plan has been created (shown in Figure 4-1 below) which covers the progression of the TRLs and key milestones up to the first orbital flight, projected for 2025. This first flight requires the majority of the spaceport to be constructed including the power infrastructure, transmission antenna, tracking and communication systems.

Before that three technical demonstrations of the key technologies are required, followed by a 1,000th/s scale free flying version that tests these technologies together for the first time. Around three years later a 1/10th scale suborbital vehicle is aimed to be tested, using a ≈70m diameter transmission antenna, potentially on the expected site of the eventual spaceport.

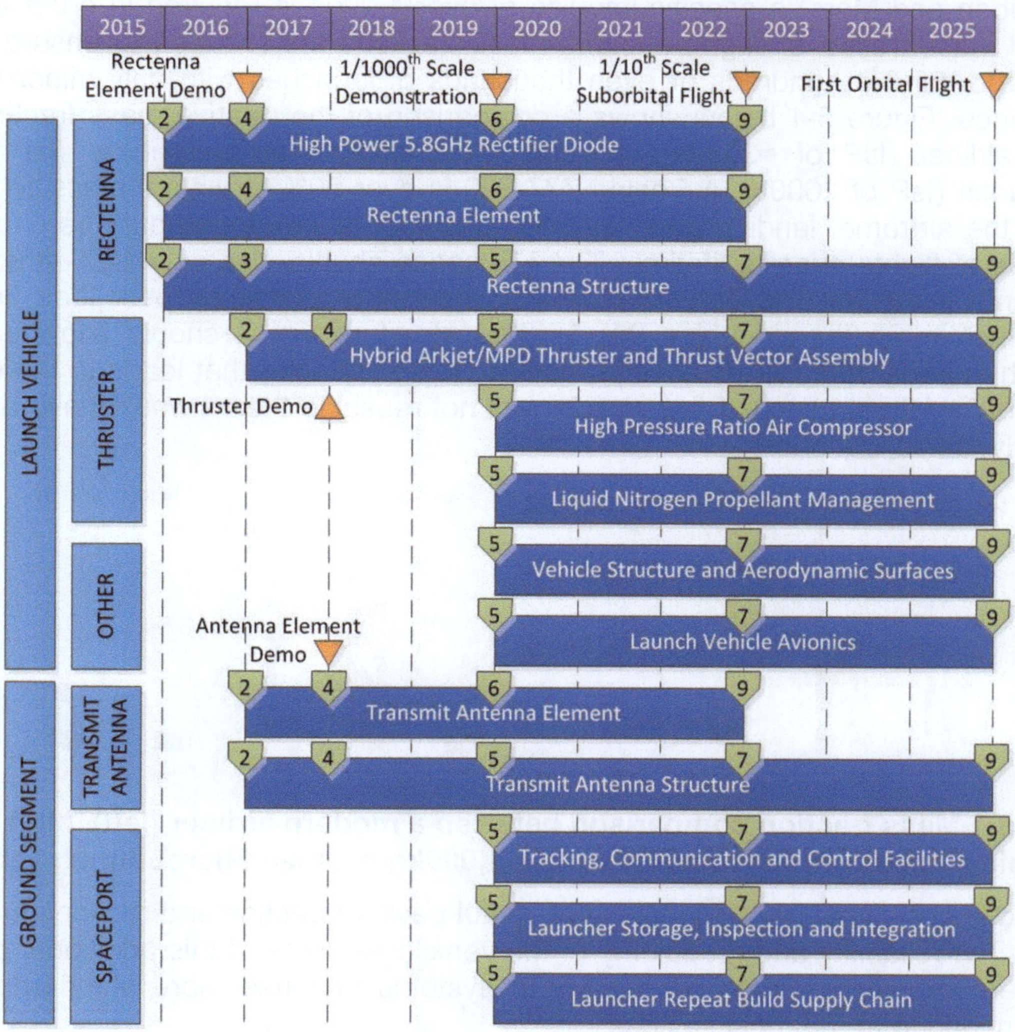

Figure 4-1 Development Timeline of the MicroLaunch Concept to first orbital flight. The progression of the technology TRLs are indicated in the green boxes and key milestones are orange triangles

5. COMPARISONS AND BENEFITS

5.1 Cost

Traditional rockets based upon chemical fuels are only on the cusp of being capable of reaching orbit. If Earth's gravity was slightly higher it would not be possible to reach orbit at all, and as it stands complete reusability is almost impossible. As with the US space shuttle, Buran, X-37 and potentially the Falcon 9 only a fraction of a chemical rocket can be recovered and reused, not the whole vehicle. This is due to the limitations of the specific impulse achievable by chemical fuels, if this can be raised above 500s then complete reusability is possible. If it is raised further then reusable single stage to orbit is possible.

However, in order to meet the two orders of magnitude cost reduction required to meet our most basic goals for space utilization (e.g. solar power from space, asteroid mining, manned bases on the Moon and Mars) a specific impulse of over 1,000s is required in order to provide high structural mass fractions. High structural mass fractions are required to ensure that airframes are rated to survive hundreds or even thousands of launches with only minor inspection and maintenance. Figure 5-1 below shows a comparison of the relative mass fractions between a modern airliner (IsP of ≈3000s), a traditional chemical launch vehicle (IsP of 450s) and MicroLaunch (IsP of 1600s). A Boeing 777-300 is over 50% structural mass on takeoff (which includes the airframe, landing gear, avionics and fuel tanks) and should be capable of 10s of thousands of flights over its lifetime. The US space shuttle was only 12% structural mass on launch and hence the five orbiters combined only flew 135 times over their 30 year history. MicroLaunch more than doubles this to 30% structure which should allow each vehicle to achieve hundreds of launches over its lifetime. It is this factor that leads to the enormous cost reductions for MicroLaunch, if the vehicle was not reusable then it would have a similar cost to chemically fueled launchers.

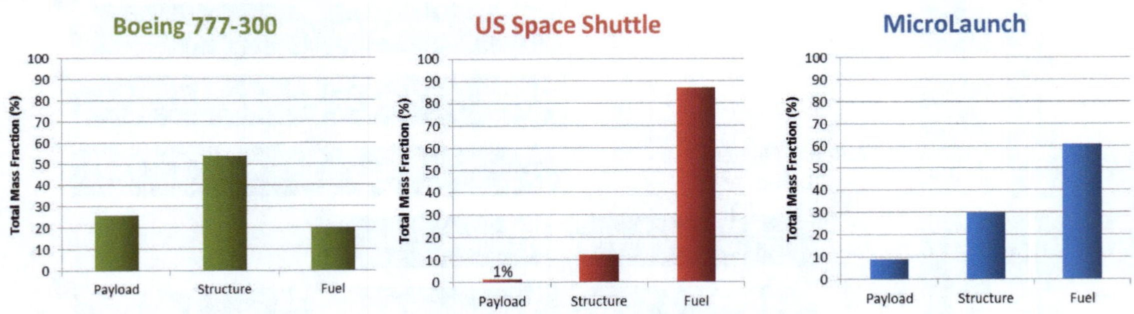

Figure 5-1 Mass fraction comparison between a modern airliner (left), high performance chemical launch vehicle (centre) and a 1,000kg payload MicroLaunch vehicle (right)

Increased specific impulse also raises the useful payload fraction and reduces fuel costs. It also improves the reliability and modularity of the vehicle as some of this additional payload can be reallocated to include more redundancy in systems and use more standardised equipment which can lead to further cost savings.

Taking this further, the cost per kilogram of all viable MicroLaunch configurations against specific impulse is shown in Figure 5-2 below. With few exceptions it can be seen that higher specific impulse leads to a lower cost. Also included in Figure 5-2 is a non-exhaustive list of currently available launchers and some notable launcher systems currently in development. It can be seen that all existing launchers have specific impulses in the range of 250s to 450s and costs from €5,000/kg to €30,000/kg. The small satellite launchers currently in development (Launcher One, Electron, Firefly) all have costs above €30,000/kg and it is difficult to see how the market will support this. Two notable launch systems in development are Skylon and the Falcon 9R. Skylon will increase its average IsP by air-breathing for a significant portion of its flight just to the point where a reusable SSTO vehicle is possible. The Falcon 9R (and potentially its larger variant the Falcon Heavy) is intended to return its first stage to the launch pad for reuse. Both the Falcon 9R and Skylon would be highly complementary to MicroLaunch as they can launch large payloads that would not be feasible for MicroLaunch due to the input power requirements.

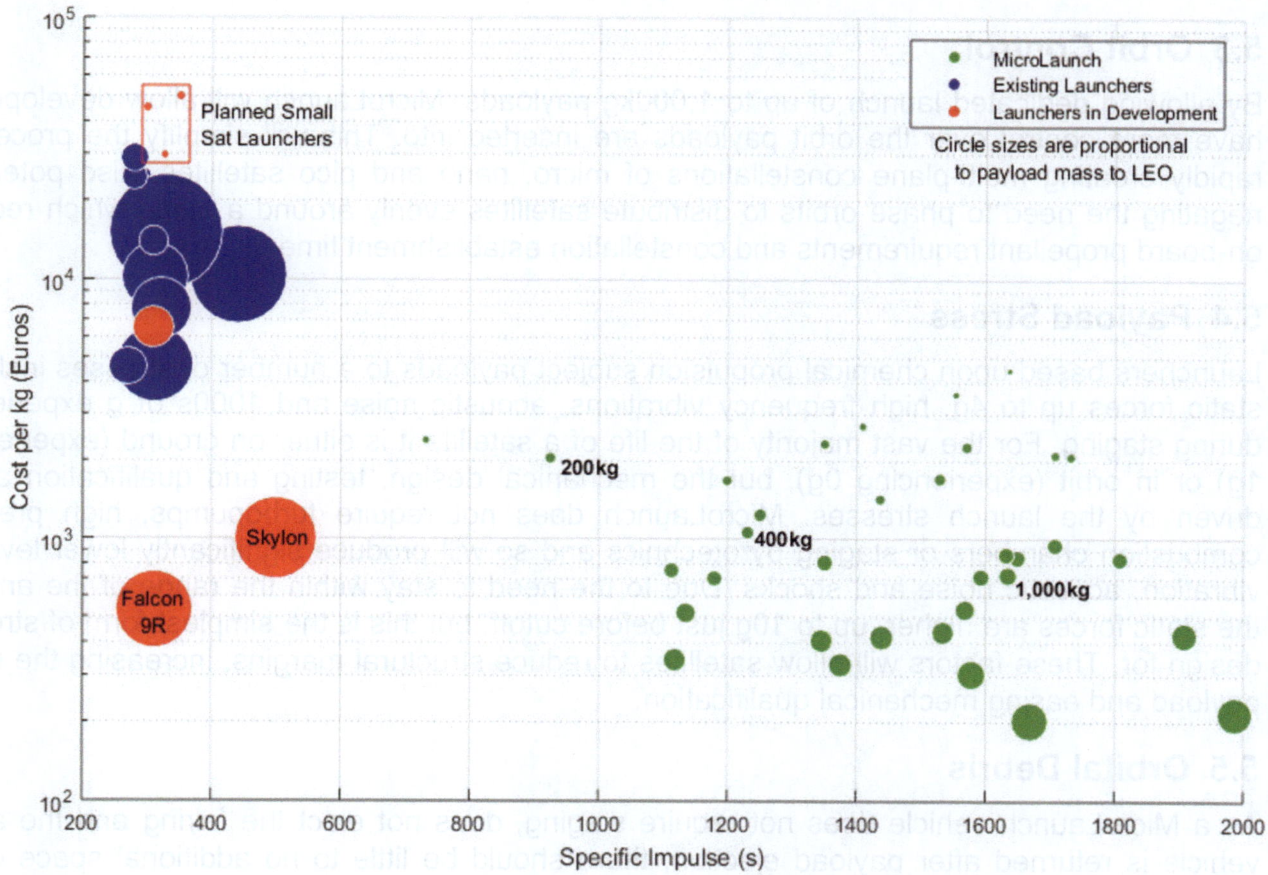

Figure 5-2 Cost per kg against primary stage specific impulse for MicroLaunch configurations (200, 400 and 1,000kg baselines highlighted) compared to other LEO launchers, both existing and in development (some parameters estimated)

The three MicroLaunch baseline configurations are highlighted, with the 400kg and 1,000kg variants being cost competitive with the Skylon and Falcon 9R. There are other configurations that could achieve a lower cost per kilogram (potentially down to €200/kg) but these were not baselined as they require unrealistically high input powers (up to 20GW in one case) which is not compatible with the current capability of the UK's national grid. This may change in the future if an effective method of grid energy storage can be deployed, but this is not seen as credible in the pre-2035 timeframe.

5.2 Turnaround Time

By reducing the size of the launch vehicle, limiting the number of payloads and utilising inert propellants the turnaround time (minimum time between successive launches) should be significantly shorter than the current state of the art. For 200kg payloads the estimated total mass of the launcher is 7,800kg which enables the use of flatbed vehicles and cranes to move the vehicle into position on the launch pad, rather than a traditional crawler, further reducing turnaround time.
With vehicle fleet sizes of 20 or more multiple launches per day are possible with MicroLaunch from a single spaceport. In this way a multi-plane EO or telecommunications constellation could be launched in timeframes measured in days rather than months.

5.3 Orbit Control

By allowing dedicated launch of up to 1,000kg payloads, MicroLaunch will allow developers to have more control over the orbit payloads are inserted into. This will simplify the process of rapidly creating multi-plane constellations of micro, nano and pico satellites, also potentially negating the need to phase orbits to distribute satellites evenly around a plane which reduces on-board propellant requirements and constellation establishment time.

5.4 Payload Stress

Launchers based upon chemical propulsion subject payloads to a number of stresses including static forces up to 4g, high frequency vibrations, acoustic noise and 1000s of g experienced during staging. For the vast majority of the life of a satellite it is either on ground (experiencing 1g) or in orbit (experiencing 0g), but the mechanical design, testing and qualification are all driven by the launch stresses. MicroLaunch does not require turbopumps, high pressure combustion chambers or staging pyrotechnics and so will produce significantly lower levels of vibration, acoustic noise and shocks. Due to the need to stay within the range of the antenna the static forces are higher, up to 10g just before cutoff, but this is the simplest form of stress to design for. These factors will allow satellites to reduce structural margins; increasing the useful payload and easing mechanical qualification.

5.5 Orbital Debris

As a MicroLaunch vehicle does not require staging, does not eject the fairing and the all the vehicle is returned after payload ejection, there should be little to no additional space debris created for every launch. This is a vital consideration to the long term usage of key orbits (particularly sun-synchronous polar LEO orbits) and in the long term reduces risks of in-orbit failure of satellites due to debris impacts.

6. APPLICATIONS

6.1 Satellite launch

In the 2025 to 2035 timeframe, the primary application of MicroLaunch is expected to be launch of small satellites up to 200kg, although the capability to launch up to 1,000kg may also be widely used depending upon how trend for small satellites evolves. The great advantage of

MicroLaunch comes in the rapid launch of constellations of small satellites to provide greater coverage of the Earth than single large satellites can provide. Currently the deployment of small satellite constellations is extremely difficult to plan as opportunities to ride share on multi plane launches to the same altitude are rare.

Several LEO mega-constellations of small satellites are currently in early development that are planning for hundreds to thousands of satellites in twenty or more planes. In all cases the launch costs are dominant and the time taken for deployment is at least a few years. MicroLaunch is capable of launching thousands of small satellites a year at orders of magnitude lower cost; making the business case for mega-constellations easier to close.

In addition to small satellites, the ability for MicroLaunch vehicles to stay in orbit for extended periods and safely land cargo would make it ideally suited for microgravity research that requires longer than the few minutes offered by sounding rockets but is too large or dangerous for the International Space Station.

6.2 Debris removal

One concern for the deployment of large numbers of small satellites is that they may rapidly lead to a runaway crisis in the form of space debris. In addition there are thousands of inactive large satellites and rocket stages already in orbit that may become hazards in the future. Microlaunch can offer a solution to this issue as its low cost and propulsive capabilities make active de-orbiting of debris technically and economically viable.

Missions can be imagined where a vehicle would drop off a new satellite to replace one failed in the same plane and then proceed to grab its defunct brethren and return it to Earth for failure diagnosis, refurbishment and eventual relaunch.

6.3 In-Orbit Maneuvers

MicroLaunch rectenna technology could be added to new large satellites to allow them to receive and convert power from a ground antenna to on-board DC power. This could be used to drive a high power radar system or an electric thruster for large orbit maneuvers (such as altitude raising or inclination changes). This allows the satellite solar panels to be sized for the average power usage case rather than the peak.

6.4 Lunar Base Resupply

If a second MicroLaunch spaceport were to be built close to the equator the low cost and high frequency of launches would make it ideal for creating a permanent supply chain to a manned lunar base. Transport tugs using solar electric propulsion and resupplied by MicroLaunch vehicles could move back and forth between Earth and Lunar orbits to provide regular deliveries of food and essential equipment that cannot be produced on the surface.

6.5 Space Based Solar Power

One of the ultimate goals of MicroLaunch is to enable the build of economically viable SBSP stations, which have the potential to be the ultimate low carbon energy source; providing solar energy 24 hours a day and beaming it to wherever it's needed. These will require thousands of launches as well as developments in microwave power transmission and rectenna technology at 5.8GHz, all of which MicroLaunch can offer. Whilst it is unlikely that a ground based transmission antenna could provide the circularisation burn needed to deliver the required components into geostationary orbit, the station could be assembled in LEO and then its collective power output used to drive an upgraded MicroLaunch thruster. A 4,000tn SBSP station could provide up to 4GW to such a thruster which is sufficient to achieve 40kN of thrust at 10,000s IsP. This would only require ≈4% of its mass as propellant to reach GEO and could do so within a week. This is compared to nearly 50% of its mass if chemical propulsion is used. Once completed, such a station would be ideal for directly powering a MicroLaunch spaceport, reducing its equivalent CO_2 emissions by orders of magnitude.

6.6 Interplanetary Transit Network

The combination of MicroLaunch and SBSP technology would unlock the full potential of beamed propulsion and enable the establishment of an interplanetary transit network. An SBSP powered spaceport on Earth could launch MicroLaunch vehicles directly into a geostationary transfer orbit. When the vehicle reaches the apoapsis of the orbit a nearby SBSP station can directly power the vehicle to allow it to circularise its orbit into a true geostationary one. There the vehicle can unload its cargo of propellant into the SBSP station and then return to Earth for the next round trip.

When enough propellant has been stored in the station, it can use its solar arrays to power an upgraded MicroLaunch MPD thruster to propel it within the space of less than a year to another planetary body such as Mars, Venus or the larger asteroids. Once in orbit the station can be used to provide ample power to any colonists to keep them and their farms warm as well as vaporize ice and rocks to make oxygen and water. This is the first step towards the interplanetary transit network where MicroLaunch vehicles adapted for longer in space can use the power of the SBSP stations to accelerate out of the orbit of one planet/moon/asteroid and decelerate into the orbit of another. This removes the need to carry large solar arrays (rectennae can handle 1,000x higher energy densities than PV cells) or heavy and politically unfavorable nuclear reactors. Additionally almost any atmospheric gas can be used as propellant and so that required for the return trips can be extracted from Venus and Mars' atmosphere in the form of CO_2.

In this way the SBSP stations become orbital spaceports where vehicles can be decelerated, docked, refueled and accelerated to the next destination. This will allow the inner solar system, where solar energy is abundant, to be completely opened up to humanity in a timeframe measured in decades rather than centuries.

In the further future, the beamed propulsion technologies developed for MicroLaunch are likely to form the basis for the first interstellar probes.

7. CONCULSIONS

Beamed propulsion is a method of ground to orbit launch for which the time is finally right for development as the technology has caught up with the market need. MicroLaunch is a concept designed with the needs of the 2025 to 2050 space industry in mind; that of orders of magnitude reduction in cost and increase in launch frequency.

A 10 year development program has been proposed to bring the key technologies from TRL 2 to 9 in time for the commissioning of the first spaceport and orbital launch. There are technological and regulatory challenges to overcome with the deployment of GW scale beamed power transfer, but the long term economic benefits are significant.

Ultimately MicroLaunch could replace chemical systems as the dominant form of space launch into LEO and open up the inner solar system for exploitation.

8. REFERENCES

[i] http://spaceflightservices.com/pricing-plans/
[ii] SF-1100-PUG-00001, Secondary Payload Users Guide, Spaceflight, Inc. Rev D March 2013.
[iii] K. Parcin, "The Microwave Thermal Thruster and its Application to the Launch Problem" PhD Thesis, California Institute of Technology, 2006
[iv] https://web.stanford.edu/group/stanfordbirds/text/essays/How_Fast.html
[v] http://srren.ipcc-wg3.de/report/IPCC_SRREN_Annex_II.pdf
[vi] http://www.energy-uk.org.uk/energy-industry/electricity-generation.html
[vii] http://www.aef.org.uk/uploads/NewsPulloutCO2.pdf
[viii] http://www.heathrow.com/file_source/Company/Static/PDF/Communityandenvironment/air-quality-strategy_summary-document_LHR.pdf

Reinventing Space Conference
BIS-RS-2015-82

Design Criteria of Remote Sensing Constellations of Small Satellites with Low Power Electric Propulsion and Distributed Payloads

Salvo Marcuccio
Università di Pisa

Stefan Gregucci
SITAEL

Pierpaolo Pergola
Università di Pisa

13th Reinventing Space Conference

9-12 November 2015
Oxford, UK

S. Hatton (ed.), *Proceedings of the 13th Reinventing Space Conference*,
https://doi.org/10.1007/978-3-319-32817-1_20

BIS-RS-2015-82

Design Criteria of Remote Sensing Constellations of Small Satellites with Low Power Electric Propulsion and Distributed Payloads

Salvo Marcuccio
Università di Pisa
Via G. Caruso 8, 56122, Pisa (Italy); +390502217211
salvo.marcuccio@unipi.it

Stefan Gregucci
SITAEL SpA
Via Gherardesca 5, 56121, Pisa (Italy); +39050967211
stefan.gregucci@sitael.com

Pierpaolo Pergola
Università di Pisa
Via G. Caruso 8, 56122, Pisa (Italy); +390502217211
pierpaolo.pergola@sitael.com

ABSTRACT

The recent explosion in proposed microsatellite missions is based on the possibility to mass-produce cheap platforms capable to deliver acceptable performance over a limited lifetime. The assumption behind such scheme is that individual microsatellites are expected/allowed to fail in reasonable numbers, the resulting degradation of constellation performance being limited due to the large population of active spacecraft. We argue that cheap platforms do not necessarily need to be seen as disposable assets, so that low cost constellations featuring a low number of microsatellites may nevertheless be capable of remarkable performance. The key technology needed to enable such feat is low power electric propulsion, whereby microsatellites are allowed to acquire and maintain precisely tuned orbital locations, compensate atmospheric drag to fly longer, and de-orbit safely at end of life. A number of such microsatellites may be fitted with an instrument each from a suite of different sensors operating in various spectral bands. The constellation would operate as an actively controlled system, with the individual instruments providing well coordinated raw data that may be processed using data fusion techniques to yield the final product. Starting from the proven performance of a currently available low power Hall thruster, we present general design criteria for constellations based on a 50 kg-class microsatellite bus. The potential benefits of such technology are outlined with respect to applications such as precision farming, urban area monitoring, and dual use land surveillance.

KEYWORDS: Electric Propulsion, Small satellite, Constellation, Distributed payload, Earth Observation

INTRODUCTION

Aimed at enhancing worldwide government, commercial, industrial, civilian, military and educational communities capabilities to support Earth Observation (EO) missions (e.g. to manage natural resources, to support agricultural practices, to provide climatic assessments, to detect and monitor natural disasters), small satellite technology offers unique opportunities to obtain high performance reducing the mission cost. Such platforms provide the opportunity to carry out Earth observation missions using small, low cost satellites, and correspondingly to reduce the cost of launch, ground stations, data distribution structures, and space system management approaches. In additional small satellites provide unique opportunities to setup affordable constellations[1]. In this respect, small satellites can realize tasks that are not practical with large satellites. In addition, the present technological readiness level reached by low power, high performance electric thrusters, like SITAEL's HT100[2], in combination with a growing number of ongoing technological advances (in particular high efficiency solar cells), make it now possible to equip a microsatellite platform with an electric propulsion system[3].

Microsatellites equipped with low power electric thrusters can enable new kinds of missions in support of Earth observation, e.g. by providing extended lifetime at low orbital altitude and therefore achieving better Ground Sample Distance (GSD) performance with small optical instruments. In particular, the use of electric propulsion makes it possible for an entire microsat constellation to be placed in orbit with a single launch, possibly as secondary payloads. Each constellation element, based on a common microsatellite bus, can be equipped with a payload chosen among a number of different instruments. Once released in the initial orbit, each individual microsatellite can maneuver autonomously to achieve its own operational orbit.

This constellation architecture is based on the implementation of a distributed array of different instruments, acting in a cooperative way to achieve optimal data collection performance. Each satellite, however, occupies an orbital location that can differ in altitude (and if necessary also in inclination or eccentricity) from the other members of the constellation, so to let each instrument operate in the most appropriate orbital conditions. If desired, different microsatellites could be given different revisit periods of the target area, in order to observe the scene in different, co-ordinated moments in time. Such schemes, which can obviously change in the degree of complexity and in the associated cost of operations, are made possible by the combination of low launch mass (enabling multiple platforms to be orbited by a single launch) and propulsion capability featured by last generation, electrically propelled microsatellites.

This study aims at identifying some design criteria for Remote Sensing Constellations of small satellites based of the distributed payload/multiple orbits concept, in order to offer the best compromise in terms of spatial, spectral and temporal resolution performance, ground coverage (from regional to global accessibility) and satellites number. As an additional output of this study, we outline the design of a microsatellite bus equipped with the SITAEL HT100 thruster and compatible with existing small Earth observation optical instruments to cover the whole range of potential remote sensing applications.

MISSION REQUIREMENTS

Orbit class selection

The orbit design starts with the selection of orbital altitude. Only Low Earth Orbits (LEOs) have been considered with an altitude lower than 1000 km. This upper limitation is imposed due to the difficulty to maintain both high resolution, limited instrument dimensions and power demand at increasing orbital altitude.

Traditionally [4,5] the lower bound on the altitude of Earth Observation missions is set at around 500 km, due to the action of drag that limits the spacecraft lifetime and would imposes severe requirements on any propulsion system that should compensate for it. The use of a low power electric thruster enables microsatellite to counteract atmospheric drag even at very low altitude; therefore in our analysis the limitation on the altitude is set at 300 km [6].

Three different LEO geometries have been considered:

- elliptical vs. circular orbits;
- equatorial vs. inclined orbits;
- Sun-Synchronous orbits (SSO).

Elliptical orbits are an attractive solution for Earth observation purposes. They offer a significant potential gain in terms of coverage: when the orbit is elliptical the satellite stays for an extended period at the apogee, so allowing for a major coverage in the corresponding hemisphere. Due to the altitude range restrictions previously set (300 to 1000 km), the maximum value of eccentricity to consider is 0.049. This is a relatively small eccentricity value and therefore such elliptic orbit offers limited coverage advantages, while it is characterized by a wide set of perturbations typical of this kind of orbit. Moreover, due to satellite altitude and velocity variations, adequate instrument performance can not be guaranteed[5]. Accordingly, the use of elliptical orbits appears not convenient in the altitude range chosen, thus only circular orbits are considered in the following analysis.

Equatorial orbits are not suitable for EO missions since these cannot observe high or even mid-latitudes regions. Inclined obits are appropriate if a specific region or latitude belt has to be observed. In these kinds of orbits, the inclination of the orbit itself is determined by the location of the region of interest. The use of inclined orbits has been proposed specially for military applications[5]. Moreover for a generic orbit of this kind the orbit plane rotation induced by the RAAN-rate causes a variation of the illumination conditions of the target sites between consecutive satellite passages; hence the impossibility to observe the same place every time in the same illumination conditions. Furthermore it is very likely that dedicated launches might be required, increasing the overall mission cost.

In conclusion, the use of circular Sun-synchronous orbits is envisaged since they allow for uniform coverage and Sun illumination conditions, high latitude accessibility, limited satellite altitude and velocity variations, high opportunity of launch as piggyback payload.

Sun-synchronous Repeating Ground Track Orbits

A Sun-Synchronous Repeating Ground Track Orbit (SSRGTO) is an orbit which provides simultaneously the capabilities of repeating ground track orbit and Sun-synchronous orbit[7]. SSRGTO orbits are well exploited for example by Landsat, SPOT and RapidEye programs[6].

Sun synchronous orbits are characterized by the combination of inclination (i), eccentricity (e) and semi-major axis (a) that guarantees to have an average regression of the line of nodes due to the Earth oblateness (J_2) equal to the Sun apparent motion around the Earth (1 deg/day).

Repeating ground track orbits are generated by the combination of perturbations on the argument of perigee, mean anomaly and RAAN so to have an integer number of revolutions after a given number of days (accounting also for the Earth natural rotation). In a repeating ground track orbit, the spacecraft returns after a given number of days on the same Earth location, thus the ground trace of the spacecraft repeats itself. The design of such an orbit requires a fixed orbital period; perturbations, however, will cause an orbital period variation. In particular, the rotation of the orbit due to Earth oblateness has to be considered. This results in an iterative process for the design of a RGTO due to the fact that the Earth oblateness effects are a function of altitude.

To design SSRGTO orbits satisfying both requirements a non linear system for a, e, i has to be solved. At first order, however, such orbits can be considered as circular, near polar and with an altitude given by the initial estimate of the altitude of a repeating ground track orbit neglecting Earth oblateness effects, according to relation 1 [7]:

$$h_o = \mu^{1/3}\left(\frac{2\pi j}{\tau_E k}\right)^{-2/3} - R_E \qquad (1)$$

where $\tau_E \cong 86164.10035$ s is the sidereal rotation period of the Earth (relative to the fixed stars), R_E is the equatorial radius of Earth, μ is the Earth's gravitational constant, j is the integer number of orbit periods completed in an integer number of k days.

OPTICAL INSTRUMENTS

Following a conservative approach, we restrict our study to state-of-the-art, flight proven instruments, excluding any new developments. To meet the a wide range of EO requirements in terms of spectral and spatial resolutions, the following options have been considered [6,8]:

- Multispectral instrument;
- Hyperspectral instrument;
- Thermal infrared instrument.

Under the assumption to design a small platform with a launch mass of less than 70 kg, only existing instruments with a mass lower than 20 kg and power requirements up to 50 W have been selected for each class. The key parameters considered for each instrument are:

- spatial resolution;
- swath width;
- number, type and width of spectral channels.

Table 1 summarizes the main instruments selected with the associated flight heritage.

Instrument Class	Instrument
Multispectral	HPT (Rising-2), HRMS (Hodoyoshi-4), IRIS (X-Sat), Mx-T (IMS-1), NAOMI (SPOT 6-7) OC (Hodoyoshi-1), SLIM-6-22 (DMC)
Hyperspectral	CHRIS (PROBA-1), COMIS (StSat-3), Phytomapper (-)
Single Thermal IR	CIRC (Alos-2), HSRS (Bird)

Table 1: Reference EO payloads considered

CONSTELLATION DESIGN

With the aim of designing a constellation based on the presence of different optical sensors, the orbital parameters analysis has been conducted separately for each kind of instrument. The entire constellation is conceptually divided in a set of sub-constellations, each based on a single common payload, and with microsatellites conveniently spaced into a given orbit plane.

Obviously, the number of microsatellites is a driver factor for the overall system cost, thus the number of satellites shall be minimized. The number of orbit planes is another design variable open to multiple choices. In terms of constellation growth and degradation, a single-plane constellation has some advantages with respect to constellations with multiple orbit planes: if a satellite fails, it is possible to re-phase the remaining satellites, with a relatively limited propellant consumption, by means of an in-plane maneuver. On the contrary, repositioning a satellite in a multiple plane constellation may be prohibitive for the high maneuver cost[5]. In addition, using a single near-noon plane the optical payloads will acquire the images with the better and the same illumination conditions[5].

Sub-constellations are aimed to independently achieve payload specific requirements exploiting individual, ad-hoc designed orbits. The entire constellation resulting from their combination offers a very high degree of completeness and versatility, aimed at allowing users to exploit images of only one sub-constellation, or to use a logic combination of some or all of them, depending on specific objectives. A convenient and fast access to space-data to many different users is therefore ensured.

SRGTO identification

Our constellation design begins with the definition of the SRGTO orbit altitude (and associated inclination), starting from the required repeat cycle. The requirements of most instruments set the Revisit Time (RT) in between 1 and 3 days[6]. However, taking into account the unique opportunities offered by the microsatellite constellation, a nominal Repeat Cycle (RC) of up to 5 days has been considered. Equation 1 is used to calculate the altitudes corresponding to the desired nominal RCs. Figure 1 shows the recurrence diagram for Sun-synchronous satellites for altitudes between 250 and 1200 km.

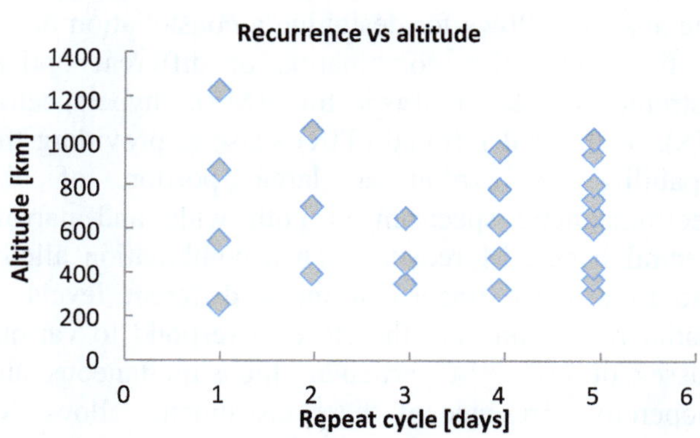

Figure 1. SRGTO altitudes for 1-5 days repeat cycles

Once the possible orbital altitudes are known, the most suitable solutions to cover a given region of interest has to be identified. For this purpose, given the instrument Field of View (FOV), the swath width (S_w) is calculated for each altitude h_o through a spherical Earth assumption[9].

To design the orbit constellation, in addition to the swath width, the dimension of the observation area perpendicular to the satellite motion direction (D_c) shall be taken into consideration. If D_c is equal or lower than the swath width S_w, only one satellite placed in a 1-RC orbit is sufficient to cover every day the area of interest. Otherwise, the area of interest could be divided into a number of strips (N_{strip}) characterized by a dimension equal to the instrument swath width S_w.

Under the assumption to cover the whole area of interest at the same time through a micro-satellite constellation, the number of strips N_{strip} corresponds to the number of orbital planes in which at least one microsatellite has to be placed. These planes are characterized by the same orbital parameters, besides the Right Ascension of the Ascending Node (*RAAN*): they will be separated by an angle $\Delta\Omega$ satisfying the relation:

$$S_w / 2 = R_E \Delta\Omega \qquad (2)$$

As an example, Figure 2 shows the number of planes required to cover at once an area of interest for a specific instrument (FOV=26.6°, SLIM-6). The number of planes is plotted as function of altitude h_o for different values D_c of the target size.

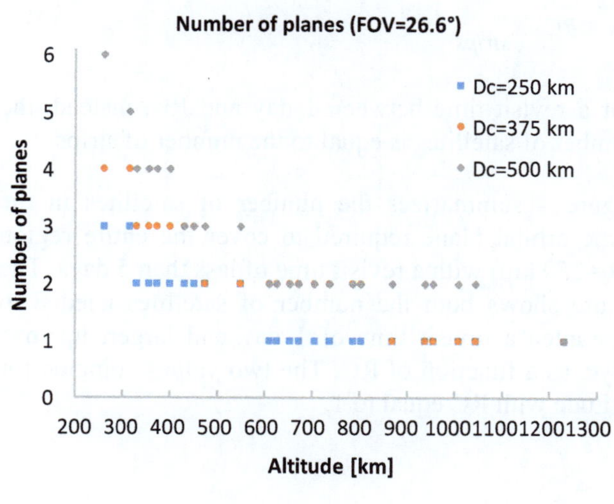

Figure 2. Number of planes for different altitudes at fixed instrument FOV

Figure 3 shows the number of planes required to cover the area of interest (assumed to be D_c = 250km), at once with instruments with different FOV. As a reference, the following values of FOV are presented: 19° (CIRC), 26.6° (SLIM-6), 42° (Phytomapper). The number of planes is plotted as function of altitude h_o .

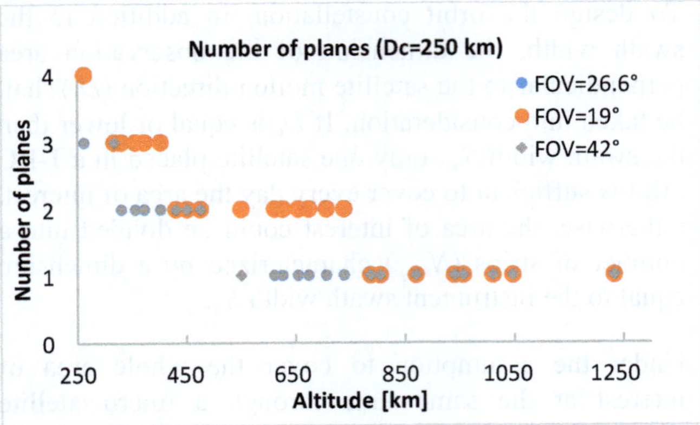

Figure 3. Number of planes for different altitudes at fixed dimension of area of interest (D_c=250 km)

This approach allows for a simultaneous coverage of the area of interest but it does not guarantee the required revisit time, except for altitude with 1-RC. To revisit the region of interest every day in any case, one or more satellites shall be placed along the same orbit. Under the assumption to cover the whole region in quasi-nadir pointing mode, this number of satellites (N_s) is a function of the repeat cycle, and of the swath width resulting by the instrument performance and satellite altitude. Accordingly, for a revisit time of 1 day the maximum number of satellites per orbital plane is:

$$Ns \leq RC \cdot N_{strips} \quad (3)$$

For a revisit time between 1 day and RC, instead, the number of satellites is equal to the number of strips.

Figure 4 summarizes the number of satellites in the same orbital plane required to cover the entire region (D_c=250 km) with a revisit time of less than 5 days. The figure shows both the number of satellites needed to guarantee a revisit time of 1 day and larger, up to 5 days, as a function of RC. The two values coincide for altitude with RC equal to 1.

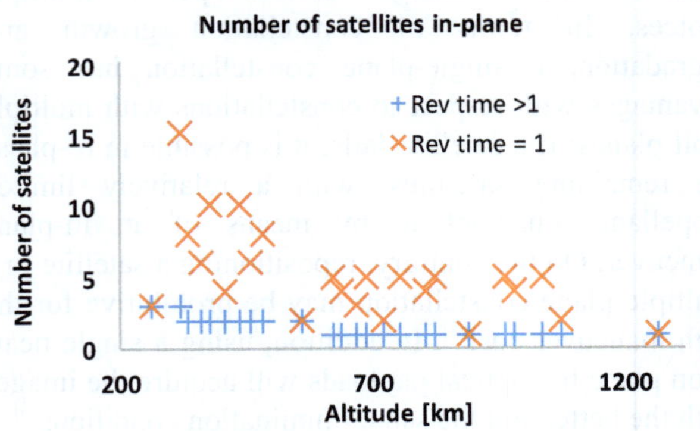

Figure 4. Number of satellites per orbital plane

CASE STUDY: TUSCANY REGION COVERAGE

A possible EO system for continuous monitoring of the agricultural activities in the Tuscany region of Italy is here analyzed as reference case. Table 2 presents the preliminary user requirements, considered as upper level constraints.

Requirements	Value
Product level	Surface reflectance and temperature.
Spectral wavelengths	Wide-band visible (VIS)/near infrared (NIR). Hyperspectral VIS/NIR/short wave infrared (SWIR). Thermal infrared (TIR).
Spatial resolution	Multispectral and hyperspectral images: 100-30 m, 30-10 m, < 5 m. TIR images from 500 m to 100 m.
Revisit time	From 2 months to 1 day.

Table 2: Preliminary user requirements

The analysis allows for designing a constellation based on the cooperative combination of different optical instruments, like multispectral (MS), hyperspectral (HS), or thermal infrared (TIR) sensors, providing the capability to exploit a large portion of the electromagnetic spectrum in both wide and narrow spectral bands. Moreover, such a combination allows also to produce images at many different levels of spatial resolution, and therefore to respond to various classes of users. In particular, the simultaneous and cooperative presence of these instruments allows for covering the whole range of spatial and spectral resolution levels potentially required in Tuscany agriculture activities. Four optical instruments were selected to provide with high spatial and spectral resolution both surface reflectance and temperature measurements.

The most promising solution is a constellation of four microsatellites each equipped with a specific different optical instrument, which acquires images in the VIS Red (R), Green (G), Blue (B) and NIR channels, and in the SWIR and TIR domains. Considering the limited extension of the Tuscany region (D_c = 210 km), RC values not larger than 3 days have been finally selected. Indeed, in the case of a small geographical coverage and for the same RT, it is convenient to stay with low values of RC, so to limit the number of satellites. Table 3 and Figure 5 summarize the constellation architecture.

Sensor	Altitude, [km]	GSD, [m]	Spectral bands	RT, [days]
MS #1	554	10 – 30	R, NIR	1
MS #2	358	2 – 5	R, G, B, NIR	3
HS	358	30 – 50	Thousands VIS, NIR, SWIR	3
TIR	554	100 – 500	TIR	1

Table 3. Microsatellites constellation architecture characteristics

Figure 5. Satellites passes over the Tuscany region

The space-born data obtained can be used for a large number of applications; e.g. land cover and use mapping, crop classification and health monitoring, soil moisture quantification, timely and located fertilization and irrigation strategies definition (precision agriculture).

The MS #1, HS and TIR instruments are expected to be able to provide a swath large enough to cover the entire region during each pass. The MS #2 instrument has been added with the aim at providing very fine spatial details suitable for add-value applications, and for very targeted observations down to the single-crop level[8]. Observations from at least one VIS/NIR and one TIR instrument are daily provided, and this allow to provide also a marginal service of disaster monitoring (floods, wild fires detection).

Finally, observations from the entire constellation are ensured two times per week, perfectly in line with agriculture and disaster monitoring requirements if also partial cloud coverage is considered.

PLATFORM DESIGN

The standard platform is sized and designed to demonstrate the feasibility of the combination of a microsatellite platform, an electric propulsion system and a set of existing small remote sensing instruments. The platform is designed according to the following requirements:

- use of off-the-shelf components to the larger possible extent;
- the whole system has to be designed to be compatible with the presence of an electric propulsion system on-board;
- overall launch mass <70 kg, including payload and propellant;
- maneuver capabilities to counteract the atmospheric drag at very low altitude and to perform orbital maneuvers.

The design is aimed at exploiting a thrusting module based on the SITAEL's HT100 low power Hall effect thruster[2]. Table 4 and Figure 6 show the main thruster performance and characteristics.

Performance	Value
Power, [W]	120-350
Thrust, [mN]	6-18
Specific Impulse, [s]	1000-1600
Efficiency	Up to 40%
Thruster Unit Mass, [g]	< 440
Thruster Envelope, [mm]	Φ 60 x 41 (I/F and cathode excluded)
Propellant	99.996% Xenon
Technology	Hall Effect Thruster, closed electron drift with extended acceleration zone

Table 4: HT100 main performance

Figure 6. HT100 thruster assembly

According to thruster power and thermal requirements, the entire design has been developed in order to make the platform completely thermally isolated from the electric propulsion device. The platform shall also be compatible with the great majority of the existing small optical instruments, and shall be able to provide in-orbit performance (e.g.: attitude pointing accuracy, payload data rate) suited to EO missions. The standard platform is based on off-the-shelf components, and is aimed at ensuring a sufficient power margin for electric thruster operation[2]. Table 4 summarizes the platform dimensions and Figure 7 shows the platform external layout.

Performance	Value
Platform dimensions, [m3]	0.5 x 0.4 x 0.5
Dry mass w/o payload, [kg]	< 40
Power generation BOL, [W]	250
Battery capacity, [Wh]	252
Payload available volume, [lit]	20
Payload available mass, [kg]	12
Payload available power, [W]	30
Fine pointing accuracy, [°]	0.025
DeltaV capacity, [m/s]	1250
Mission lifetime	Up to 5 years
Launch compatibility	VEGA, DNEPR
Communication	X-band downlink (up to 100 Mbit/s) S-band uplink

Table 5: Platform performance

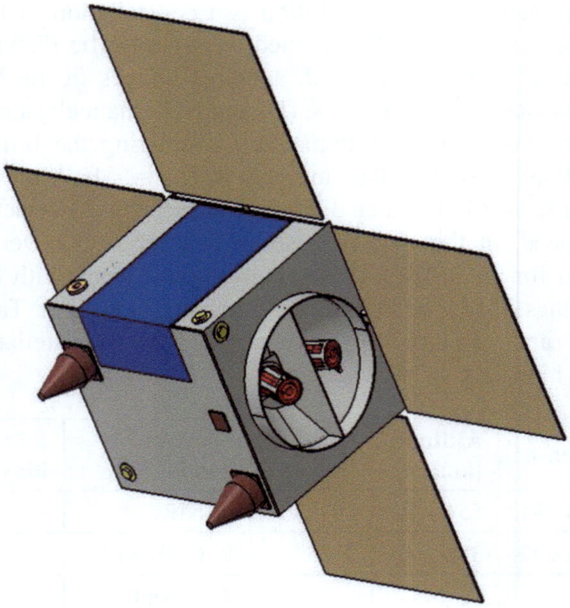

Figure 7. Platform design (*payload vane* in blue)

Two HT100 thrusters in cold redundancy are considered. The thrusting module is completed by appropriate plume shields and by an internal monolithic titanium tank. Four deployable solar panels and an additional body mounted solar array allow for generating up to about 250 W at satellite begin of life (with a reduction of about 10% at the end of life). This power level, rather high for a microsatellite and enabled by the recent technological developments in terms of high efficiency solar cells, is fundamental to operate the HT100. The attitude determination and control system relies on four redundant reaction wheels coupled to a pair of star trackers to provide a very fine attitude pointing accuracy during thrusting, target acquisition or data transmission.

The design proposes the exploitation of coarse sun sensors and magnetic torques to perform coarse attitude control during acquisition or safety mode phases. Magnetic torques take care of momentum dumping too. The platform design is completed by two redundant X-band antennas, and by two Li-Ion secondary batteries aimed to provide a total storage capability of 252 Wh[3,7]. This storage capability is aimed at providing the possibility to perform altitude maintenance ignitions also during eclipse periods. This eclipse thruster ignition capability is aimed to perform very fine station keeping maneuvers, and to allow for, limited electric thruster ignitions in favor of platform thermal control.

The preliminary design resulted from this analysis offers high versatility to the payload in terms of available volume, mass and power, and high performance in terms of data transmission, and pointing accuracy.

Figure 8 illustrates the overall platform logic architecture.

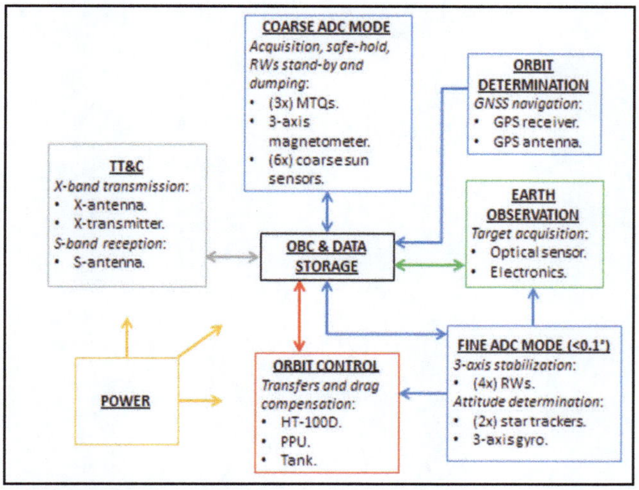

Figure 8. Platform logic architecture

CONCLUSIONS

The paper presents a simple analytic approach for the preliminary design of a small satellite constellation. The approach requires the definition of the size of the region to be observed, of the frequency required and of the payload characteristics and returns the number of satellites and of orbital planes required for a complete every day coverage.

As an applicative case, the design of a Tuscany region agriculture support mission is presented. Starting from upper level user requirements and considering the performance of existing small optical instruments, the analysis results in a constellation based on four microsatellites, each equipped with a different optical instrument (multispectral, hyperspectral and thermal infrared) responding to specific spatial and spectral performance.

In order to guarantee very frequent revisit, microsatellites are placed in SSRGT orbits from 358 km to 554 km. Each microsatellite is equipped with two low power Hall effect thrusters, to provide orbital maneuvering capability and drag compensation for station keeping.

The versatility ensured by the presence of the electric thruster, the consequent capability of optimally and simultaneously exploiting different optical sensors, and the large compatibility of the platform with the great majority of existing small optical sensors, make the proposed constellation able to easily respond to requirements coming from a variety of different users.

Finally, the total mass and the overall dimensions of the microsatellites are such that the entire constellation can be launched in a single shot with any of several low cost launchers.

References

1. Sandau, R., "Status and trends of small satellite missions for Earth observation," *Acta Astronautica,* Vol. 66, Issue 1-2, Jan.-Feb. 2010, pp. 1-12.

2. Ducci, C. Oslyak, S. Dignani, D. Albertoni, R. and Andrenucci, M., "HT100D performance evaluation and endurance test results", IEPC-2013-140, *Proceedings of 33rd International Electric Propulsion Conference*, October 2013, Washington DC.

3. Mission Design Division Staff, "Small Spacecraft Technology State of the Art", *NASA/TP-2014-216648***,** California, 2014.

4. Krueger, J and Selva, D., "Spacecraft and Constellation Design for a Continuous Responsive Imaging System in Space", AIAA 2009-6773, *Proceedings of the AIAA Space 2009 Conference*, Pasadena, California.

5. Spera, P. and Gallon, A. "Constellation Orbit Design Criteria for a Dual Use EO system," RTO MP-61, *Proceedings of the RTO SET Symposium on Space-Based Observation Technology*, Samos, Greece, 16-18 October 2000.

6. Gregucci, S. Pergola, P. Papa, A. and Marcuccio, S., "Electric Propulsion Microsatellites in a Versatile Constellation for Remote Sensing Applications," IAA-B10-1102, *Proceedings of 10th Symposium of Small Satellites for Earth Observation,* IAA, Berlin, Germany, 2015.

7. Wertz, J., Everett, D., and Puschell, J., *Space Mission Engineering: The New SMAD*, Space Technology Library, Microcosm Press, 2011.

8. Moran, S. 2000. "New imaging sensor technologies suitable for agricultural management, Remote sensing in agriculture", *Remote Sensing in Agriculture*, *Aspects of Applied Biology* 60, 2000.

Feasibility study of LTA launch system for micro and smaller satellites

Andrea Testore

13th Reinventing Space Conference

9-12 November 2015

Oxford, UK

S. Hatton (ed.), *Proceedings of the 13th Reinventing Space Conference*,
https://doi.org/10.1007/978-3-319-32817-1_21

Feasibility study of LTA launch system for micro and smaller satellites

Andrea Testore
School of Aerospace, Transport and Manufacturing, Cranfield University, UK
Via S. Michele 14, Pino T.se, Turin, Italy; +39 340 8770829
andrea.testore@icloud.com

Nowadays, space companies are investing more and more in smaller satellites aiming to reduce projects costs and complexity. Despite the use of Pico/Nano and Micro-sats has considerably increased throughout the last few years, a reserved launch vehicle is still missing. This paper analyses both the economic and technical feasibility of a reusable stratospheric launch pad based on lighter-than-air (LTA) technology. The possibility to lift the launch pad up to stratospheric altitude with the use of airship technology has been investigated. Around 20-25 km of height the rocket ascension is less affected by some typical losses, such as air drag and atmospheric loss. Thus, stratospheric launch can ensure higher performance of the rocket, higher mass ratio, and less fuel is required. The above mentioned configuration seems promising for a small satellite launcher. Being tailored to service the smaller spacecraft it will be able to provide a much better and more reliable service with respect to the *piggyback* solution. Thanks to atmospheric condition in which the rocket will operate, LTA design would not only ensure advantages but it will be also able to provide a larger launch window availability with respect to the ground launch station. Indeed, thanks to its capability to fly above the jet streams it would be possible to ignore weather condition that usually provokes ground launch delay and abort. Moreover, it would be able to ensure launch operation to a wider range of orbit with respect to any kind of ground launch pad. Starting from the market analysis (section 1), the future need of a dedicated launch system for small satellites has been proved. This analysis identified the payload mass requirement. Since the production and development cost evaluation of this system is based on the size of the airship, a preliminary environment analysis and mass breakdown is required. For this reason section 2 and 3 are focused on these aspects. In order to evaluate the economical feasibility of the project, the final section will outline the cost comparison between Airship Assisted Launch System (AALS) and traditional launch facilities. If feasible, AALS could ensure valuable advantages in small satellites launches. The serviceable orbits and the launching window availability are just two among the most important benefits that air launch based on LTA technology could provide.
KEYWORDS: airship, low cost launch system, Airship Assisted Launch System, Lighter Than Air, small satellite, feasibility study.

1.Market Analysis: Customers, Competitors, and Opportunities

Thanks to an analysis of the market trend since January 2000 and the competitors right now available in the sector the main mission requirements have been identified. This analysis took into account all the satellites launched with overall mass lower than 500 kg. All the studies outlined throughout this section have been performed considering satellite mass category (pico, nano, micro, mini), mission purpose (civil, military, commercial), and satellite owner. All this information has been useful to better describe the evolution of smaller satellites market throughout these 15 years. Both customers and competitors statistical analysis has been performed using *SpaceTrak™* database (i.e. online catalogue developed by *Seradata,* collecting satellites and launches information).

Customers: Satellites Trend and Purposes.

Throughout the last decades the market share of nano and micro satellites has dramatically increased gaining more and more importance not only in civil but also in commercial and military missions. The increment of the use of nano-satellites is well shown in figure 1.

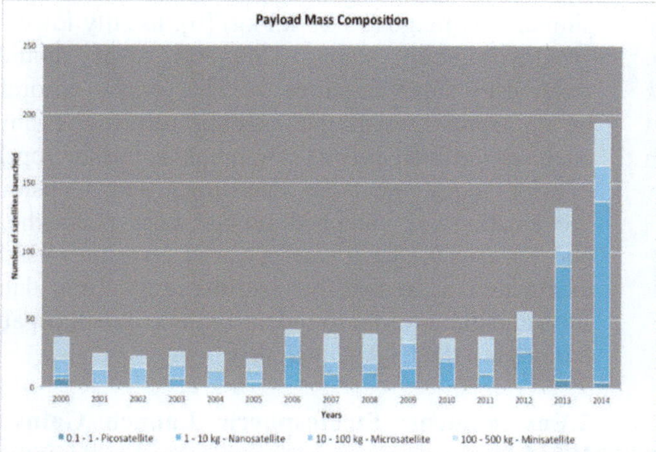

Figure 1: Satellites launched for each mass category since Jan 2000 (data source: [1])

Throughout 2013 the launches of this kind of systems have been increased around 300% with respect to 2012. A further increment of 160% has been registered in 2014 [1]. This clearly shows the health of this market share. Although only one company has produced a huge part of these satellites, with its EO constellation *Planet Labs* proved the high feasibility level of nano-sats based commercial mission. Furthermore, thanks to this constellation, commercial missions became the first purpose for nano-sats in 2014. With 94 3U cubesats activated last year *Planet Labs* produced, for commercial purpose, more than 70% of the nano-sats launched in the same period. Since this trend is expected to continue in the next years, the market share importance of this type of satellites will lead to a whole new amount of possible customers. A further proof of the relevance of this field is the interest that companies such as Airbus, Qualcomm and Virgin Galactic are having in this kind of technology. *OneWeb* project is expecting to provide worldwide Internet service launching 648 micro-satellites (≈150 kg). Many companies, also not directly related to space industry, such as Coca Cola are investing huge assets in this

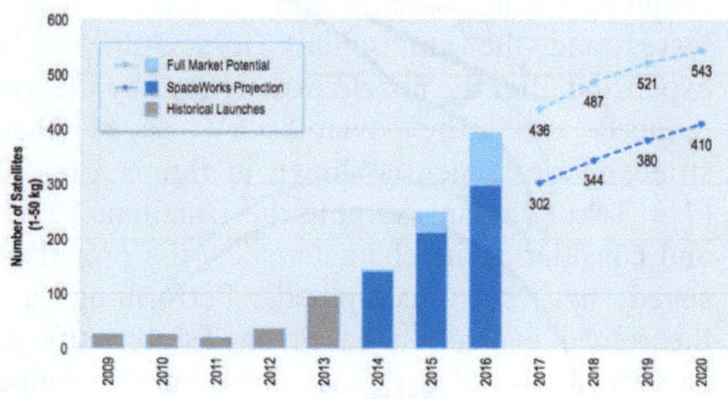

Figure 2: SpaceWorks market forecasts (source: [4])

project. In June 2015 the project received $500M [2] from various investors. This gives a further proof of the value that satellites with mass lower than 200 kg will have in the near future. For this reason the AALS has been sized to ensure service to satellites up to 200 kg. American Federal Aviation Administration (FAA) in June 2015 registered 8 future missions that would lunch 329 satellites with mass lower than 200 kg within 2019 [3]. Moreover, *SpaceWorks*, American consultancy firm, foresees an increasing trend throughout the next 5 years of the use of satellites with mass lower than 50 kg, as shown in figure 2 [4]. Considering all these forecasts, it is quite clear how this market share will acquire more economical importance in the near future ensuring a large amount of possible customers. Figure 3 shows the customers prevision for the next three years. In order to draw this preliminary forecast the market composition has been analysed. Considering the commercial mission planned and just 15% of the overall civil missions planned for the near future, 100 satellites have been identified as possible customers starting from 2016. The ratio of civil mission requiring

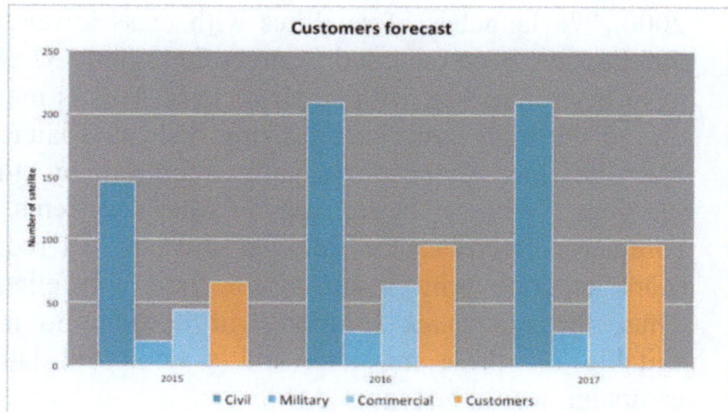

Figure 3: Customers forecasts for small launcher service (based on [3] [4])

dedicated launch has been evaluated considering the number of satellites launched during the last 15 years by gvernments and national space agencies. Throughout these decades 43% of the satellites with mass lower than 500 kg and civil purpose have not been launched for educational purpose. For this reason it is fair to expect that these satellites will have more demanding launch requirements. In order to be conservative not the whole cross-section has been considered anyway. This preliminary consideration has been valuable to identify the rocket payload requirement. A capability of 200 kg has been considered a capability sufficient enough to service main future commercial missions.

Competitors: Small Rocket Market.

Although the smaller satellite market share is acquiring more importance, a prominent dedicated launcher for satellites up to micro/mini category is still missing.

Because of their small size and weight these systems are especially suitable for small rocket and, thus, they could take important advantages in air launch design. This kind of satellites is usually launched as *piggyback* and rarely dedicated launches are performed. The most used launchers for this purpose are listed in table 1 with

Table 1: Small launcher employment since January 2000 (data source: [1])

Launcher	Satellite launched	n of launch	Pico	Nano	Micro	Mini	Civil	Commercial	Military
Falcon 1	8	5	0	3	3	2	5	1	2
Pegasus XL	13	12	0	0	11	2	11	1	1
Shavit	5	5	0	0	0	5	0	0	5
Shtil	1	1	0	0	1	0	1	0	0
Start 1	3	3	0	0	0	3	1	2	0
Volna	2	2	0	0	0	2	2	0	0
Tot	32	28	0	3	15	14	20	4	8

the amount of satellites launched grouped by mass and purpose category. Using *SpaceTrak™* database it has been possible to identify the market size. Since January 2000, 774 launches of satellites with mass lower than 500 kg have been recorded. Among these, only 32 (4%) have been launched with small rockets. This is mainly due to the high cost characterizing dedicated launches and the low precision usually required by these missions. With a larger use of these systems for commercial purpose the satellites owners will require more accurate orbit insertion for their constellation. Thus, the *piggyback* solution will become no more suitable for these missions and a dedicated launch campaign would be required [5].

Although the services available up to now are few, they are quite heterogeneous. Both ground based launches and air launch are currently available or they will be available in the next few years (Virgin Galactic LauncherOne, Zero2Infinity Bloostar). Cost per kilo of all these services are shown in figure 4. As it is clear

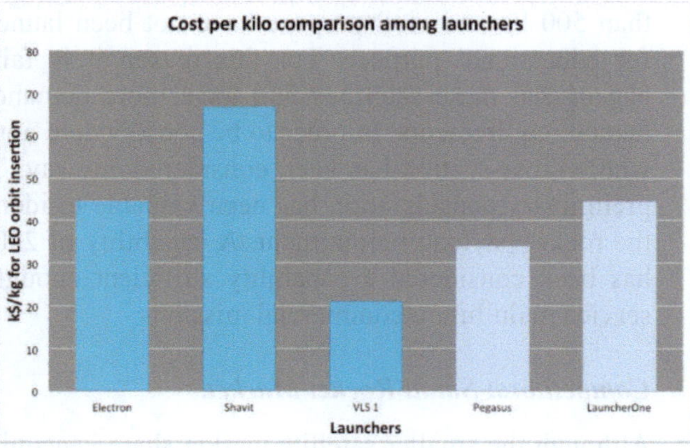

Figure 4: Cost per kilo characterizing the most used or in development small launchers (light blue: air launch, turquoise: traditional launchers)[9]

from the image, air launch systems are characterized by similar cost range (35-45 k$/kg) [6], usually lower than ground launch systems. This cost range should be ensured by small satellites launcher system in order to be competitive with the existing services. From the analysis outlined in this paragraph a further top-level mission requirement has been identified. Considering the existing air launched rocket mass (LauncherOne and Pegasus XL) a lift-off mass of 20 tonnes has been exstimated a reasonable assumption. This data is fundamental in order to indentify the payload capability required to the LTA launch pad.

2.Environment: Stratospheric Launch Gains and Wind Issues

Launching from high altitude can ensure several advantages. The rarefied air can guarantee better rocket performances and, thus, a reduction of many types of losses. Moreover, performing the launch from stratosphere ensures a minor influence of wind within the launch operation. Nevertheless, reaching stratospheric height with 20 tonnes of payload is challenging and it requires innovative design and technology.

Atmosphere: Gains Achieved Launching from High Altitude.

Thanks to the relevant decrease in the atmospheric density with the height, the intensity of forces such as air drag dramatically decreases. Considering three different scenarios the great advantages achievable thanks to stratospheric launches has been proved. Traditional air launch systems (i.e. Pegasus XL) release the rocket around 10 km reducing to almost one third the ΔV required to overcome the air drag. At stratospheric altitude the velocity that the rocket has to provide in order to overcome this loss is just 7% of the one required by a ground launched vehicle. Although reducing drag loss is the primary benefit of launching from high atmosphere, it is not the only source of gain achieved thanks to stratospheric launch. The gravity losses and the atmospheric losses due to exhaust expansion after the nozzle experienced by the rocket are reduced too. The overall ΔV saved thanks to stratospheric launch is shown in figure 5 (≈ 700 m/s) [7]. Taking as an example the Brazilian VLM rocket and considering its characteristics, the propellant mass saved has been computed. Performing a purely theoretical evaluation based on Tsiolkovsky equation and VLM overall dry mass (16 tonnes) 3 tonnes have been calculated as saved propellant mass. For this reason stratospheric launch is considered a very promising option for LTA launch pad design.

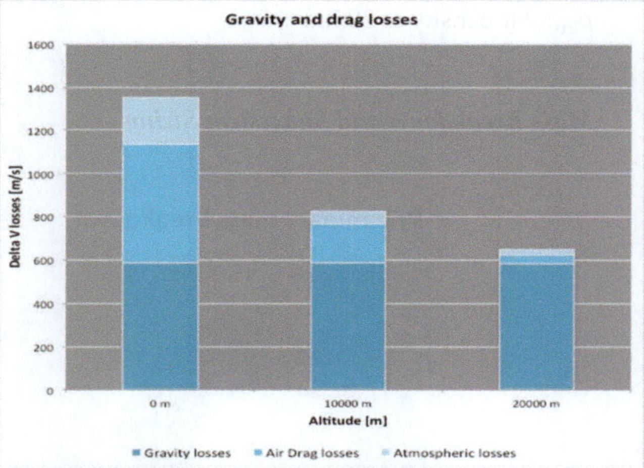

Figure 5: ΔV required to overcome different type of losses for sea level 10 km and 20 km launch (computed using [7] and VLM data)

Winds: an Issue Deeply Affecting Launch Operations.

Wind can force to postpone or even abort launch operations. Also the largest and more powerful launchers can be deeply affected by wind slower than 20 m/s. Rockets such as Atlas V and Falcon 9 cannot be launched with wind speed faster than 15-17 m/s [8] [9]. Launch sites located at latitude between 30 and 60 degree both north and south are affected by the Jet Streams. These are strong and continuous wind situated around 10 km of altitude mainly affecting these latitudes during winter months. They thus dramatically reduce launch availability from these spaceports during winter months. Flying over these winds the launching capability would be considerably enhanced and it would be possible to ensure a more reliable service to the customers. Figure 6 clearly shows this high-speed wind areas located at 12 km of altitude (within Ferrel Cells area) during winter months. In order to ensure this higher launching capability the LTA launch pad has to

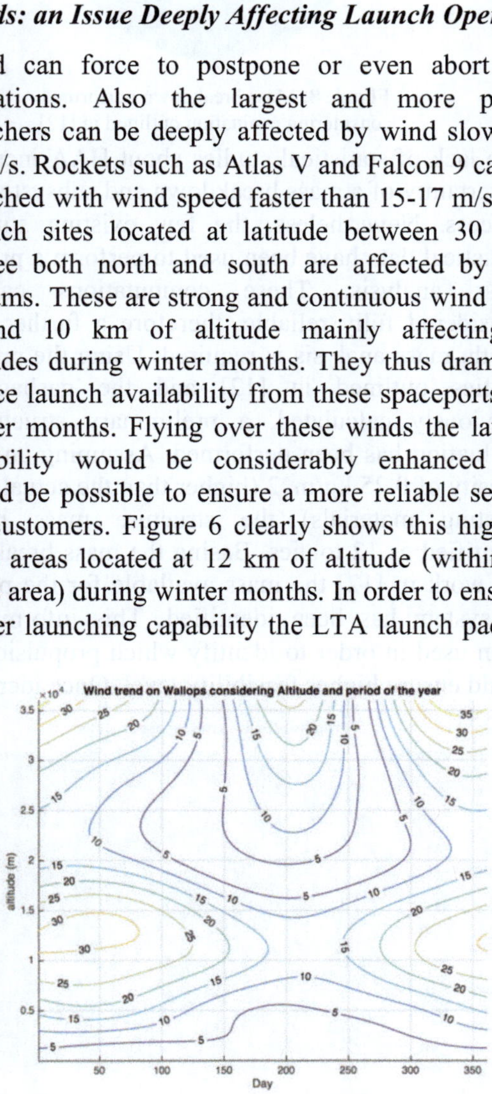

Figure 6: Statistically evaluated wind trend over Wallops spaceport considering both altitude and period of the year.(computed using *hwm07* MATLAB® function)

be able to actively fly through these Jet Streams. For this reason the mission scenario requires a propulsion system to safely reach the launch location. Thus, only airship design has been taken into account throughout this research. Thanks to the presence of propulsion subsystem the launching capability can be dramatically increased. Moreover the launch site precision, and thus the orbit that can be serviced, would get further enhancement. Finally, AALS launch altitude has been identified considering the statistical wind trend computed thanks to HWM07 model. At around 20 km of altitude the wind usually reduces its intensity ensuring a reliable altitude to perform launch operations. Considering the *wind knee* and the air density, 20 km has been identified as the most suitable altitude to perform good launch operations. This altitude has been set as the last top-level requirement and further used to size the airship envelope.

3.Airship Design: Envelope Sizing and Mass Breakdown.

Table 2: AALS top-level requirements used to study the airship sizes and mass

AALS Requirements	Value	Unit
Operational Altitude	20 000	[m]
Payload Capability	>20 000	[kg]
Cruise Speed	20	[m/s]

Once identified the payload capability required to lift the rocket and the operational altitude the preliminary design of the airship has been performed. Market and environment analysis previously outlined has been useful to recognize the mission top-level requirements fundamental to identify the main characteristics that the AALS should have. These are summarised in table 2. The operational altitude of 20 km has been chosen not only for the great reduction in losses achievable from that altitude but also for the low wind speed characterizing this layer. Considering both the propulsion subsystem weight and the launch capability, the cruise speed requirements has been set as 20 m/s. Once analysed the air launch competitors, 20 tonnes has been identified as reasonable payload capability requirements for the airship.

Lifting Gas and Envelope Sizes.

LTA technology is based on few possible gases. Hydrogen and Helium are certainly the ones with better performance. Both these gases have a lifting capacity higher than 10 N/kg and can ensure the buoyance required by LTA designs. Unfortunately, both these gases have relevant disadvantages. Although Hydrogen

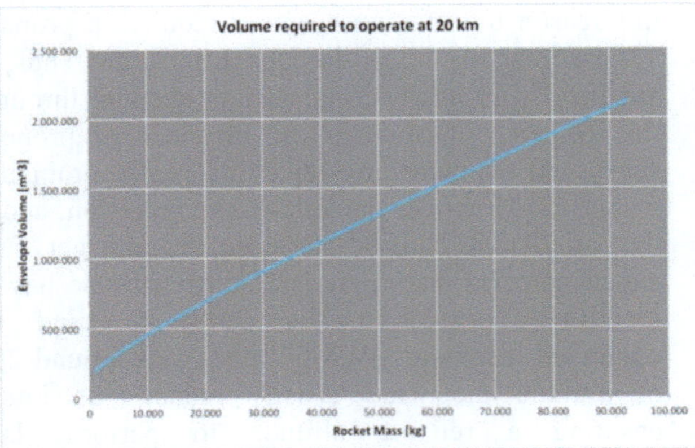

Figure 7: Envelope volume required for pressure height of 20 km vs rocket mass (computed considering prolate ellipsoide and [11])

is the lightest, cheapest, and the most common LTA gas on the planet, its high flammability reduced its use since Hindenburg disaster in 1937. On the contrary, Helium is much safer than hydrogen and is currently the most used lifting gas. Nevertheless, Helium is far rarer than Hydrogen and its high demand, due to its use in chip production, makes its price steadily increase since last decades [10]. Lately Hydrogen based airships have been considered as an option again. Lower price and possibility to use fuel cell as power source are very interesting factors for airship industry. Nevertheless, in order to be conservative, airship sizing has been executed considering the more expensive and less well-performing Helium gas. This choice ensures a valuable margin both in economic and technical aspects of this feasibility analysis. Starting from the payload mass requirements (20 tonnes) and performing an iterative mass breakdown, the volume of Helium required has been computed around 53 000 m^3. Once identified the volume of Helium required the envelope sizes have been computed. To calculate these sizes, both atmospheric conditions at operational altitude and pressure requirements imposed by the non-rigid gasbag have been studied. The envelope structure for High Altitude Airship (HAA) cannot be developed with rigid design because of mass issues. The non-rigid design, thus, imposes pressure gradient limits between the inner and the outer environment of the gasbag. Once computed the envelope volume, imposing a fineness factor $f = 4$ in order to reduce the air drag intensity [12], the sizes and surface of the envelope have been computed.

$$V_{20} = \frac{V_{sl} * \rho_{sl}}{\rho_{20}}$$

V_{20}: Helium Volume required at 20 km
V_{sl}: Helium Volume required at sea level
ρ_{sl}: Air density at sea level
ρ_{20}: Air density at 20 km

Mass Breakdown and Subsystem Sizing

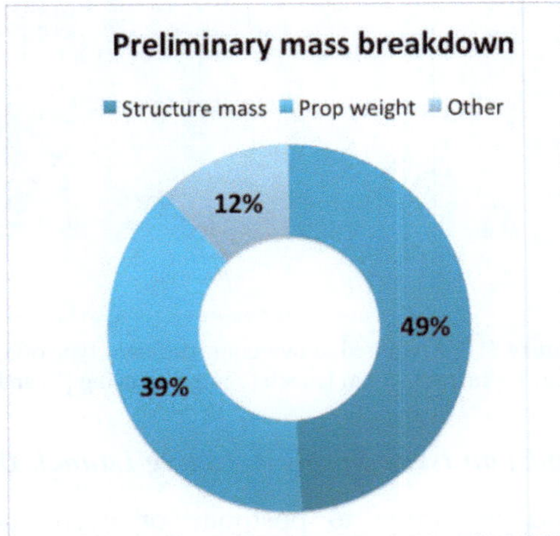

Figure 8: Mass breakdown performed considering evaluation outlined in [12]

The lack of statistical studies about HAA increases the uncertainty of a mass breakdown and subsystems sizing process. Nevertheless, the few existing studies and statistical data have been used to perform a preliminary mass analysis. These computations cannot be considered fully reliable, therefore a further, more in depth, mass analysis is required. Using the preliminary studies outlined in [12] and the gasbag surface previously calculated, a preliminary structure mass evaluation has been performed. Assuming an envelope density of 0.25 kg/m^2 (higher than the actual value for existing materials) the structure mass has been identified as 12 tonnes. Basing the mass breakdown on the work in [12] the mass available for the propulsion subsystem has been identified. This information has been used in order to identify which propulsion system could ensure higher feasibility level. Once identified the

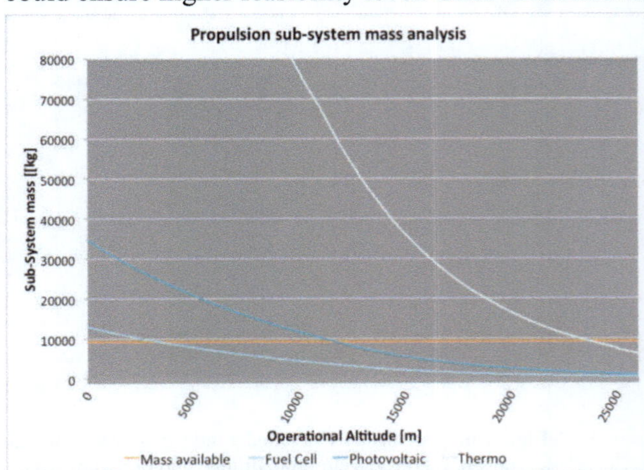

Figure 9: Fully operational altitude for three different type of power subsystem considering the air drag value and mass limitation (computed using [18] and air drag eq.)

mass available for the propulsion subsystem, different options have been considered taking into account both engine and power source efficiency. Considering the cruise speed requirement (introduced in table 2) and air drag produced by the envelope, the power and thrust requirements for different altitudes have been estimated. This analysis allowed to compute the minimum operational altitude ensured by each different propulsion subsystems. This analysis is shown in figure 9. Thanks to its higher power density, fuel cell based power subsystem proved its better performances. Fixed the mass available for the power and propulsion subsystem, fuel cells are able to ensure the required cruise speed starting from a 3400 meters of altitude. On the contrary, in order to ensure 20 m/s as cruise speed using photovoltaic power system, the airship has to previously reach lower air density (12 km). This could imply a not fully workability near Jet Stream layer. Using the iterative process, outlined throughout Appendix A, the main AALS characteristics have been computed and are summarized in table 3.

Table 3: AALS main characteristics evaluated starting from top level requirements considering statistical studies and using iterative process (Appendix A)

AALS parameters	Value	Unit
Envelope Volume	725 362	[m^3]
Envelope Length	281	[m]
Envelope Diameter	70	[m]
Dry Mass (40% margin)	33 273	[kg]
Lifting Capablity	54 991	[kg]
Serviceable Payload	21 717	[kg]
Air Drag (sea level)	170	[kN]
Power required (sea level)	3.4	[MW]

4.Cost Analysis and Comparison.

Using Cost Estimation Relationships (CER) [13] corrected for HAA design both development and maintenance costs for AALS have been evaluated. This estimation has obviously preliminary value, therefore a more in depth cost analysis is required to provide more reliable data.

AALS Development and Maintenance Cost.

This preliminary analysis has been based on 5 parameters used in CER model in order to obtain a cost estimation. The most important parameters are certainly airship dry mass and cruise speed. These two parameters combined with production rate (rp), number of prototypes (Np) and difficult factors (Df) [13] have been used to compute the costs characterizing this project. The costs do not include research and development of a dedicated propulsion subsystem. Because of the specific environmental conditions the cost of a dedicated propulsion system for the airship is characterised by the following equation:

$$10^9 + 2 * (2251 * 0.043 * T * 243.25 * M_{max} + 0.096 * T_i - 2228 * Np * Ne)$$

Where T is the thrust required, M_{max} is the maximum Mach number, T_i is the turbine inlet temperature, and Ne is the number of engines. The altitude range at which AALS has to operate implies the constant initial addition of 1 billion dollars [13]. This constant value due to the high complexity of engine design process would increase the overall AALS development cost of over 300%. It has been, thus, decided to use existing engines in order to reduce development cost. Airplanes such as *Pathfinder* and *Perseus B* have previously proved the suitability of both electric and Brayton based propulsion subsystem [14]. The overall development cost has been computed around 320 FY2015M$ while operation and maintenance costs are listed in table 4. According to [15], these expenses can be considered part of the *Launch Site Facility Costs*. It usually representing the 8% (≈ $ 1M) of the overall fly away costs. Thanks to the simplicity of the AALS with respect to a traditional launch site, it is reasonable to expect a reduction of this entry within the overall fly away cost. Thanks to AALS moving capability a

Table 4: Per mission costs related to developmet, production and maintenance operations of AALS (computed using CER model [13])

	Costs
Development	$155M
Production	$168.3M
Total	$323.3M

Costs	Per hour	Per mission
Maintenance cost block		$6 055
Spare parts and repl. hour	$1 116	$4 857
Spare parts and repl.sortie		$1 954
Helium replicement		$1 313.05
Total per mission		$14 179.05
Personnel	N	Mission cost
UAV Operators	2	$922.20
Manager	2	$2 610.00
Op. Engineer	1	$578.55
Maintenance	2	$565.50
Total per mission		$4 676.25
Total per mission		$ 18 855.30

reduction in *Range Costs* can be expected as well. In order to obtain quantitative cost predictions, further and more in-depth costs evaluations are required.

Traditional Launch Pad Development and Maintenance Cost.

In order to provide a more in depth comprehension of the actual competitiveness of AALS with respect to traditional launch systems, development and maintenance costs of the ground spaceports have been analysed. Launch pad complexity and development cost are deeply related to the overall rocket thrust. Higher is the rocket thrust, more complex the deflector and the system are. Moreover, the launch facilities cost is also related to the physical sizes of the rocket. Since the relation between building height and cost is not linear [16], higher is the building, more relevant are the costs. Figure 10 shows, the maximum and minimum cost expected for a traditional launch pad development as

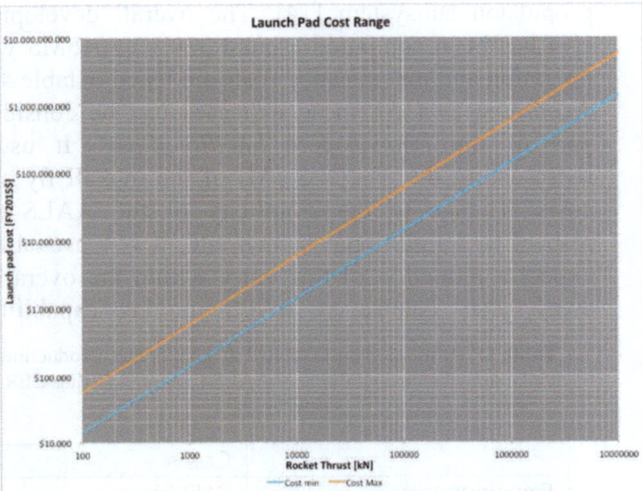

Figure 20: Launch facilities cost forecast considering overall rocket thrust [17]

function of the rocket thrust [17]. Remembering that the cost computed thanks to this relation can just give a preliminary evaluation, and that often the final cost of the facility development is much higher than the estimated, it is now possible to compare the AALS and traditional launch pad cost. Fixing the payload capability (200 kg), the mean development cost of a traditional ground facility has been identified around $129M. In order to compute this value Israeli *Shavit* rocket has been used. The development and production cost characterizing a traditional launch pad is less than 41% of the cost involved by AALS development (≈$320M). Completely different is the ratio considering only AALS production costs. Because of the big lack in experience for HAA and Heavy Lift Airship, the development is characterized by very high costs. On the contrary, ground launch sites can count on a valuable heritage level. Analysing only production cost, figure 11, it is clear how the two costs are completely comparable. This provides a promising preliminary feasibility evaluation. Moreover, analysing the maintenance cost of a traditional launch site, the scenario is even more encouraging. Because of the sea proximity that many spaceports require for launch corridor and safe range requirements, the metal corrosion is a relevant issue. Furthermore, the acidic nature of the exhaust gases and the presence of alumina particles further increase this issue. Analysing the maintenance costs related to KSC launch and transportation platforms a higher cost related to the traditional launch facilities than AALS has been identified. Since the almost null interaction that the rocket must have with the airship for safety reasons, the maintenance costs are much lower compared to traditional launch pad. Figure 11 and 12 show these comparisons.

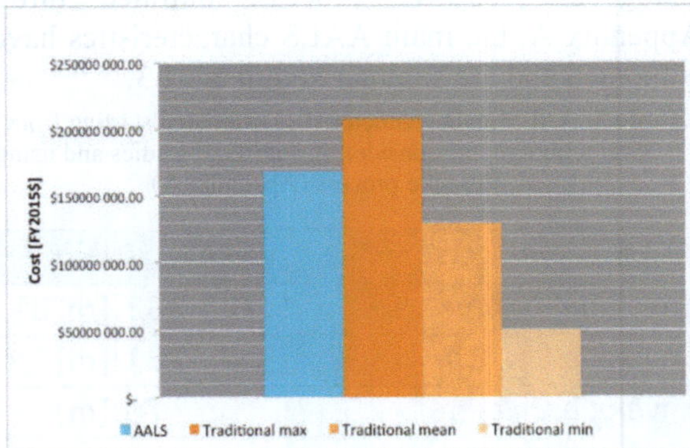

Figure 11: Production launch pad cost tailored for 200 kg payload (computed thanks to [13] and [17]).

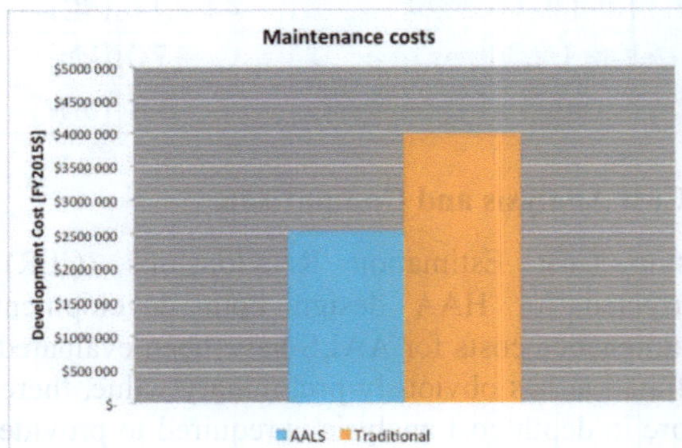

Figure 11: Maintenance costs for 6 launches per year for AALS and traditional launch pads (computed thanks to [13] and [19])

Discussion and Conclusion.

Starting from the market trend identified during the first section of this research, the value of a launch system dedicated to small satellites has been identified. Up to now, dedicated vehicles serviced just a small amount of the smaller satellites launched. A commercial and civil missions growth, with their high precision requirements, could consequently lead to an increment of dedicated launch demand. Indeed, *piggyback* service cannot ensure a service reliable and punctual enough to perform commercial and military missions [5]. Starting from this consideration the payload mass requirement has been identified around 200 kg. This requirement has been decided considering both the average mass of the commercial satellites launched up to now and the LTA technology limitations. Higher payload mass would, indeed, involve a higher overall rocket mass. Higher rocket mass would involve higher airship mass and thus a larger amount of lifting gas. Larger gas volume implies larger envelope and thus higher thrust and power required to satisfy the operations compliancy. The use of Hydrogen rather than Helium could dramatically cut the production and maintenance costs. The inevitable leaking of lifting gas would imply a recurrent cost per mission in order to refill the gasbag. Since Helium can cost up to 5 times with respect to the Hydrogen the use of this gas in such large amount could deeply influence the overall production and maintenance cost. Throughout the last few years higher safety standards have been studied in order to renew Hydrogen based airships. Using this design with fuel cell powered propulsion system would ensure a much lower cost, 24/7 operation capability, and lower *fully operational* altitude. Although rocket ignition certainly adds important issues on hydrogen based AALS, the use of this technology would certainly increase the commercial feasibility level of this project. Higher profitability certainly attracts a larger amount of financing helping further research and analysis.

Moreover, the environment analysis identified 20 km as the most suitable operational altitude. At this height the air density is low enough to ensure much higher rocket performances. Furthermore, this altitude is characterised by low wind speed and it, thus, could ensure a safer location for launch operations. Imposing 20 km as main mission requirement the envelope volume required to ensure this pressure height has been computed. Gasbag size is certainly the main issue concerning the technical feasibility level. Figure 13 shows the dimension of an AALS system with respect to the largest airship ever built, USS Akron. Although the economic feasibility is certainly the most important problem within this project, also design and structure has to be studied more in-depth. Considering payload mass (rocket) and operational altitude has been possible to identify a preliminary cost estimation for AALS development and maintenance. The overall system cost has turned out to be deeply influenced by development expenses, and, thus, higher than an equivalent capability ground launch facility. The high development cost is due to big lack in heritage level related to airship-based systems. A cost reduction would be fundamental in order to ensure a higher feasibility value.

A further increment in feasibility level could be achieved designing the AALS for multiple purposes. Throughout the last few years long endurance airship missions have been carefully analysed thanks to their lower cost than traditional satellite missions. Designing AALS with both photovoltaic and fuel cell powered electric propulsion system could ensure its use for surveillance or geo-engineering focused missions. These two fields have been deeply analysed throughout the last decades. Heavy lift transportation can be a further application for AALS. A payload capability of 20 tonnes is very attractive for transportation applications. Unfortunately, the 20 km pressure height requirement implies a large envelope size and, thus, a low manoeuvrability at lower altitude reducing the efficiency of this system for traditional transportation purpose. A modular envelope design could reduce this limitation ensuring a higher suitability for traditional airship mission. A more in depth configuration study is therefore required in order to better understand the possibility to ensure a broader applications range.

The economical evaluation performed up to now does not suggest the project is unfeasible. Although the development cost is much higher than the traditional

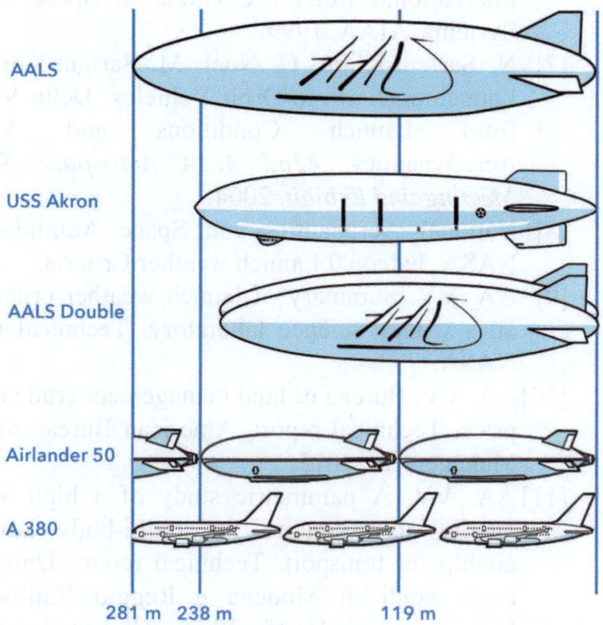

Figure 13: Artistic representation of both mono and double –envelope AALS sizes with respect to other flying vehicles

launch facilities, larger orbit range, wind tolerance and slightly lower maintenance cost look promising prospective for a feasible product. The service quality that AALS could provide is, indeed, much higher than traditional small ground launch rocket. Even considering traditional air launch options, the overall gain in launch performances achievable with AALS worth a more in depth analysis.

As conclusion, although the development cost deeply affects the overall feasibility of the project, with a wise design related to additional applications as well, the AALS could be a powerful launch system. For these reasons a further and more in-depth analysis of the configuration seems valuable to better understand the possibilities allowed by this innovative system.

References

[1] Seradata, http://www.seradata.com, July 2015.

[2] Daniel Thomas. OneWeb Secures 500M Dollars from Virgin, Qualcomm, Airbus and Bharti. *Finacial Times*, June 2015.

[3] AA. VV. 2015 Commercial Space Transportation Forecasts. Technical Report, Federal Aviation Administration, FAA, 2015.

[4] Elizabeth Buchen. 2014 Nano/Microsatellite Market Assessement. 2015.

[5] M. N. Sweeting. Space at Surrey: Micro-Mini-Satellites for Affordable Access to Space. *Air & Space Europe*. Vol. 2, number 1. 2000.

[6] J. B. Hopkins S. J. Isakowitz, J. P. Hopkins Jr. International Reference Guide to Space Launch Systems. AIAA, 1999.

[7] N. Sarigul-Klijin, C. Noel, M. Sarigul-Klijn. Air Launching Earth-to-Orbit Vehicles: Delta V gains from Launch Conditions and Vehicle Aerodynamics. *42nd AIAA Aerospace Science Meeting and Exhibit.* 2004.

[8] National Aeronautics and Space Administration NASA. Falcon 9 Launch weather Criteria.

[9] AA. VV. Summary of launch weather criteria for atlas v/mars science laboratory. Technical report, NASA.

[10] AA. VV. Bureau of land management crude helium price. Technical report, American Bureau of Land Management, 2015.

[11] AA. VV. A parametric study of a high altitude airship according to the multi-body advanced airship for transport. Technical report, Universita` degli Studi di Modena e Reggio Emilia, The University of Hertfordshire Higher Education, Southern Federal University, Engys Ltd., 2012.

[12] R. S. Pant M. I. Alam. An improved methodology for multidisciplinary design analysis and optimization of high altitude airship configurations. *In 10th International Airship Convention, Friedrichshafen,* 2015.

[13] Brandon Suarez Greg Keogh David W. Keith Justin McClellan, James Sisco and Jay Apt. Supplemental online material: Cost analysis of stratospheric albedo modification. *Environmental Research Letters*, 2012.

[14] Anthony Colozza. High altitude propeller design and analysis overview. Tech- nical report, Federal Data Systems, 1998.

[15] T. Kuennen, H. D. E. Knorreck, M, W. Elson, S. J. Buckley, Reducing the Cost of Launching Small Satellites: A Perspective From the Field, *18th Annual AIAA/USU Conference on Small Satellites*. 2004.

[16] Brian S. Gulliver G. Wayne Finger, David L. Keller. Launch site infrastructure cost trends. In *AIAA SPACE 2007 Conference and Exposition*, 2007.

[17] Heinz Hermann Koelle. Handobook of Astronautical Engineering. McGraw-Hill Book Company, 1961.

[18] J. R. Wertz, Space Mission Engineering: The New SMAD. Space Technology Libray , Microcosm Press, 2011.

[19] M. R. Kolody, P. E. Hintze, W. Li, L. M. Calle. A 20 Year Lifecycle for Launch Facilities at the Kennedy Space Center. NASA.

Appendix A

As follows the iterative process implemented to preliminary size the Airship Assisted Launch System.

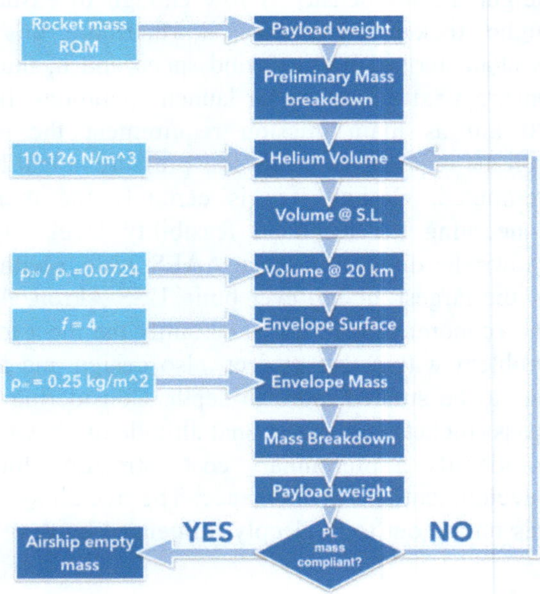

Reinventing Space Conference
BIS-RS-2015-76

Could reusable air-launch break the space access paradigm?

David J. Salt
Telespazio VEGA Deutschland GmbH

13th Reinventing Space Conference
9-12 November 2015
Oxford, UK

S. Hatton (ed.), *Proceedings of the 13th Reinventing Space Conference*,
https://doi.org/10.1007/978-3-319-32817-1_22

Could reusable air-launch break the space access paradigm?

David J. Salt[1]
Telespazio VEGA Deutschland GmbH
Europaplatz 5, D-64293 Darmstadt, Germany; +49 (0)6151 8257-0
dave.salt@telespazio-vega.de

ABSTRACT

Although the performance and operational benefits of air-launch have been recognised since the dawn of the Space Age, few operational systems have been fielded to date and none have generated any significant customer demand. This lack of success is a consequence of two basic factors: 1) performance limitations of the air-launch platform or 'zero-stage', especially if existing aircraft are considered, which thereby limit the addressable market; 2) insufficient advantages over ground-launch systems with similar payload performance when expendable rocket stages are used. This paper assesses the potential of a small, fully reusable, subsonic air-launch system to overcome both of these limiting factors and, in doing so, shows how such a system could break the current space access paradigm (i.e. space access is expensive because the demand for it is small, but the demand is small because space access is expensive). This 'disruptive' potential is assessed in terms of three key system aspects – performance, operations and economics – and thereby highlights the significant and unique benefits of a small, fully reusable, subsonic air-launch system.

KEYWORDS:

ELV	=	**Expendable Launch Vehicle**
IRR	=	**Internal Rate of Return**
kg	=	**kilogram**
LEO	=	**Low Earth Orbit**
Mg (t)	=	**Metric tonne**
Mn	=	**Mach number**
R&D	=	**Research & Development**
RLV	=	**Reusable Launch Vehicle**
TSTO	=	**Two Stage to Orbit**
UHNWI	=	**Ultra High Net-Worth Individuals**

1. INTRODUCTION

THE frequency and complexity of space activities have evolved relatively slowly over the last three decades and rates of growth have much reduced in comparison with the first two decades of the Space Age, which began over half a century ago.

The significant reduction in rate of growth resulted from changes in both the political and economic drivers of space activities [RD.1], especially those involving human spaceflight. More importantly, prospects for future growth appear restricted by one fundamental constraint; our limited access to space due to the continued high cost and risk of launching into orbit.

The aim of this paper is to provide an understanding of the critical issues that constrain our access to low Earth orbit (LEO) and, by highlighting the key factors that drive them, identify one possible approach that may represent the 'path of least resistance' for breaking the current paradigm. It does this by:

- describing the nature of today's space paradigm and the issues that constrain it;
- discussing the dilemmas of reusability and their relationship to technology and markets;
- identifying the benefits/limitations of air-launch and how it favours fully reusable systems;
- assessing launch market sizes and 'elasticity';
- outlining a launch system concept that is designed to address the most promising market;
- analysing the associated business case in order to identify key requirements for success;
- indicating how such a launch system could stimulate development and growth of in-space infrastructures that will enable the exploration and utilization of space-based resources.

Given this understanding, the reader should be better able to consider the question that forms the title of this paper: could reusable air-launch break the space access paradigm?

2. THE CURRENT SPACE ACCESS PARADIGM

Current space activities range from pure science missions through to civil and military applications like communication, navigation and observation systems. Nevertheless, growth and evolution in all these areas is limited by a few key factors:

- government priorities and constraints;
- competition from terrestrial alternatives;
- low market 'elasticity' (i.e. lower prices stimulate only limited market growth);
- launcher cost/availability/reliability.

The first factor is important because the growth of space activities is still dominated by government programmes, both civil and military. Communication satellites represent the nearest thing to a truly

[1] Senior Consultant, Telespazio VEGA Deutschland GmbH, Europaplatz 5, D-64293 Darmstadt, Germany.

commercial market sector, but government funding still underpins much of their basic R&D while the second and third factors have placed significant restraints on their growth and evolution, as witnessed by the problems of commercial ventures like Iridium, Globalstar, ICO, SkyBridge and Teledesic.

To put the situation into perspective, Figure 1 shows a breakdown of the global space industry's annual revenue, which was $330 billion in 2014. Note that that the 'Commercial Infrastructure and Support Industries' segment, which includes satellite manufacture and launch, is responsible for only around one third of this revenue (i.e. $128Billion) and is comparable to that from direct-to-home TV services (i.e. $93Billion), which represents three-quarters of the 'Commercial Space Products and Services' segment.

The relatively small size of current space activities may be better appreciated by comparing it to other global business sectors. One example is Wal-Mart [RD.3], which was founded in 1962 and had a revenue of $476 billion in 2014; thus singlehandedly managing to outgrow the entire world space industry because it services vastly bigger markets and numbers of people.

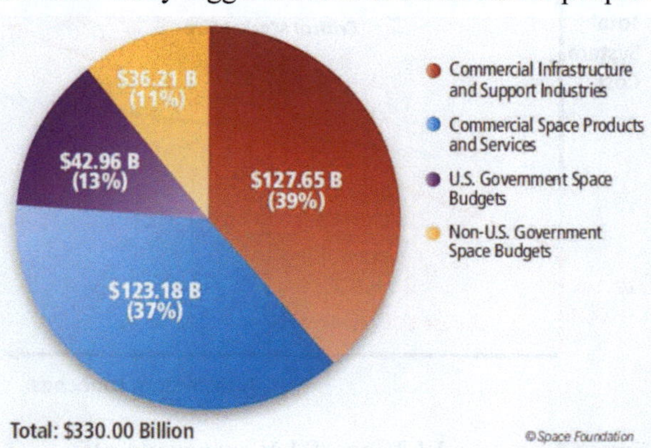

Figure 1. Global Space Activities, 2014 [RD.2]

Though a wide range of future space-based activities and associated markets[2] have been discussed for many decades (e.g. space manufacturing facilities, solar power satellites) their realisation has been limited by a few key factors:

- large investment requirements;
- operation and utilization cost uncertainty;
- market demand and 'elasticity' uncertainty;
- launcher cost, availability and reliability.

Given these circumstances – and in the absence of a major government imperative, equivalent to that which justified Apollo (i.e. the Cold War) – it has become clear to many that the current paradigm will not lead to any significant growth of space activities within the foreseeable future.

One way to change the current paradigm could be to stimulate existing or new markets by reducing specific launch costs by an order of magnitude to below about $1000/kg to LEO, the point where significant growth in all market sectors is expected to be triggered. However, a major problem with this approach is that current markets have very poor 'elasticity'[3] (i.e. lower prices stimulate only limited market growth), as illustrated in Figure 2. This not only stifles any incentives to reduce current commercial ELV launch prices, but also undermines the business case for developing far more cost-effective launch systems. It also highlights the root cause of our current paradigm that *'space access is expensive because the demand for it is small, but the demand is small because space access is expensive'*.

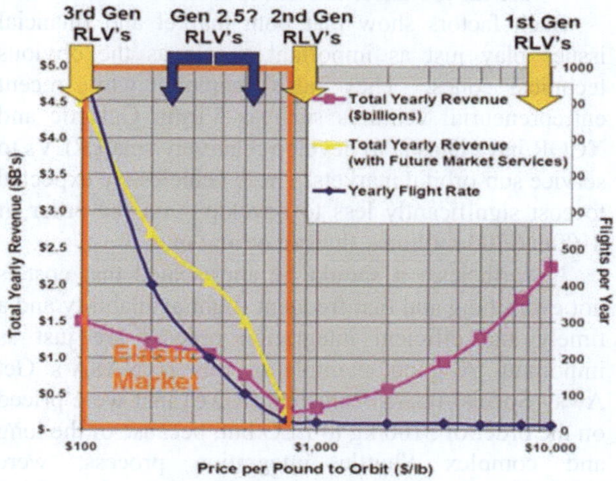

Figure 2. Market Elasticity Issues [RD.6]

Nevertheless, global competition is now forcing commercial launcher companies to reduce their specific launch costs by employing advanced manufacturing methods to build cheaper expendable vehicles. A few, most notably Space-X, are also working to introduce some degree of reusability into their expendable designs. However, no one is currently developing a fully reusable launch vehicle for reasons that are discussed in the next section of this paper.

3. THE REUSABILITY DILEMA

It has long been recognised that the only way to achieve significant improvements in space access is via fully reusable launch vehicles (RLVs) instead of expendable launch vehicles (ELVs), because they offer:

- major reductions in marginal costs, as expensive components tend not to be discarded after use;
- better amortisation of investments, as costs can be spread across more users;
- higher reliability and safety, due to the intrinsic value of the vehicle.

---

[2] For example, the Commercial Space Transportation Study (CSTS) performed a comprehensive review in 1994 of all current and foreseeable markets [RD.4]

[3] Space market elasticity is difficult to estimate due to the relatively small size and low diversity of current markets, though studies such as the CSTS [RD.4] and the NASA ASCENT Study [RD.5] have derived tentative estimates.

Unfortunately, government efforts to field such systems have, to date, either missed many of their original goals (i.e. Shuttle), or been outright failures (X-33/VentureStar, X-34, etc.). Moreover, commercial efforts to develop such systems have been hampered because their development costs are difficult to justify against potential markets. For example:

- many studies estimate it will cost $10-20 billion to field an operational system;
- existing markets are insufficient to justify their development because they have limited growth and elasticity (i.e. lower prices stimulate only limited market growth);
- new markets that could justify their development are far too uncertain and speculative.

Such factors show that both market and financial issues play just as important a role as the obvious technical ones. They also explain why recent entrepreneurial ventures such as Virgin Galactic and XCOR have chosen to develop relatively small RLVs to service sub-orbital markets. These vehicles are expected to cost significantly less to develop – on the order of $100m-$200 million – than an orbital one.

Nevertheless, it should be appreciated that cost is not everything and that frequent flight availability and a timely and efficient integration process are just as important. A good example of this is NASA's Get Away Special (GAS) canisters [RD.6] that were priced on the order of $100/kg to LEO but, because of the long and complex Shuttle integration process, were undersubscribed so that many GAS canisters were filled with ballast. The service was eventually discontinued after the Columbia accident.

3.1 Cost Factors

The cost of operating mature transportation systems (e.g. railroads, trucks, ships, airlines) tends to be a function of their fuel cost; typically between two and five times the fuel cost. Today, the operating cost of an ELV is well over a thousand times the cost of its fuel/propellant and is why many believe that a mature space transportation system has to be reusable.

Unfortunately, due to the nature of the rocket equation, the goal of reusability creates a fundamental dilemma as it requires a design that both minimizes margins, but maximizes reliability. In the case of the US Space Shuttle, the designers were forced by Congress to minimise development costs so much that it resulted in operations costs that were higher than those of the ELVs it was supposed to replace.

Reusability increases development costs directly by requiring more robust structures and propulsion, plus the addition of systems for recovery (TPS, landing gear, etc.) and maintenance (access ports, interfaces, etc.). It also increases development costs indirectly because all of this additional mass decreases payload performance, which can only be recovered by increasing the vehicle's size/mass. The need for additional testing at all levels (i.e. component, unit, system, in-flight) to verify both safety and reliability also adds significantly to the vehicle's development costs.

Reusability also increases operational costs due to additional equipment/facilities/personnel to both return the vehicle back to the launch site and then perform all necessary refurbishment/maintenance. These particular aspects are critical to the RLV's economic viability as they must be less than the ELV's production cost in order to ensure the RLV can be in any way competitive.

3.2 Cost Trades

Taking all of these issues into account, both positive and negative, the conditions under which RLVs become more cost-effective than ELVs are not obvious. One way to visualise this trade-off is presented in Figure 3, which assumes that RLV development costs are far larger than those of the ELV, but that their operation costs are far less on a per flight basis. Here, the 'total system cost' is the accumulated sum total of both development and operations costs and grows with the number of flights performed.

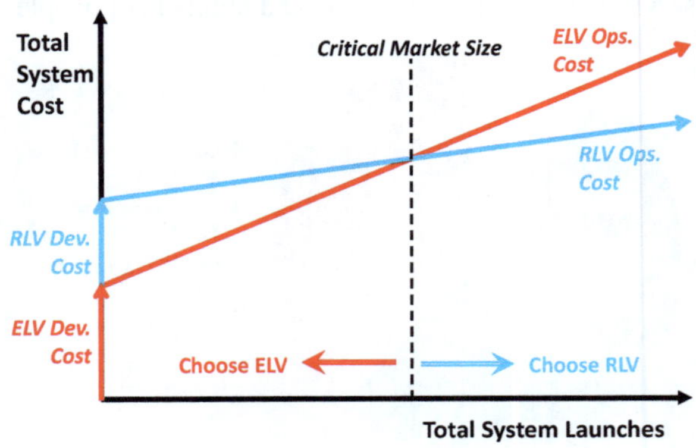

Figure 3. ELV–vs –RLV cost trade-off

The basic message of this type of plot is that the choice is very dependent upon the number of flights (i.e. payloads) that the system is expected to perform, which in-turn depends upon the size of the market it is able to both address and capture.

Most launcher development decisions have been based upon this type of assessment (N.B. the Shuttle was sold by assuming 25-30 flights per year over 12 years [RD.7]) and, based upon current market projections, the consensus seems to be that a launch system will need to perform more than 30-60 flights per year in order to justify choosing the RLV design option.

Many studies have examined these trade-offs more generally to assess the nature of these 'break-even' conditions and have concluded that the RLV is most sensitive to its nonrecurring development cost while ELVs is most sensitive to the learning curve that can applied to reduce its recurring production costs.

The fundamental question for the RLV is whether amortizing the construction cost over multiple flights is worth the increase in cost in most other categories.

Economic models [RD.8] suggest it does not appear to be worthwhile at launch rates less than about 100 times the current rate. Therefore, a significant increase in demand for space launch should be seen as a great opportunity for RLVs to become economically viable.

4. THE POTENTIAL FOR AIR-LAUNCH

The cost of developing and operating a new launch vehicle can be estimated using parametric relationships, which use historical data to indicate cost trends with respect to some parameter or sets of parameters such as vehicle dry mass, engine performance, numbers and types of subsystems and so forth. One general feature of such estimates is that larger/heavier vehicles tend to cost more and goes some way to explaining why launch vehicle designers try to minimise mass margins, since larger margins will result in a larger vehicle for the same payload performance.

The relationship between mass and cost is clearly more critical to RLVs because of their inherent design overheads. Therefore, any way of relaxing the impact of mass margins will prove more advantageous to RLVs than to ELVs and is the reason why air-launch may represent an extremely effective means of achieving a radical reduction in the cost of space launch.

4.1 Brief History of Air-Launch

The idea of air-launching a rocket has a long history that dates back to the early 1950's when *rockoons*, which were sounding rockets launched from helium balloons. These allowed the rocket to achieve a higher altitude so that it did not have to move under power through the lower and thicker layers of the atmosphere. Unfortunately, they had some serious disadvantages because the balloon could not be steered and so both the launch direction and the region where it fell were not easily to control. Possibly the most successful was the USAF's Project Farside, which launched six vehicles in late 1957 though only two reached their target altitude of just over 2000km.

The first aircraft launched rockets were primarily developed as anti-satellite (ASAT) weapons. The first of these was Project Pilot, which was an attempt by the Naval Ordnance Test Station (NOTS) at China Lake to orbit a 1kg payload in response to Sputnik. The vehicles, named NOTS EV-1 (NOTSNIK), were solid rockets launched by a Douglas F-4D1 Skyray and ten were flown in mid-1958, though none were successfully tracked to orbit. Similarly, a Bold Orion missile, which was air-launched from a B-47 Stratojet on 19th October, 1959, against the Explorer 6 satellite. However, this was a limited test and it was not until 13th September, 1985, that an F-15A launched an ASM-135 ASAT destroyed the Solwind P78-1 satellite flying at an altitude of 555 km. Since then, the only operational air-launched rocket has been Pegasus, which was developed by the Orbital Sciences Corporation as a commercial satellite launch vehicle and first flown on 5th March, 1990, with 42 launches to date.

Most air-launch concepts carry the rocket external to the launch vehicle, either on top or under the fuselage or wing, as shown in Figure 4. However, a few concepts

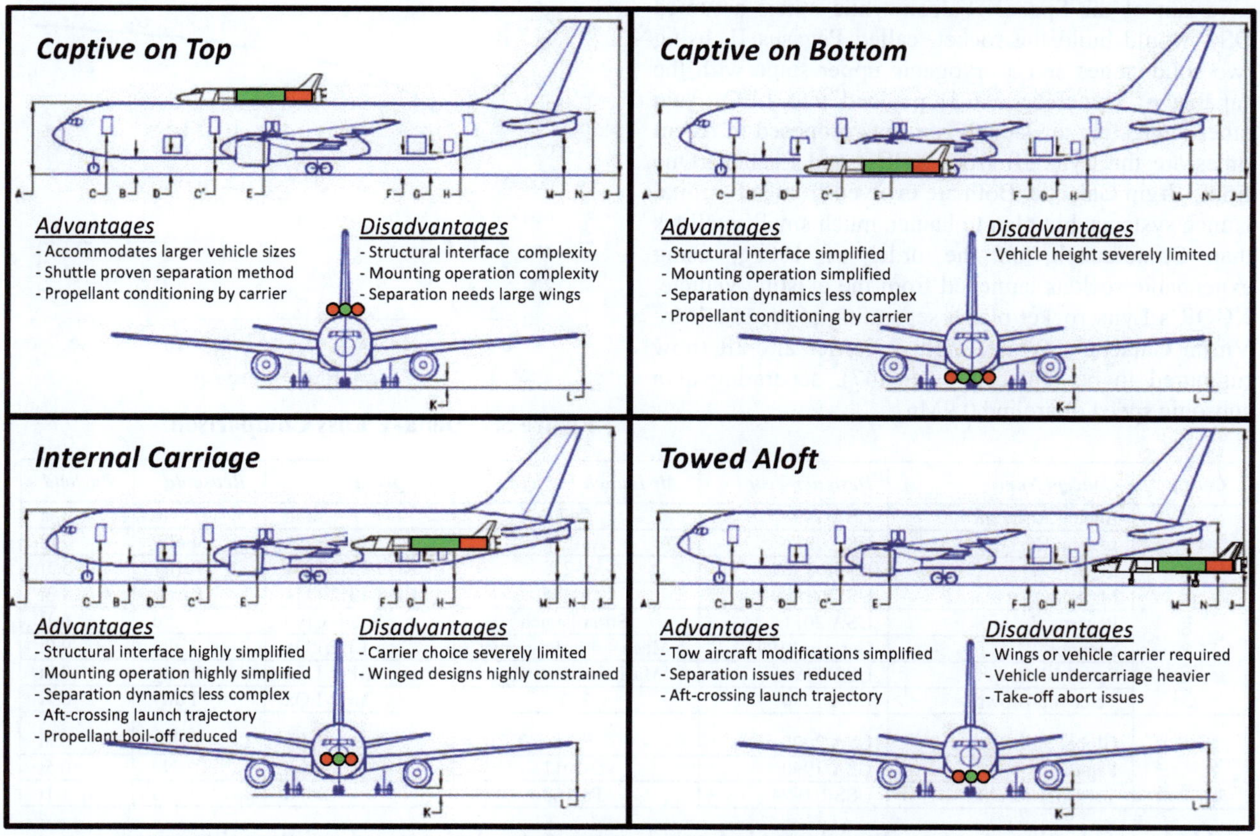

Figure 4. Subsonic Air-Launch – Rocket Vehicle Mounting/Interface Options

have proposed carrying the rocket inside the fuselage and 'extracting' it during launch via drag chutes, which also provide stability during the subsequent free-fall phase, before igniting the rocket motor. The USAF tested air launching a Minuteman ICBM from a C-5A Galaxy transport aircraft on 24th October 1974, but this concept was never pursued. However, the AirLaunch LLC performed significant demonstration tests in 2006 of a very similar concept called QuickReach for the DARPA/USAF FALCON programme, which launch a liquid ELV from a Boeing C-17A. Similarly, the Air Launch Aerospace Corporation proposed an air-launched system capable of placing satellites into LEO using the AntonovAn-124 "Ruslan", though this was never developed.

Other concepts proposed in the mid-1990's have envisaged towing the launcher behind an aircraft (i.e. Astroliner, proposed by Kelly Space & Technology) while others have envisaged in-air fuelling of the launcher in order to reduce take-off mass (i.e. Black Horse, proposed by Pioneer Astronautics). Neither of these approaches were ever pursued beyond the conceptual design stage, though Kelly did perform tow tests of an F-106 jet behind a C-141 cargo aircraft in early-1998 under a NASA SBIR award.

The most recent air-launch concept to attract serious attention has been the Stratolaunch Systems proposal in 2011 to build a massive aircraft by combining the wings and fuselage of two Boeing 747 airliners. However, the exact nature of the launch vehicle is still not specifically defined, with initial speculation suggesting it would be a variant of the SpaceX Falcon while others assumed OSC would build the rocket, called Pegasus II, using two solid-stages and a cryogenic upper stage with the capable of launching a 6.1t payload into LEO. Two other interesting air-launch concepts proposed in recent times are the Lynx III from XCOR and LauncherOne from Virgin Galactic. Both are evolved from sub-orbital launch systems, but plan to launch much small satellites than Stratolaunch, on the order of 100kg, using expendable rockets launched from the existing vehicle: XCOR's Lynx rocket plane, separating at around 4Mn; Virgin Galactic's WhiteKnight 2 carrier aircraft (now rumoured to be replaced by a 747), separating at a subsonic speed of around 0.9Mn.

4.2 Benefits of Subsonic Air-Launch

There have been numerous studies of air-launch concepts and Table 1 provides an overview of a very small, but representative selection of them. The interest in such concepts arises because air-launch offers both unique and important benefit with respect to vehicle performance and operations, plus it offers several paths to improve these benefits in an evolutionary manner.

i) Performance Benefits

Rocket operations above the dense atmosphere reduce significantly both drag and gravity losses. It also allows for a significant increase in engine specific impulse (Isp) by allowing the use of a larger expansion ratio nozzle, which is constrained at lower altitudes because over-expanded nozzle flows suffer destructive instabilities. Theoretically, the latter problem can be overcome by using some sort of altitude compensating nozzle, though the additional mass and complexity tend to cancel out any performance benefit.

Figure 5 illustrates the delta-V losses encountered by a rocket as a function of launch altitude for both sea-level and the 10km case, which represents a subsonic air-launch. It shows a major reduction in velocity losses and, more specifically, that these losses represent around 20% of the ideal ascent delta-V for a typical sea-level launch (i.e. 7.7km/s to LEO), but only around 10% of an air launch.

Another small but positive benefit of air-launch is that the launch point may be chosen to match the

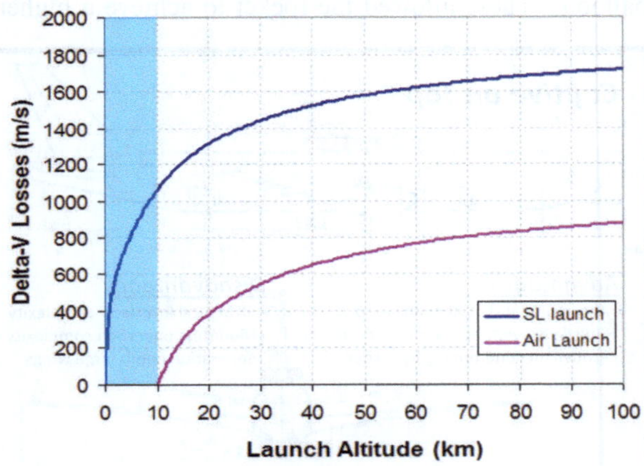

Figure 5. Delta-V Loss Comparison

Config.	*Concept Name*	*Designer/Year*	*Air-launch Vehicle*	*Propellant*	*Reusable*	*Payload*
Captive on Top	Boeing AirLaunch	USA/1999	747	Solid	No	3.4t
	Interim HOTOL	UK/1991	An-225	LH2/LOx	Fully	7.0t
	MAKS-M	USSR/1989	An-225	RP-1/LH2/LOx	Partly	5.5t
	MAKS-OS	USSR/1989	An-225	RP-1/LH2/LOx	Partly	8.3t
	Pegasus II	USA/2011	Stratolaunch	Solid+Cryo	No	6.1t
	Saenger II	Germany/1991	Mach 4.4 turbo-ramjet	LH2/LOx	Fully	9.0t
	Spiral 50-50	USSR/1965	Mach 6 turbo-ramjet	RP-1/LOx	Partly	10.0t
	Teledyne-Brown	USA/1986	747	LH2/LOx	Fully	6.7t
Captive on Bottom	Global Strike Eagle	USA/2006	F-15	Solid	No	0.3t
	Pegasus	USA/1990	L-1011	Solid	No	0.5t
	Yakovlev HAAL	USSR/1994	Tu-160	Solid	No	1.1t

Table 1. Selection of external carriage Air-Launch concepts (excludes towed or internal carriage)

inclination of the target orbit. This not only allows for maximum exploitation of Earth's rotation (~400m/s for equatorial orbit), it also reduces trajectory losses by reducing or even removing the need for plane changes to achieve the target orbit.

ii) Operational Benefits

Air-launch offers the only realistic way to operate a space launch system from existing airfields, including the possibility of someday operation out of major civil airports. This is because the launch aircraft uses air-breathing propulsion as opposed to a pure rocket, which enables an enormous reduction in noise during take-off due to the reduced exhaust velocity. However, concepts that use a supersonic military jet will never be as 'quiet' as those that use a subsonic transport and will also be penalised because of their much reduced payload capacity, which will likely be at least one order of magnitude less.

Using an existing military or commercial aircraft also means that the air-launch system can build upon this vehicle's inherent safety, reliability, maintainability and availability. Moreover, these will be extremely valuable if rapid and/or frequent launch is one of the primary system requirements. In addition, it leads to a launch system whose elements are all processed and operated horizontally, which helps to streamline the maintenance and launch workflow as it simplifies access to the vehicle.

As already mentioned, air-launch offers the possibility to choose the launch point to match the inclination of the target orbit. An additional, but extremely important benefit is that the launch point can be 'tracked' so that the launch window for rendezvous with an orbiting target can be widened significantly. This not only improves operational flexibility but, as already stated, also reduces the need for plane changes to achieve the target orbit and so has the potential to reduce the size of the upper stage by reducing the on-orbit propellant requirements.

Figure 6 presents a schematic of the operational profile of a generic subsonic air-launch RLV and also shows another operational advantage of this concept, which is that it can use the launch aircraft to ferry the rocket back to the launch site if it should have to land at an alternate. More importantly, it also highlights the potential to use the cruise phase to either harvest liquid oxygen in-flight [RD.1], or to transfer it from a tanker aircraft in order to reduce the rocket's mass at take-off dramatically and thereby increase its payload performance. Separating oxidiser and fuel during the take-off also reduces significantly the associated risks.

Another capability that is not obvious from the figure, but could have very important operational benefits is the ability to fly the launch vehicle up-range so that the 1st stage booster of any TSTO RLV can return directly to the launch site after staging, thus avoiding the need to fly or glide back up-range. Requiring the booster to fly-back up-range is a very constraining problem for ground launched TSTO RLVs because it either:

- limits the staging to around 3Mn at 30km altitude to ensure the booster has sufficient 'energy height' to glide back to the launch site;
- forces the booster to carry extra propellant in order to perform an up-range boost-back manoeuvre;
- forces the booster to carry an additional air-breathing propulsion system in order to fly back up-range;
- requires an additional landing site down-range of the launch site as part of the basic infrastructure.

For an SSTO RLV, it also means an aborted launch could fly-back directly to the launch site should the abort occurred sufficiently early in the mission. Thus,

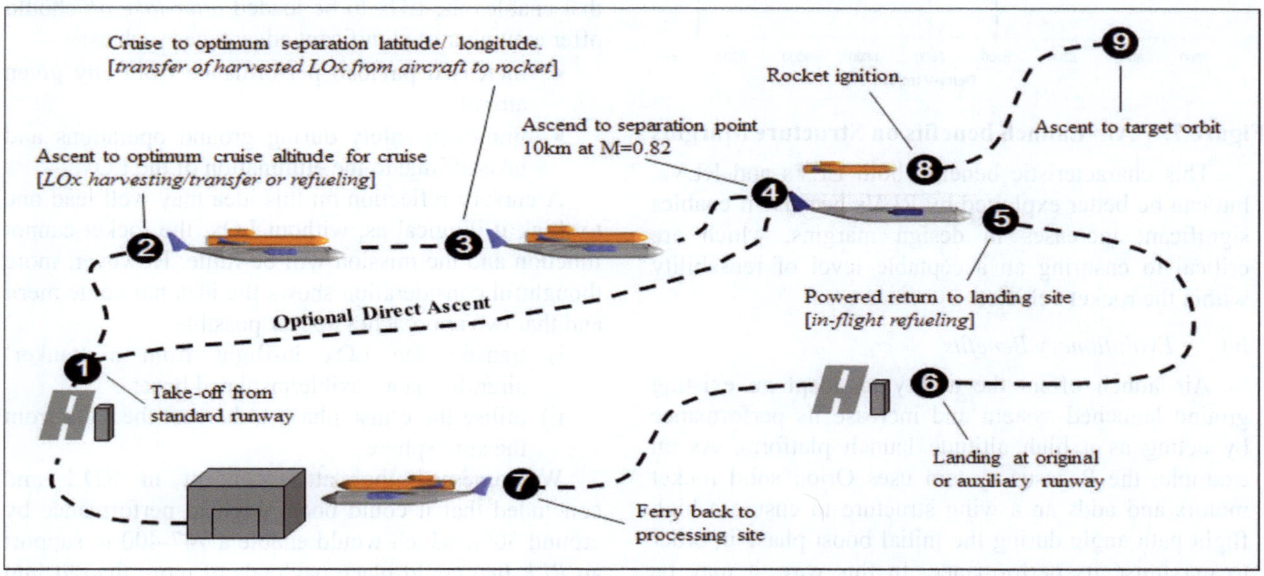

Figure 6. Subsonic Air-Launch Operations [*optional in-flight LOx harvesting/transfer or in-flight refueling*]

air-launch also increases the number of abort options and so improves both safety and operational robustness.

iii) Design Margin Benefits

The effective reduction in delta-v shown in Figure 5 has the synergistic effect of reducing key component design requirements for both ELVs and RLVs:

- Rocket engine thrust/weight can be smaller because the low initial trajectory angle does not have large gravity losses;
- Rocket engines can lower combustion chamber pressure to reduce pump complexity and stress;
- Propellant reserves can be reduced as most meteorological uncertainties are below the launch altitude.

However, other benefits are more specific to RLVs;

- Wing area can be smaller as they do not need to lift the gross weight at low subsonic speed;
- Wing bending structure need not be designed for gross weight take-offs or gust loads;
- Undercarriage weight is reduced significantly as it only supports the vehicle's landing mass;

One other significant benefit of air-launch results from the exponential nature of the rocket equation and is illustrated in Figure 7. This shows how the reduction in effective mission delta-v will also reduce, by a factor of about eight, the impact of any increase in margins on structures that scale with the propellant load.

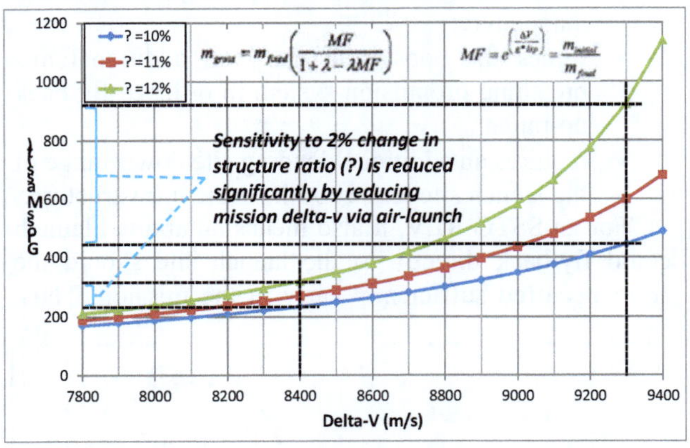

Figure 7. Air-Launch benefits on Structure Margins

This characteristic benefits both ELVs and RLVs, but can be better exploited by RLVs because it enables significant increases in design margins, which are critical to ensuring an acceptable level of reusability within the rocket vehicle.

iv) Evolutionary Benefits

Air-launch offers the ability to adapt an existing ground launched system and increase its performance by acting as a high altitude launch platform. As an example, the Pegasus system uses Orion solid rocket motors and adds an a wing structure to ensure a high flight path angle during the initial boost phase in order to maximise its performance. In this way, it may be possible to evolve an existing sub-orbital launcher into an orbital launcher, or at least improve its payload performance.

As increasing vehicle size tends to increase both development and operational costs, air-launch could offer an importance path for commercial ventures. However, the orbital payload performance of any air-launch concept is fundamentally limited by the aircraft's carrying capacity and, more specifically, its maximum take-off mass. Currently, the world's largest operational aircraft is Russia's An-225, but this is a one-off design, based upon a heavily modified An-124, which is likely to be an impractical option for an air-launch system. Commercially available options include the Airbus A380 and the Boeing 747-400, though the former is relatively new and so is very expensive. Table 2 gives an overview of the relevant performance of the most likely candidate aircraft and includes a rough estimate of the maximum payload mass (taken from RD.9) they could deliver to LEO if used as the basis for an air-launch system supporting an *expendable* rocket.

Candidate Aircraft	*External Mass (Mg)*	*Max. P/L to LEO (Mg)*
An-225	200	13.8
A380-800F	120	7.8
747-100 SCA -911	109	7.0
747-400F	140	9.1
Dual-fuselage C-5	350	23.7
Stratolaunch Carrier	120	6.1

Table 2. Candidate Aircraft for Air-Launch

The gross mass of any launch vehicle that uses liquid oxygen (LOx) as an oxidiser will be dominated by the amount of LOx it must carry. Typical oxidiser/fuel ratios of 5.2 and 2.3 respectively for liquid hydrogen (LOx/LH2) and kerosene (LOx/RP-1) fuelled rockets mean that the LOx will account for more than half the launch vehicle's gross mass at take-off. Therefore, it is reasonable to think any design approach that enables the LOx to be loaded *after take-off* should offer a number of significant advantages such as:

- increased payload performance from any given aircraft;
- improved safety during ground operations and take-off due to the elimination of the LOx.

A cursory reflection on this idea may well lead one to think it illogical as, without LOx, the rocket cannot function and the mission will be futile. However, more thoughtful consideration shows the idea has some merit and that two approaches appear possible:

i) transfer the LOx in-flight from a 'tanker' aircraft, via a flexible/insulated hose;
ii) utilise the cruise phase to harvest the LOx from the atmosphere.

We assessed the latter concept in RD.1 and concluded that it could boost payload performance by around 30%, which would enable a 747-400 to support an *RLV* that could place payloads of more than 9t into LEO.

5. ASSESSING MARKET SIZE & ELASTICITY

Details presented in the previous sections suggest that a subsonic air-launched RLV may offer the 'path of least resistance' towards radically reducing the cost of access to space because it is an approach that:

- reduces the mission delta-v, which reduces the rocket's mass to launch a given size of payload;
- accommodates the additional margins needed for reusability in exchange for very tolerable performance penalties;
- allows the use of existing and relatively low cost aircraft that can support high flight rates and require relatively few modifications;
- offers ways to evolve the system through small, incremental, development steps.

Although these qualities are very encouraging, they must also enable the RLV to both address and then capture the sort of markets that will be needed to justify the substantial investments required to build it.

5.1 Addressable Markets

As noted earlier, communication satellites currently represent the nearest thing to a truly commercial market sector, but their elasticity is rather poor. New market sectors, such as Earth observation and navigation, have grown substantially over the past few decades, but still represent only a tiny fraction of the market sizes needed to justify development of an RLV. Meanwhile, the market for launching very small satellites, such as cube-sats or nano-sats, into LEO appears to be growing rapidly and is fostering the development of a large number of small expendable launch systems. Unfortunately, the growth potential of this sector remains highly uncertain and may ultimately be limited by concerns over their potential to increase the problem space debris.

Until recently, we considered comsats to be the only real and addressable market that could justify a commercial RLV development, presenting an outline of the business case scenario in RD.1. However, a recent internet article [RD.10] highlighted new and intriguing evidence for a more substantial and addressable market: human passenger flights to LEO. This market is sometimes referred to as 'space tourism' and has been recognised for many decades, though very few studies have analysed it in any real depth [RD.4 & RD.11]. Nevertheless, there is now real-world evidence that a number of people are both willing and able to spend a substantial fraction of their wealth on a trip into LEO since, to date, seven people have each paid $30M or more to visit the International Space Station.

What makes this market special is that many people world-wide share the desire to travel into space, but are restricted primarily by their wealth, which suggests that the market elasticity may be quite positive (i.e. numbers will increase substantially as ticket prices drop). An initial assessment of this elasticity was made by t/Space [RD.12] in 2005, but was effectively forgotten until the recent internet article, which re-assessed the results with respect to more recent surveys of ultra-wealthy people, like that from Wealth-X [RD.13]

5.2 Passenger Market Size & Elasticity

Taking the identified analyses and data sources as a starting point, plus another survey of ultra-wealthy people by Credit Suisse [RD.14], we built upon and then expand the analysis as follows.

Wealth statistics for the world's Ultra High Net-Worth Individuals (UHNWI) were tabulated, as shown in Table 3, within a spreadsheet and functions derived to define the cumulative wealth pool within a specific set of wealth tiers or levels, as shown in Table 4.

Wealth Tier	# of Individuals	Total Wealth	Average Wealth
>$1B	2,325	$ 7,291,000,000,000.00	$ 3,135,913,978.49
$750 - 999M	1,295	$ 1,075,000,000,000.00	$ 830,115,830.12
$500 - 749M	3,590	$ 2,464,000,000,000.00	$ 686,350,974.93
$250 - 499M	9,335	$ 3,530,000,000,000.00	$ 378,146,759.51
$200 - 249M	14,580	$ 3,170,000,000,000.00	$ 217,421,124.83
$100 - 199M	25,400	$ 3,660,000,000,000.00	$ 144,094,488.19
$50 - 99M	63,120	$ 4,775,000,000,000.00	$ 75,649,556.40
$30 - 50M	91,630	$ 3,760,000,000,000.00	$ 41,034,595.66
$10 - 30M	682,775	$ 13,655,500,000,000.00	$ 20,000,000.00
$5 - 10M	835,950	$ 6,269,625,000,000.00	$ 7,500,000.00
$1 - 5M	14,930,000	$ 44,790,000,000,000.00	$ 3,000,000.00
$0.1 - 1M	366,340,000	$ 201,487,000,000,000.00	$ 550,000.00
$0.01 - 0.1M	1,265,000,000	$ 69,575,000,000,000.00	$ 55,000.00
<$0.01M	3,248,000,000	$ 16,240,000,000,000.00	$ 5,000.00

Table 3. UHNWI Distribution Statistics [RD.13]

Wealth Level	Number Individuals
> $1B	2325
> $750M	3620
> $500M	7,210
> $250M	16,545
> $200M	31,125
> $100M	56,525
> $50M	119,645
> $10M	894,050
> $5M	1,730,000
> $1M	16,660,000

Table 4. UHNWI Cumulative Wealth Pool by Tier

These numbers were then factored to account the following findings from the Futron study:

- the fraction of net-worth an individual would be prepared to pay for a ticket (1.5%, 5%, 10%,);
- the likelihood that any UHNWI would purchase a ticket at a specific price point;
- the fraction sufficiently fit to fly (61%);
- the additional fraction who would fly if training were in the US, instead of Russia (24%);
- the additional fraction who would fly if training were reduced from 6 to 1 month (50%);

One additional factor that was accounted for in this analysis was that, on average, only around 25% of the UHNWI wealth is held in cash, which may be a very relevant factor if this is seen as a luxury purchase and not an investment.

These arrays of values were then plotted to show how the distribution of addressable customer (Figure 8) and addressable market value (Figure 9) varies with each RLV ticket price point. The resulting trends illustrate how the market's size and value may respond to a reduction in ticket price and suggest that the market to fly humans into LEO holds the potential for a significant degree of elasticity.

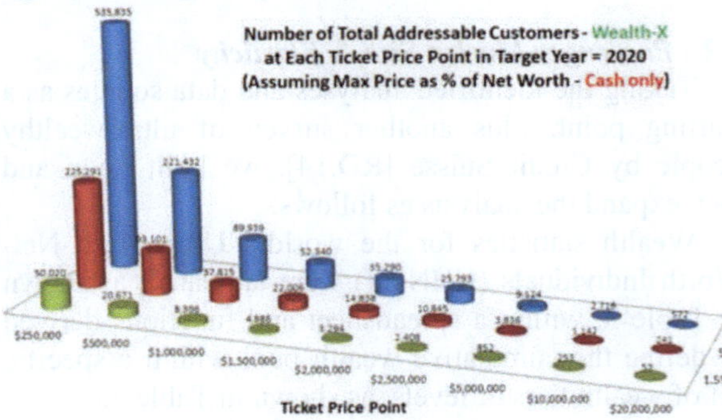

Figure 8. RLV Customers – vs – Price Point

Figure 9. RLV Market Value – vs – Price Point

A cursory examination of these plots reveals the following trends and features:

- elasticity is essentially linear above the $10M per ticket price point;
- significant/geometric growth begins as we drop below the $10M per ticket price point;
- growth below the $2M per ticket price point becomes exponential.

Note that the analysis assumed market conditions starting in 2020AD as this was thought to be the earliest starting date for such a venture. The values used were therefore scaled in accordance with estimated annual growth rates, which were 5.3% for UHNWI wealth and 4.4% for the UHNWI population.

Due to the number of assumptions and the limited nature of the population survey, great caution must be taken when interpreting these results. However, given their very encouraging nature, we decided to take the analysis one step further to investigate the business case for an RLV designed specifically to address this market.

6. AN RLV CONCEPTUAL DESIGN

There are two fundamental requirements for RLVs designed to enable human passenger flights to LEO:

- The RLV's payload performance must be big enough to carry at least one human, plus associated life support systems (e.g. space suit), which is assumed to be around 250kg;
- The aircraft's payload performance must be big enough to carry the fully loaded RLV mass, which is assumed to be more than 50t.

For the purposes of this analysis, these requirements lead to a conceptual RLV design capable of launching two passengers (i.e. 500kg) into LEO from a 767-300.

The choice of aircraft was based upon the details for a range of existing aircraft presented in Table 5, which took most of its numbers from RD.15, with operations costs expressed per flight hour (FH) and maintenance costs expresses per flight cycle (FC). Note that the operational cost of the SR-71 were included as a basis for estimating RLV operations costs.

Aircraft	$(m)/Unit [new]	Value Retention (% after 10 yr.)	$(m)/Unit [after 10 yr.]	Ops. $(k)/FH	Maint. $(k)/FC	Payload (t)	Range (km)
B747-400	260.0	23	60	25.5	10	107.8	4091
B777	300.5	45	135	8.4	6	104.3	8689
B767-300	188.0	35	66	9.1	6	51.8	3113
A330-200	232.2	52	121	5.9	6	68.1	5900
SR-71	—	—	—	165		—	—

Table 5. Aircraft Costs & Performance

The same spread-sheet model developed for previous RLV analyses [RD.1] was adapted to derive vehicle mass and performance characteristics for the current vehicle design and mission assumptions, which are listed in Table 6. The scaling rules applied to these models are outlined in Table 7.

RLV Design & Mission	
1	Baseline mission delta-v to 400km LEO = 7820 m/s
2	Delta-v loss: 1750 m/s from sea-level; 850 m/s from 10km
3	Existing rocket engines (e.g. Merlin 1C & RL10A-4-2)
4	Oxydised/Fuel ratio: 2.28 for LOx/RP; 5.24 for LOx/LH2
5	Isp: 450s @10km for LOx/LH2; 300s @10km for LOx/RP
6	Current available structural materials (i.e. TRL 6+)
7	TPS mass: 5% Booster dry mass; 20% Orbiter dry mass
8	Wings + Empennage + body flap: 7% dry mass

Table 6. Air-Launch Model – Assumptions

Wing & TPS Mass: Scales directly with materials factor (S) and the change, with respect to the baseline, in the sum of Fuselage, Tank, Systems, and Engine masses (Ms3 + Ms4 + Ms5 + Ms6).
Fuselage Mass: Scales directly with materials factor (S) and the change, with respect to the baseline, in the propellant tank mass (Ms4).
Tank Mass: Scales directly with materials factor (S) and change, with respect to baseline, in propellant mass (Mf) raised to the power of 2/3.
Systems & Engine Mass: Scales directly with the change in the propellant mass (Mf), with respect to the baseline.

Table 7. Air-Launch Model – Scaling Rules

Another basic assumption for this concept was that it would be a two stage to orbit (TSTO) design using kerosene (RP-1) for the 1st stage fuel and hydrogen (LH2) for the 2nd stage. This combination was selected to ensure a relatively compact size for the 1st stage to enable mounting the rocket under the aircraft and so reduce drag losses during both the combined flight and separation phases [RD.16].

The mass budget and payload performance of the vehicle were modelled for a range of separation speeds by splitting the baseline mission delta-v between the two stages, but accounting the full delta-v loss only on the 1st stage. The resulting payload performance of this concept was then estimated for a range of separation speeds to select the optimal RLV design point, which is presented in Table 7 in terms of a breakdown of its mass budget and mission delta-v.

Separation Mach number (Mn) = 10			
Materials density scaling factor (S) [%]	1.00		1.00
TSTO Booster Details		TSTO Orbiter Details	
Specific Impulse (Isp) [sec.]	300	Specific Impulse (Isp) [sec.]	450
Rocket equation factor (R=Exp(dV/Isp/g)	2.8228	Rocket equation factor (R=Exp(dV/Isp/g)	3.5685
TSTO Gross Mass (MTg=MBp+MBs+MBf) [kg]	52095	Orbiter Gross Mass (M0g=MOp+MOs+MOf) [kg]	10010
Booster Dry Mass (MBs=SUM(MBs1:MBs6)) [kg]	8445	Orbiter Dry Mass (MOs=SUM(MOs1:MOs6)) [kg]	2260
Wings Mass (MBs1) [kg]	645	Wings Mass (MOs1) [kg]	259
TPS Mass (MBs2) [kg]	463	TPS Mass (MOs2) [kg]	458
Fuselage Mass (MBs3) [kg]	1824	Fuselage Mass (MOs3) [kg]	592
Tank Mass (MBs4) [kg]	1888	Tank Mass (MOs4) [kg]	713
Systems Mass (MBs5) [kg]	797	Systems Mass (MOs5) [kg]	220
Engines Mass (MBs6) [kg]	2827	Engines Mass (MOs6) [kg]	278
FSSC-16 Defined Propellant Mass (MBf) [kg]	33640	FSSC-16 Defined Propellant Mass (MOf) [kg]	7205
Booster Payload (MBp=MOg, Orbiter Gross Mass) [kg]	10010	Resultant TSTO Payload (MOp) [kg]	545
Booster delta-V loss (LdV) [m/s]	850	Orbiter delta-V loss (LdV) [m/s]	---
Booster delta-V (BdV) [m/s]	2204	Orbiter delta-V (OdV) [m/s]	5616
		TSTO System Details	
		Total Mission Delta-V [m/s]	8670
		TSTO Dry Mass (MTs=MBs+MOs) [kg]	10705
		TSTO Gross Mass (MTg=MTs+MBf+MOf+MOp) [kg]	52095

Table 7. TSTO+767-300 Performance & Mass

Based upon this design point data, a rough estimate of the vehicle's physical size was made in order to visualise the combined aircraft/rocket configuration, which is presented in Figure 10.

The details in the Figure 10 also highlight several interesting features of this design:

- the aircraft's ground clearance has been raised 0.4m (red lines) by increasing oleo fluid/gas, as suggested in RD.16, to give additional room for mounting the rocket below the 767 fuselage;
- the 2nd stage rocket is nested inside the 1st stage rocket to minimise the combined stack's height;
- the tanks were sized with a maximum diameter of 1.5m to ensure sufficient ground clearance below the aircraft;
- the LOx (green) and RP-1 (red) tanks were split to both shorten each stage and better distribute their mass, while a single LH2 (light blue) tank was used in the 2nd stage;
- the slight reduction in tank length/volume when slush LH2 (dark blue) is assumed.

Based upon this cursory assessment, we concluded that the conceptual design appears technically feasible. Nevertheless, we did not address the complexities of any pull-up/separation manoeuvre that may be needed for safe release from the aircraft and to optimise rocket ascent trajectory, especially if a high separation flight path angle of around 30deg. [RD.17] is required.

7. THE BUSINESS CASE ANALYSIS

Justifying the commercial development of a subsonic air-launched RLV requires more than just an assessment of the vehicle's design, operations and performance. It also requires an assessment of the associated costs and, more importantly, the revenue that it can be expected to generate from selling its services to commercial customers.

7.1 *Business Model Assumptions*

Development and operating costs can be based upon past estimates, but will be highly uncertain. However, they can be used to bound the analysis and so indicate the range of values required to justify any investment.

Our assessment of the potential market for human passenger flights to LEO can be used to construct a business model spread-sheet that generates an Income Statement and a Cash Flow Statement for any given scenario. This then enables the performance of the venture to be assessed and, more importantly, provides insight as to the most important parameters and their sensitivity to changes in key business assumptions.

From the investors' point of view, the key is to get an acceptable return on any investment. A common yardstick to measure this is the internal rate of return (IRR), which is defined as "the rate of return at which the present value of the cost of the investment and the present value of the future income stream equate" – in simplistic terms, this is somewhat akin to the annual interest rate of a savings account. For high risk aerospace investments, the IRR has to be 20-30% for such projects to merit serious consideration. Another parameter of interest is the end-of-year (EOY) cash balance, which gives a good indication of the level of cash assets a company is generating and, more importantly, allows the payback period – the time needed to recoup the initial investment – to be assessed.

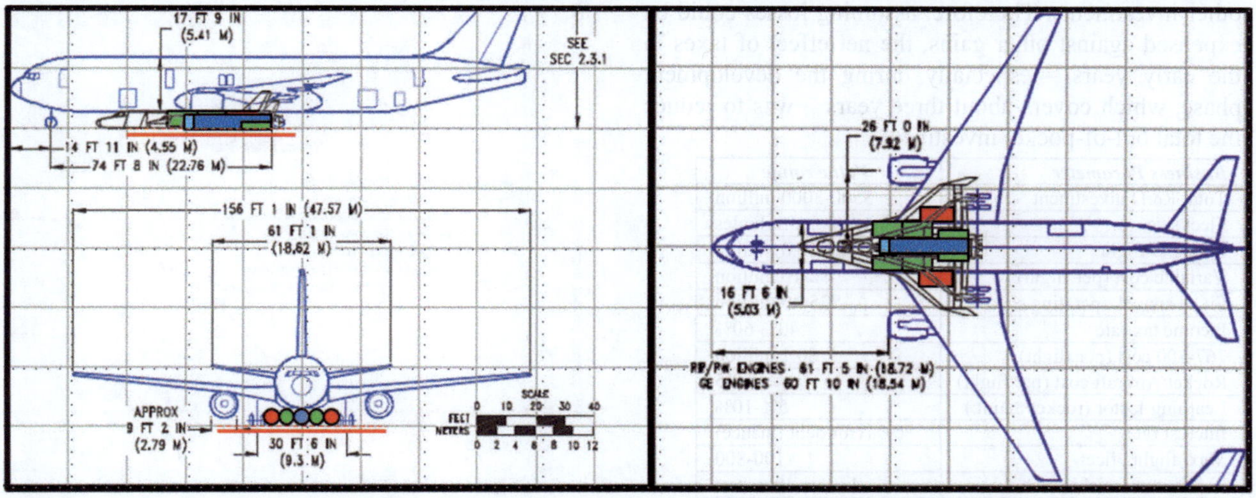

Figure 10. Subsonic air-launched RLV configuration with 767-300

There are three or four fundamental parameters that drive the results: the available market; the cost of services (development and direct plus indirect operations cost); the revenue that can be generated by selling services at a given price per flight; and the annual number or flights. Other factors such as depreciation, taxes, amortisation and insurance generally have a relatively minor impact on the final result. Therefore, in order to simplify the analysis in the face of so many unknown or ill-defined values, a number of shortcuts or approximations were applied.

a) All up-front investment was expensed (i.e. put down as business expenses) in the same year it was applied. Strictly speaking, investments related to flight hardware and other capitalised equipment should be depreciated over their expected lifetime, however, as no useful breakdown is available here, they were expensed as they were incurred.

b) Depreciation was not accounted for since it has only a marginal effect upon taxable income – it may change a 20% IRR into a 23% IRR, but not much more – and only occurs after the assets are paid for and in use.

c) Vehicle insurance, which could have been addressed by including at least one additional vehicle as an added expense (i.e. "self-insurance" against hull replacement), was simply taken as a nominal cost of $0.2M per flight against third party liability.

d) Interest was taken at a nominal annual rate of 10%, though this can vary and should be put to zero if the venture can be funded entirely by equity rather than debt, as assumed here. A more reasonable estimate for an all debt scenario could be 12-13%, which is essentially what it cost before taxes to borrow money at a corporate level in the US during the late 1999s, though this would have had minimal impact on the results.

e) Tax, which was accounted after interest and before the net income, was written-off when the venture incurred losses in the early years – in other words, it got a "tax credit" which could either be used to offset future gains, or shared amongst the investors to offset gains in other investments. Therefore, assuming losses could be expensed against other gains, the net effect of taxes in the early years – especially during the development phase, which covers about three years – was to reduce the total out-of-pocket investment.

Business Parameter	***Value range***
Total R&D investment	$500-3000 million
Fleet size	3 operational vehicles
Price per flight	$0.25-10.0 million
Variable cost (per flight)	$2-10 million
Fixed annual operating cost	$25-40 million
Income tax rate	40%-60%
767-300 cost (per flight)	$0.1 million
Rocket/Aircraft cost (per flight)	5-15
Learning factor (rocket maint.)	5%-10%
Interest rate	10% (for debt finance)
Max. flights/fleet	100-800
First commercial launch	4 years after start

Table 8. RLV business model parameters

In addition, the financial and operational business parameters shown in Table 8 were assumed in order to bound the business model and investigate its sensitivity against changes in the baseline assumptions. Note that the price per flight was only allowed to vary up to a maximum of $10M, as it was assumed that competition from other commercial ventures (e.g. Space-X flying Dragon II) would be offering flights at around this price point within the 2020AD timeframe.

We considered a staged development of the business scenario that incorporated four key phases:

1a) *NASA flights*, which span the 1st and 2nd year of operation, with a ticket price of $20M;

1b) *Pathfinder flights*, which span the 1st and 2nd year of operation, with a ticket price of $10M;

2) *Pioneer flights*, which span the 5th and 8th year of operation, with a ticket price of $10M/seat;

3) *Initial Operation flights*, which span the 3rd and 4th year of operation, with a ticket price of $5M/seat;

3) *Routine Operation flights*, which span the 9th and 12th year of operation, with a ticket price of $1M/seat.

The key feature of this phased scenario is that it enabled us the examine the benefits of having NASA as the initial customer, similar to the approach being taken for Commercial Crew Development (CCDev).

7.2 The R LV Business Case

We ran the spread-sheet with various combinations of the business scenario and model parameters in order to identify the most critical parameters/assumptions. The most important results of this 'sensitivity' analysis are presented in Figure 11, which shows the evolution of IRR and EOY cash balance over an eleven year period from the venture's start.

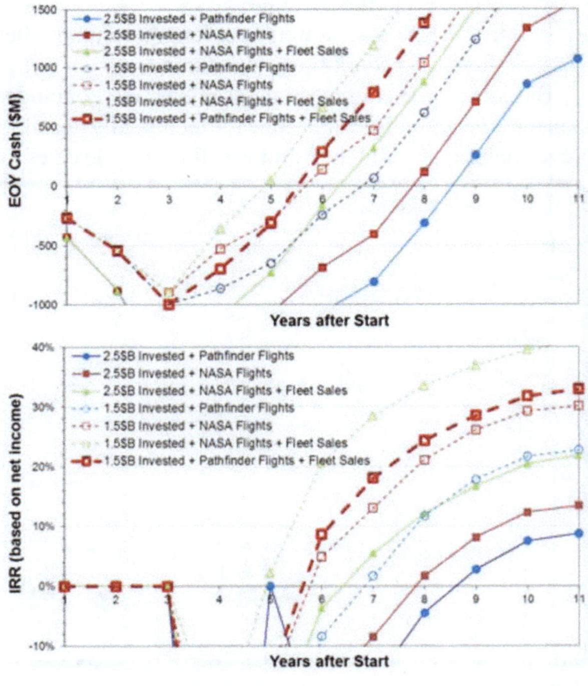

Figure 11. RLV business base sensitivity analysis

The plots show that fleet sales to third parties was also included in the business model and arose from the need to replace/update the fleet after it had performed its designed number of flight cycles. Having observed that the single operator flight rates were capturing only a small fraction of the potential passenger pool (i.e. less than 20% of the 5%NW pool), we wanted to investigate the value of selling additional RLV fleets to third-parties operators who could address the remaining passenger pool fraction.

Based upon these results and assuming that an IRR above 20% is needed to justify the initial investment in the venture, it is clear that an investment requirement greater than $2500 million *would be unacceptable* with respect to current scenarios. However, it *would become very acceptable* if the requirement was reduced down to $1500 million or below.

7.3 Observations on the Business Case

Clearly, this business case analysis is far too crude to judge the true commercial viability of such a venture. However, given these results, the general conclusion is that there are some good reasons for thinking that a fully commercial air-launched RLV venture may prove to be successful, particularly if its investment requirements can be kept around the $1500 million mark and the initial ticket price can be kept around $10 million. ***The major caveat here is that a development cost of $1500 million may be extremely low for vehicles with such a payload performance, based upon current launcher development experience.***

One important observation here is that the business case can be improved significantly if some degree of leverage can be applied to reduce the initial investment. One obvious way to achieve such leverage would be to develop key elements of the system through a separate venture or business phase. DARPA's XS-1 initiative may provide just such a leverage, while a business venture to service the nascent sub-orbital market may represent another. Whether these would be practical or sufficient to leverage development of an orbital RLV has yet to be determined. However, there are a number of real-world examples, both current and past, that may justify this approach, for example:

- SpaceX leveraging their NASA contracts to support development of the Dragon capsule;
- Boeing leveraging their USAF contracts for the KC-135 to support development of the 707.

Whatever the form of the leverage, this analysis serves to underscore the value of building up any space launch business in a series of small steps rather than one giant leap.

8. FUTURE POTENTIALS

This analysis has built upon previous attempts to show how a subsonic air-launched RLV could break the current space access paradigm by fostering new markets while also stimulating the existing ones. It suggests that if the business case can be successfully executed for a relatively small vehicle (e.g. 500kg to LEO) it may be possible to leverage the development of larger vehicles that could address other lucrative markets sectors such as geosynchronous communication satellites [RD.1 & RD.18].

The synergy of such ventures is even more important because the previous work also suggests that an RLV with a relatively modest launch performance of between 4-6t into low Earth orbit could be capable of supporting the majority of current and future launch demands by forming the key element of a fully reusable space transportation infrastructure.

8.1 Logistics & Crew Transportation

Beyond individual scientific satellites, primarily in polar orbits to support Earth observation missions, the International Space Station (ISS) currently represents the only significant market in LEO that needs frequent and routine transport services. They are currently supported by a fleet of both government and commercial vehicles, which are listed in Table 9. As can be seen, a number of them have an injected mass into LEO that appears compatible with the payload performance of an air-launched RLV. This suggests that ISS logistics resupply may be potential market for any commercial venture, especially as two of these vehicles are already built and operated by commercial companies that have secured commercial resupply contracts with NASA. Note that the reusable X-37B, which has a mass of just under 5000kg, could also represent another potential LEO payload though its military nature may make this possibility somewhat more unlikely.

ISS Servicing Vehicles	***LEO Mass (Mg)***
Soyuz (Government – Russian)	7200
Progress (Government – Russian)	7200
ATV (Government – European)	20200
HTV (Government – Japanese)	19000
Dragon (Commercial – SpaceX)	6000
Cygnus (Commercial – OSC)	4500

Table 9. ISS servicing vehicles mass in LEO

Future commercial LEO space stations, like those planned by Bigelow Aerospace, represent another potentially lucrative market because they are predicated upon the availability of routine and frequent launch services. Like the ISS, they will also require the transportation of crew and so demand a demonstrated level of safety much greater than that needed for cargo re-supply. However, such levels should be more easily achievable via a fully reusable launch vehicle because its inherent value will demand better operational contingency options in addition to a crew escape system.

8.2 Propellant Depot Resupply

A future LEO mission that may prove far more lucrative than those already mentioned is as the first leg of a space transportation infrastructure that consists of a set of operational nodes and transfer vehicles, namely:

- Space stations and human-tended experimental platforms;
- propellant depots to support missions both in, around and beyond LEO;
- short-range orbit manoeuvre vehicles (OMVs) to capture and transfer payload in and around LEO;
- long-range orbit transfer vehicles (OTVs) for travel to/from GEO and lunar orbits;
- OTVs fitted with legs and throttlable engines for lunar surface descent/ascent missions.

Details of one such space transportation architecture are shown in Figure 12, taken from RD.19, which also presents the delta-v required to reach each node and the representative masses of each of the key elements. Interestingly, the dry mass of many of these elements falls within the payload launch performance of an air-launch RLV using ACES. However, a more important point to note is that the majority of each element's mass is propellant.

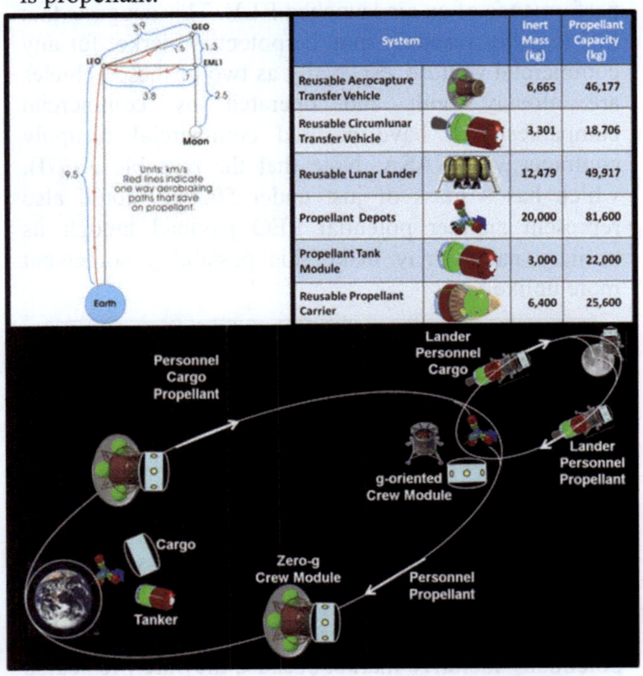

Figure 12. LEO-Lunar transport architecture

Analysis of the launch requirements for the build-up and operation of such an infrastructure [RD.19] show that the vast majority (~80%) of the mass launched into LEO is propellant. This is very significant because propellant can be infinitely subdivided and so would be the ideal payload for a small RLV capable of supporting both rapid and frequent launch and rendezvous missions. It therefore suggests that most of this architecture could be either launched and/or serviced by a subsonic air-launched RLV.

8.3 Commercial GEO Operations

Currently, the largest and most lucrative commercial launch market sector is the delivery of geostationary communications satellites (GEO comsats) into geosynchronous transfer orbit (GTO), with a perigee height of ~200km and an apogee height of ~ 36000km.

The commercial business case for developing a subsonic air-launched RLV to address the GEO comsat market sector has already been addressed by RD.1 and RD.18, so we will only present a very brief overview of the approach.

An analysis of typical comsat mass characteristics is presented in Table 10 and indicates that the majority have a beginning of life (BoL) mass ~35% below their launch mass. This is because a significant fraction of their launch mass is propellant that they use during their transfer burn from GTO to GEO. More importantly, it suggests that any vehicle capable of delivering a 4t payload into LEO could service the majority of currently planned GEO comsats if some sort of kick-stage were available on-orbit to perform the LEO to GEO transfer.

Payload Envelope (Stowed) ~7.3m x 3.8m x 3.4m

NOTE: Liquid apogee motor on all Boeing 702 & 601 series; Solid apogee motor on all Boeing 376 series

Spacecraft Name	Spacecraft Model	Launch Mass (lbs)	Launch Mass (kg)	BOL Mass (kg)	GTO to GEO Mass Fraction	RLV Flights (4t to LEO)
Anik F2	Boeing 702	13029	5910	3805	0.6438	5
	Generic	13000	5897	3796	0.6438	5
Anik F1	Boeing 702	10384	4710	3015	0.6401	4
Galaxy IIIC	Boeing 702	10604	4810	2835	0.5894	4
	Generic	12000	5443	3346	0.6148	4
Astra 1H	Boeing 601	8157	3700	2480	0.6703	3
Astra 2A	Boeing 601	7994	3626	2470	0.6812	3
Astra 2C	Boeing 601	8058	3655	2200	0.6019	3
	Generic	9000	4082	2658	0.6511	3
	Generic	6000	2722	1579	0.5800	2
Astra 2D	Boeing 376	3186	1445	824	0.5702	1
Astra 3A	Boeing 376	3318	1505	874	0.5807	1
	Generic	3000	1361	783	0.5755	1

Constraint

Table 10. Typical GEO ComSat mass characteristics

The key to servicing these markets with such a small reusable launcher is, therefore, the on-orbit assembly of a kick-stage capable of delivering the comsat directly into GEO, as illustrated in Figure 13. Such an operation would demand a rather special set of vehicle performance characteristics, namely the ability to perform:

- orbital rendezvous and docking;
- in-orbit propellant transfer or assembling sets of plug-in propellant modules;
- multiple launches within a short time period (e.g. a few days) to avoid effects of atmospheric drag, if low altitude orbits are used.

Such a vehicle would require an evolution of the basic orbital vehicle's capabilities, but the upgrades to enable rendezvous and docking are not considered too major a technological challenge since they have already been demonstrated successfully by both Japanese and US spacecraft (i.e. ETS VII and Orbital Express).

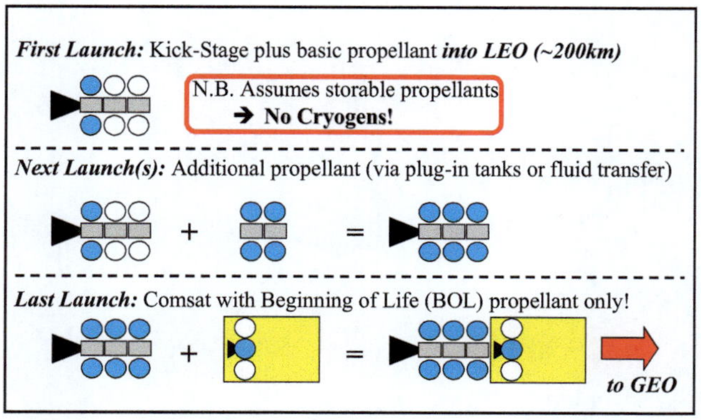

Figure 13. Orbital Assembly Scenario (Std. Comsat)

However, it is very unlikely that GTO customers would be willing to risk their satellites being launched in this manner until its operational complexity had been thoroughly proven, even if the launch price was half that of existing ELVs!

9. CONCLUSIONS

This paper has tried to show that there may be good reason to believe that the current space access paradigm could be broken by the development of a small subsonic air-launched RLV. Moreover, it has indicated that such a vehicle could be developed commercially and thereby stimulate the development of a new market sector serving human transportation to LEO.

Once this capability is fielded successfully, it is likely to generate further commercial ventures that may reach out from LEO and enable exploration of the Moon and beyond, stimulating both the growth and evolution of space activities that could provide significant benefits to all mankind (e.g. solar power satellites to provide the global population with clean, sustainable and cost competitive energy).

Clearly, many of the steps involved may be difficult to realize; in fact, this new space paradigm may prove to be unachievable because of fundamental constraints that have yet to be discovered. So, although there is good reason for cautious optimism, it would be better to regard such steps as experiments within a process of Darwinian evolution rather than the milestones of some overarching space programme.

REFERENCES

[1] "Salt, D.J. "Could a subsonic air-launched RLV enable a paradigm shift in space operations?", AIAA-2014-1897 SpaceOps 2014, Pasadena, California, 5-9 May 2014

[2] Space Foundation publication "The Space Report 2015" URL: http://www.spacefoundation.org/programs/research-and-analysis/space-report/

[3] "Wal-Mart Annual Report, Eleven-Year Financial Summary", p.26/27 URL: http://stock.walmart.com/files/doc_financials/2015/annual/2015-annual-report.pdf

[4] "The Commercial Space Transportation Study, 1994" URL: http://www.hq.nasa.gov/webaccess/CommSpaceTrans

[5] "NASA ASCENT Study Final Report", Futron Corporation 2003 URL: http://www.futron.com/pdf/resource_center/reports/ASCENTFinalReport_V1.pdf

[5] Andrews J.E., "RLV Design Issues for Future Commercial Space Applications" AIAA 2000-5104, AIAA Space 2000 Conference and Exhibit, Long Beach Convention Center, Long Beach, California, USA, 19-21 September 2000.

[6] NASA Get Away Special Wikipedia page URL: http://en.wikipedia.org/wiki/Getaway_Special

[7] "Mathematica Economic Analysis of Space Shuttle System" NASA-CR-129570 URL: http://ntrs.nasa.gov/archive/nasa/casi.ntrs.nasa.gov/19730005253.pdf

[8] "Wertz, J.R. "Economic model of Reusable vs. Expendable Launch Vehicles", IAF Congress, Rio de Janerio, Brazil Oct.2-6, 2000

[9] "Horizontal Launch: A Versatile Concept for Assured Space Access", NASA SP 2011-215994 URL: http://ntrs.nasa.gov/archive/nasa/casi.ntrs.nasa.gov/20120000791_2012000836.pdf

[10] Goff, J. internet blog article at URL: http://selenianboondocks.com/2015/06/space-tourism-musings/

[11] Futron document "Space Tourism Market Study", October 2002 at URL: http://www.futron.com/upload/wysiwyg/Resources/Reports/Space_Tourism_Market_Study_2002.pdf

[12] t/Space presentation "Final Briefing", March 2, 2005 URL: http://exploration.nasa.gov/documents/reports/cer_final/tSpace.pdf

[13] Wealth-X "World Ultra Wealth Report 2014" URL: http://www.wealthx.com/wealthxubswealthreport/

[14] Credit Suisse "Global Wealth Databook 2014" URL: http://economics.uwo.ca/people/davies_docs/global-wealth-databook-2014-v2.pdf

[15] IATA Maintenance Cost Task Force, "Airline Maintenance Cost Executive Commentary (FY2013 data) - November 2014" URL: https://www.iata.org/whatwedo/workgroups/Documents/MCTF/AMC-ExecComment-FY13.pdf
Airline Maintenance Cost Executive Commentary

[16] Sarigul-Klijn M., et al, "Selection of a Carrier Aircraft and a Launch Method for Air Launching Space Vehicles", AIAA 2008-7835, AIAA Space 2008 Conference & Exposition, 9-11 September 2008, San Diego, California

[17] Sarigul-Klijn N., et al, "Air-Launching to Orbit: Effects of Launch Conditions and Vehicle Aerodynamics", Journal of Spacecraft and Rockets, Vol.42, No.3, May-June 2005

[18] "Salt, D.J., & Lindroos, M, "The Business Case for Small RLVs", TN200, Issue 3, 20th January, 1999 (deliverable item under ESA Contract No. 970801).

[19] Bienhoff D. "From Importing to Exporting: The Impact of ISRU on Space Logistics" AIAA 2011-7112, AIAA Space 2011 Conference & Exposition, Long Beach, California, 27-29 September 2011 URL: http://arc.aiaa.org/doi/abs/10.2514/6.2011-7112

First Steps Towards the Kingston Space Shot: Low Altitude Test Vehicle

Dr. Adam M Baker

Kingston University London

Dr. Malcolm Claus

Kingston University London

S. Hatton (ed.), *Proceedings of the 13th Reinventing Space Conference*,
https://doi.org/10.1007/978-3-319-32817-1_23

First Steps Towards the Kingston Space Shot: Low Altitude Test Vehicle

Adam M Baker, Malcolm Claus
Kingston University London
Roehampton Vale Campus, Friars Avenue, London SW15 3DW, United Kingdom.
Tel. +44 (0)208 417 4722
a.baker@kingston.ac.uk

ABSTRACT

Kingston University London students supported by sponsors are working towards the most ambitious educational space activity the UK has ever seen: a low cost Space shot or rocket launch to beyond the 100km Karman line, with vehicle recovery.

The Kingston rocket launch aims to contribute to the UK civil space strategy 'Access to Space' element, the National Space Technology Strategy's Access to Space roadmap, and it is hoped it will act as an inspiration to a new generation of scientists and engineers.

The first step in a staged development programme began in summer 2015, with the design of the low altitude test vehicle and initial testing of its hybrid rocket engine propulsion unit.

Kingston University's School of Aerospace & Aircraft Engineering MEng class have been given a target of designing a vehicle capable of reaching an altitude of 25km (80000ft) that can be fully recovered for multiple uses, and to conduct an initial test launch in the summer of 2016.

Design work coupled with engine static tests at the KURocketlab began in July 2015. The vehicle will be designed around an engine that will demonstrate the full capability of the KU Rocketlab small space propulsion test facility. The engine and rocket Preliminary Design Review is planned to take place immediately prior to RISpace 2015.

Subject to support from existing and new sponsors who are assisting the student project team, the intention is to commence build by the end of 2015, conduct a system testing in early 2016 and be ready for launch by summer 2016. Success will be the first step on the road to a low cost sounding rocket capability and ultimately, with industrial and academic partners, improved UK access to space.

KEYWORDS: structure hybrid bipropellant engine testing instrumentation sounding rocket

INTRODUCTION

ACCESS TO SPACE AT KU

Kingston University London students supported by sponsors are working towards the most ambitious educational space activity the UK has ever seen: a low cost Space shot or rocket launch to beyond the 100km Karman line, with vehicle recovery. Underlying this long term aim

- The Kingston rocket launch aims to contribute to the UK civil space strategy 'Access to Space' element, the National Space Technology Strategy's Access to Space roadmap, and it is hoped it will act as an inspiration to a new generation of scientists and engineers.
- Kingston's rockets aim to market test interest in the low cost sounding rocket market which the UK left some years ago with the conclusion of the Skylark programme. Europe has limited capability here except for Rexus / Maxus rockets using non European solid propulsion, and developments in HTP hybrids at Nammo.
- Early flight swill act as a stepping stone towards flight test of a liquid oxygen hybrid rocket engine under development at KU with support from Newton Launch Systems.
- Kingston also aims to explore low cost structural approaches which have the potential to be scaled up to a future UK small satellite launch vehicle.

STRUCTURAL CHALLENGES FOR SMALL VEHICLES

Problem 1: inert mass fraction rises (and absorbs payload fraction) as gross liftoff mass decreases.

A small launcher can be defined in coarse detail starting with a minimum payload to orbit. Assuming this to be nominally 100kg, and using historical data for payload mass fractions of between 0.5-1% indicates that a gross liftoff mass of 10-20t is likely using conventional, low cost subsystems. Subsequently this mass will need to be validated using realistic inert mass fractions and ultimately a detailed model. However a major challenge for small vehicles is the Cube-Square law: as rocket size decreases, the physical volume scales as the cube of the linear dimension (say, diameter), whereas the physical quantity of material (e.g. propellant tank wall) varies with the square of the linear dimension. Hence the ratio of inert material to contained material such as propellant increases as scale decreases – the impact being that the inert mass fraction rises as the size of the vehicle falls, and the vehicle must be oversized to deliver the required deltaV. Inert mass fraction is defined as where 'm' refers to mass.

$$f_{inert} = \frac{m_{inert}}{m_{prop} + m_{inert}}$$

Further, the drag loss for small vehicles can be considerable, as boundary layers transitions and base area take up a much larger portion of the external surface, relative to contained propellant. Again this requires an increase in contained deltaV to reach the same (orbital) velocity. Efficient approaches to minimise structural mass without driving development or recurring cost too high are needed.

Problem 2: very little data on small rockets flying supersonically and in particular using cryogenic fluids is available. Most amateur rockets of any reasonable size use solid motors or nitrous oxide hybrids, most sounding rockets use solid motors, and the majority for which data is available do not fly in excess of Mach 1 for extended periods to enable an accurate drag profile to be computed.

For credible small vehicle designs to be developed and ultimately costed, testbeds must be built to generate flight test data with cryogenic oxidisers, assess structural approaches, provide instrumented data to allow accurate trade offs between the many possible propulsion and structure options to be made, and to evaluate other low cost systems including recovery, avionics and stage separation.

ACCESS TO SPACE AT KINGSTON

Kingston undergraduates and postgraduates have been researching low cost approaches to access to Space since 2012, as well as developing a hands-on test facility for a range of small chemical rocket engines and related components such as tanks.

Launcher structural approaches

In July 2014 a group of 5 M Eng students concluded their 9 months study into structural options for a small (50-100kg payload to orbit), 3 stage launch vehicle, comparing pressure stabilised, composite and semi-monocoque approaches. A trajectory profile and flow analysis for a vehicle with a stage diameter transition between 2nd and 3rd stages, and established a maximum Mach # of 0.95 at Mach 1.3.

Further , the group conducted a detailed analysis of a semi-monocoque partially pressure stabilised 3 stage vehicle . This used semi-monocoque load bearing propellant tanks supported by a number of longitudinal, circumferential and (1st stage base) crossbeam composite members.

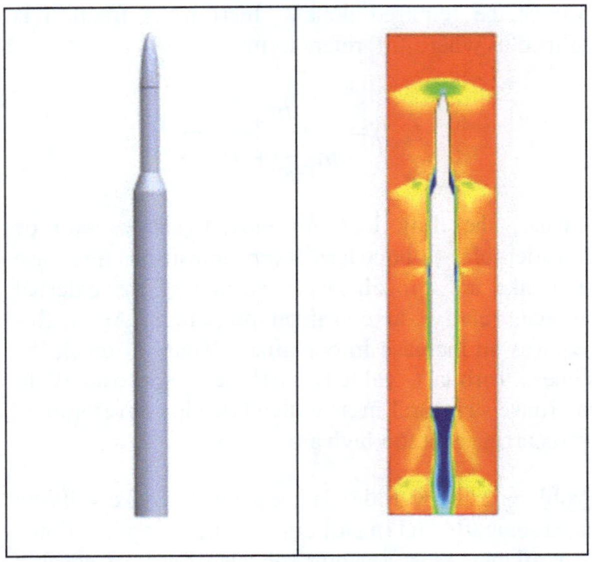

Figure 1 – Preliminary vehicle design and high speed flow model *(Kingston University – M Edwards)*

A notable result was an initial gross liftoff mass estimate for the vehicle of 11.1t, with an inert mass fraction at vehicle level of 0.125, varying between 0.11 for stage 1 to 0.29 for stage 3.

The vehicle concept design is shown below:

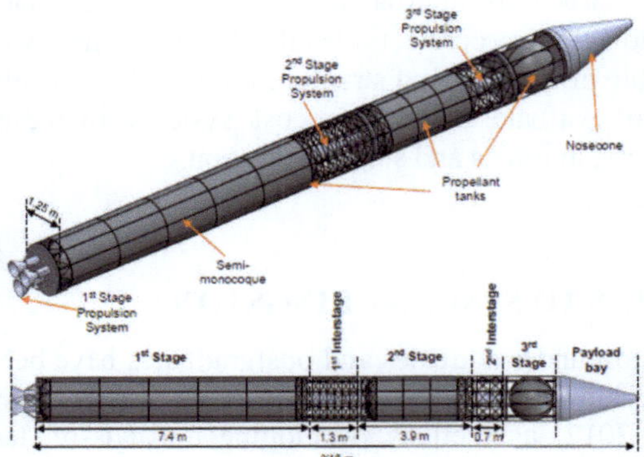

Figure 2 – 3 stage small satellite launcher concept design *(Kingston University – L Karaveckas)*

The third stage received less focus than the larger booster and second stage, which was part of the reason for the higher mass fraction. However it was noted that a 2 stage vehicle, although likely having a higher overall mass than a 3 stage design, would have larger stages which might be able to achieve lower values of inert mass and at lower cost, which was noted as an area for further research.

LOx Hybrid engine preliminary design

A small (300N-1kN thrust) liquid oxygen oxidiser / HDPE fuel engine has been designed built and a fluid feed system assembled through 3 successive undergraduate projects. This engine is shown in cross section below, and with the injector end closure and graphite nozzle end closure shown in enlarged form:

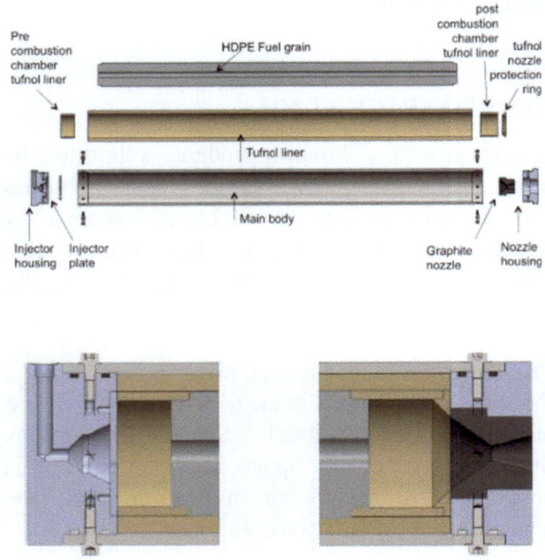

Figure – LOx hybrid rocket engine CAD model shown in cutaway *(Kingston University – Ben Edgley).*

A Liquid oxygen Fluid feed system has been designed to take LOx from a minitank supplied by BOC, to transfer it to an intermediate 25litre capacity low pressure insulated vessel able to fit into the test lab, and then to remotely transfer the cryogenic fluid into a high pressure aluminium alloy run tank . The fluid system, shown below, is designed to support gaseous nitrogen pressurisation at up to 60Bar gauge, to provide a gaseous oxygen ignition / high voltage spark ignition system, and a remote purge / Lox dump capability. Valves and plumbing are provided by Swagelok and are remote actuated using pneumatic pressurisation:

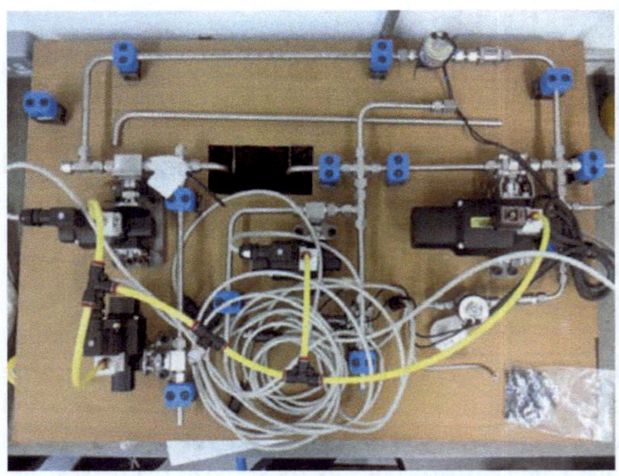

Figure – LOx hybrid rocket engine fluid control system *(Kingston University – Alex Pickard)*

The engine and a steel frame test stand designed to support it and a number of other smaller hybrid engines is shown below

Figure – LOx hybrid rocket engine components *(Kingston University – Tim Hodgkinson, Mat Bestelink, Chinonso Ezekwe, Alex Pickard)*

The image above shows also shows modular HDPE fuel grains, left; a Tufnol (cotton / phenolic) chamber liner , right, and an experimental SiC coated graphite nozzle fabricated by Archer Technicoat Ltd under a grant for the CEOI-ST / UK Space Agency National Space Technology Programme, centre.

A number of development steps are planned in 2015 to reach the first test firing, including

- Hydraulic testing
- Plumbing degrease and pressure test
- Cold flow with liquid nitrogen
- Full system test of valves under remote control with gaseous N2 at low pressure.
- Ignition tests.

Tank development

Kingston University has also begun exploring low cost propellant tank manufacture to ensure that flight vehicles are able to deliver the required propellant at suitable pressures and low mass. In house design and manufacture of cylindrical tanks with flat end caps as the lowest cost solution has produced initial hardware which is due to be hydraulically tested shortly, an example 3 litre capacity tank in Al alloy with a design MEOP of 75Bar gauge is shown below:

Figure – Prototype N2O and LOx oxidiser tank for 200mm diameter test vehicle *(Kingston University - Chris Walker, Ben Edgley)*

Further, Kingston University has been working with a local engineering consultancy, Geodome, to explore applicability of modified off-the-shelf pressure vessels from the water and fire safety industries. Polymer lined composite, and bespoke composite wrapped stainless steel lined tanks have been explored, and destructively tested. Example propellant tanks based on a glass fibre hoop overwrapped thin stainless steel liner (8 litre water capacity) demonstrated in increase in operating pressure from 12Bar gauge to over 50Bar, with a leak before break failure at 80Bar pressure.

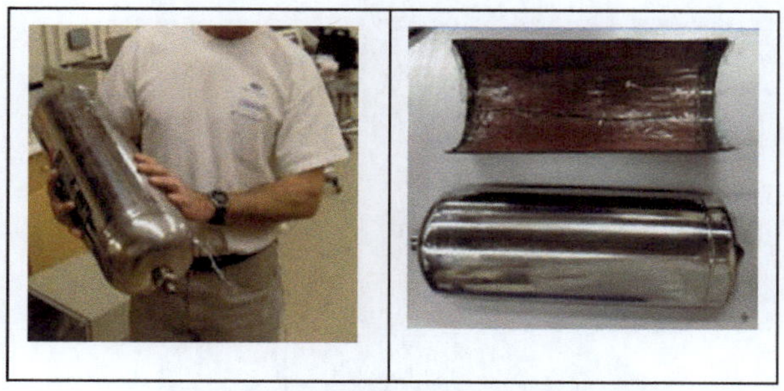

Figure – stainless steel fire extinguisher tank liner and overwrapped sectioned composite tank *(Garvey Aerospace, Geodome)*

The images above show an example tank liner derived from a water containing stainless steel fire extinguisher, designed to operate at low pressure , was structurally analysed by Geodome and then overwrapped cured and tested by Geodome using facilities at Kingston University and with the assistance of a team of final year M Eng students. Further optimisation of this approach to provide an alternative lightweight tank option for flight vehicles is expected with CEOI-ST / UKSA grant assistance in 2016.

TEST PROGRAMME

Rocket Lab – instrumented testing

Kingston university students have since 2011 designed, built and validated a small scale chemical propulsion test laboratory at the Roehampton Vale campus in SouthWest London. This facility is designed to allow undergraduate and graduate students to acquire practical skills in rocket engine design, manufacture and test. The test facility can accommodate hybrid and bipropellant engines up to several hundred N thrust using green propellants (e.g. N_2O, O_2 oxidisers and hydrocarbon fuels), with full fume extraction and remote observation and control capabilities. The lab has recently undergone a major renovation and Health & Safety review to ensure that undergraduates under supervision can acquire practical experience in rocket engine testing. The rocket lab and sound / blast proofed test cell (dimensions 2 × 3.5 × 2m) are show below:

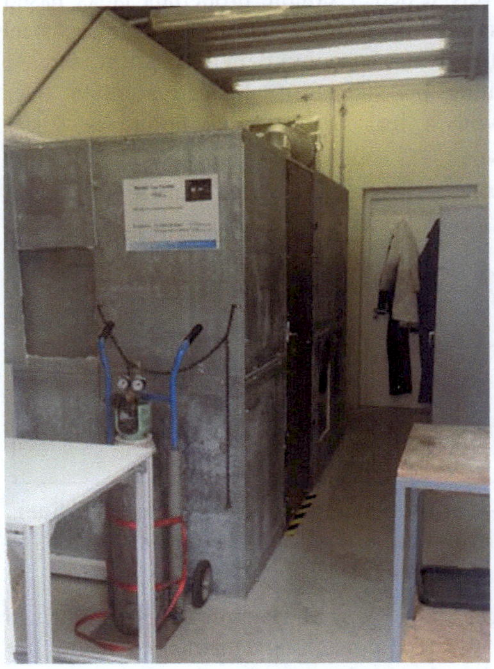

Figure – Kingston University Rocket Lab test cell (gas extraction at rear of cell)

A data acquisition unit provided by Airborne Engineering Ltd under Kingston SEC faculty support is able to monitor 16 temperature channels, 8 pressure channels, multi axis loads up to 1.5kN with kHz sampling frequency, and control automated valve sequences. A pulse counting capability for future flow measurement is also built in. Recent test on a small gaseous oxygen / PMMA hybrid rocket engine with a vortex injector has enabled the data acquisition system which runs on a Linux platform to be calibrated and customised. The current test stand setup is shown below.

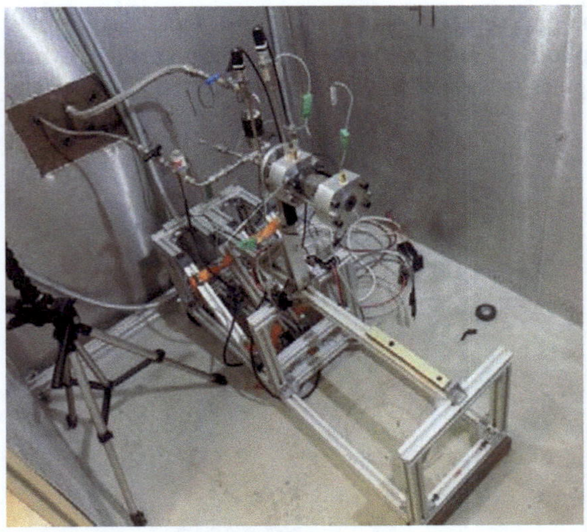

Figure – Instrumented test setup for small 50N class GOx / PMMA hybrid *(Kingston University - Wajahat Ahsan)*

The Test stand is currently set up to monitor oxidiser delivery fuel grain and test cell ambient (fan exit) temperature pressure at multiple points in the flow include Pc, and thrust using a Sensor Techniques Ltd platform load cell rated at up to 500N.

The image below shows a GoPro frame capture from a vortex injection firing of the small GOx / PMMA hybrid attached to the test frame above. The oxidiser injection is to the left and the plume is to the right. The fuel grain was burnt until failure in this case, showing a considerably reduced Oxidiser to fuel O:F ratio compared to that expected for an axial flow firing.

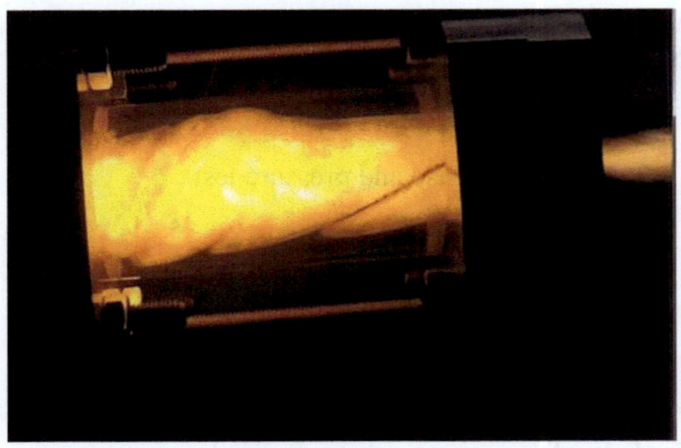

Figure – Vortex present in small hybrid PMMA fuel grain *(Kingston University – Stuart Watson).*

The Rocketlab will shortly begin hot firing a series of small bipropellant engines developed through PhD and MSc research, where vortex chamber cooling and SiC coated ablative nozzles will be evaluated for their potential contribution to low cost chemical propulsion.

Figure – Wajahat Ahsan with his vortex flow small hybrid rocket engines, designed built and tested during this 9 month 3rd year MEng individual project.

FLIGHT VEHICLE PROGRAMME

Concept design – LOx hybrid

The project brief agreed with the current M Eng group of 9 students is to develop a concept design and trajectory prediction for the High Altitude Test Vehicle HATV under guidance from Newton Launch Systems. HATV is a single stage sounding rocket capable of reaching 25km altitude with a LOX / HDPE hybrid rocket engine, which will be derived from the engine mentioned earlier in this paper, once a full static test programme to characterise it has been completed. The HATV will be not be built during this project which runs between July 2015 and May 2016, but is to be the subject of further collaboration between Newton & Kingston in 2016.

The HATV design is being completed now, and the outline design specification is as follows:

- 5kN thrust LOX / HDPE engine using 5 chambers
- 4m length and 175-200mm diameter.
- Gross liftoff mass 65kg
- Inert mass 25-26kg (inert mass fraction 0.38-0.4). The majority of the structure mass is in the engine chamber casings which will be the subject of further detailed analysis.
- Predicted altitude using a range of in house and open source / commercial tools 20km+.
- A range of launch sites in Scotland including the Mull of Galloway / Stranraer recently used by Celestial Mechanics for a high altitude 2 stage launch are under consideration. Water recovery options are under consideration.
- Thrust vector control may be needed to enable the design altitude to be reached and range safety requirements to be met. This is the subject of ongoing research at Kingston.

The image below shows the current design which is a semi monocoque using a lightweight skinned spaceframe based on composite tubing. Sections in blue represent the LOx oxidiser tank and nitrogen pressurant tank.

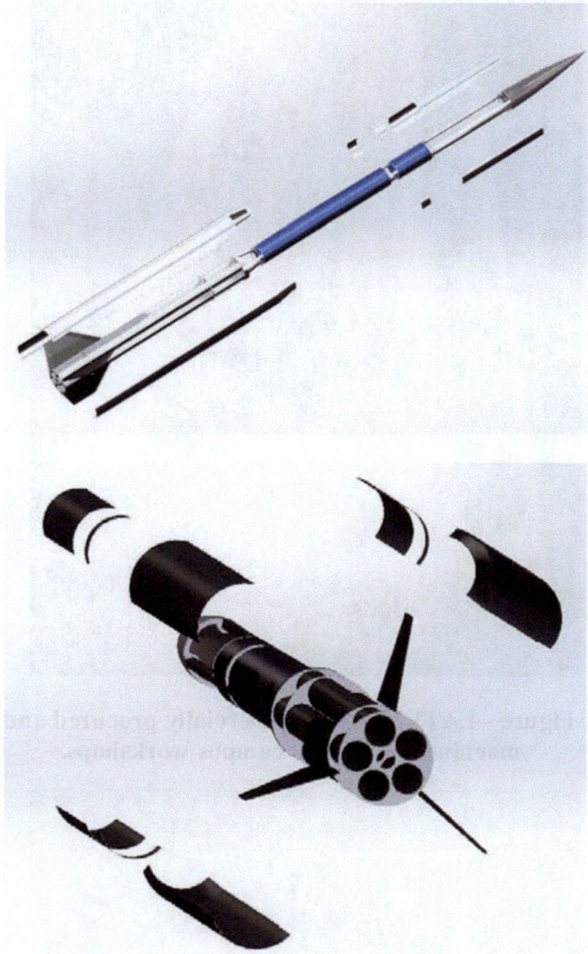

Figure - Kingston / Newton high altitude test vehicle concept design (Kingston University – Aidan Nicholls)

Flight vehicle – N2O hybrid

Following conclusion of the HATV concept design which is aimed at determining system level feasibility, the group will design, build and test a low altitude vehicle which will fly on a small (up to 1kN thrust) nitrous oxide of N2O / HDPE hybrid rocket engine. This engine is to be tested in the KU Rocketlab early in 2016, and the majority of the parts have already been designed and built . The LATV is a scaled version of HATV and although it does not have an altitude target, the ability to launch it to a sufficient altitude to test all the rocket systems and provide experience for the future HATV construction is desirable. A land based launch in southern England with the support of the UK Rocketry Association is being sought.

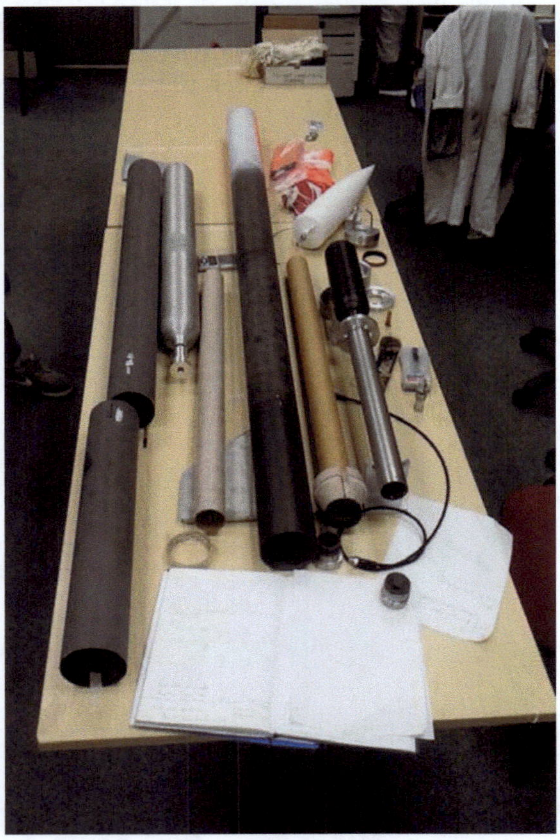

Figure - LATV parts , commercially procured and machined at KU RV campus workshops.

Figure – Kingston University M Eng group with PhD student Linas Karaveckas discussing recovery options for LATV.

The images above shows a number of the parts already assembled including off the shelf oxidiser tanks, carbon fibre body tubes and engine components. Many of these are likely to be used for an early 2016 flight test of critical systems including a flight hybrid rocket engine, avionics / payload and recovery systems.

CONCLUSIONS

Kingston University London and its RocketLab are aggressively pursuing an 'Access to Space' agenda. Through the development of a functioning rocket laboratory investigations into some of the problems associated with low cost access to space are being investigated. The KU Rocketlab has been functioning for 5 years but only recently began beginning to meet its goals of supporting both teaching and research (group and individual projects) This current academic year there are several projects investigating rocket/space propulsion including

- Small sounding rocket systems.
- Fully instrumented small chemical propulsion laboratory
- Vortex flow hybrid& bipropellant rocket engines
- N2O and LOx hybrid rockets
- Low cost composite and monolithic metal propellant tanks
- Mas efficient structural design approaches for small rocket vehicles

ACKNOWLEDGMENTS

Kingston University would like to acknowledge the generous support of the faculty of Science Engineering & Computing who have provided the test lab, control room and additional storage space, and funded a bespoke Airborne Engineering Ltd data acquisition system for rocket engine testing, as well as providing invaluable Health & Safety insight. The authors are particularly grateful to John MacBean, Kevin Attree, Helen Sharman and Peter Barrington.

Stuart Watson, Jack James Marlow and the Kingston University Rocket Propulsion Group (RPG) were instrumental in providing the data and images for this paper and the authors are grateful for their help.

Kingston University would also like to thank its sponsor Newton Launch Systems Ltd and Martin Heywood in particular, who have provided consumables and materials plus invaluable technical advice to a range of undergraduate and postgraduate projects over the last 3 years.

NEWTON Launch Systems

Reinventing Space Conference
BIS-RS-2015-32

Enabling Solutions for Small Satellite Space Access

Andreas Jonsson
RUAG Space Sweden AB
ASJ-vägen 9, SE-581 11 Linköping, Sweden
+46 13 482 8409 andreas.jonsson@ruag.com

Magnus Engström
RUAG Space Sweden AB
ASJ-vägen 9, SE-581 11 Linköping, Sweden
+46 13 482 8402 magnus.engstrom@ruag.com

13th Reinventing Space Conference
9-12 November 2015
Oxford, UK

S. Hatton (ed.), *Proceedings of the 13th Reinventing Space Conference*,
https://doi.org/10.1007/978-3-319-32817-1_24

1 INTRODUCTION

The Small Satellites market is growing and is expected to grow even further in the coming years. If the announced satellite constellations are included in the prognosis, we will see an unprecedented amount of satellites to be launched in the coming years.

To make this increase in launched satellites happen, and maintaining a healthy business case for the satellite operators - there is a large increase in need for cost-effective, flexible and easy to use launcher systems for satellite integration.

New satellite platforms as well as new launchersare being developed, to address the increase in launcher needs as well as cost-effective solutions for satellite launches.

The launch rates foreseen considering both classical single satellites as well as large constellations announced from OneWeb, SpaceX and others, calls for special measures with respect to interface standardization, logistics and simplified integration sequences to optimize cost and lead time for launched satellites. In addition to this there is a need for replacement strategies once the constellation matures.

To begin solving this increase in number of satellites one has to look at how the S/C will be dispensed once launched.
This forces us to carefully consider dispenser and separation systems as well as integration and logistics solutions that can handle these challenges.

2 DISPENSER SYSTEMS

The classical dispenser system consists of a two main systems, i.e. the separation system and the dispenser structure. In these systems the mechanical forces from the launch and the later separation is handled by the structure, whereas the separation velocity and rotation is the handled by the separation system.

For all launches the mass is critical and there is a need to optimize dispenser structure, separation system as well as the satellite in order to maximize the effective S/C payload mass and volume.

Another issue that arises for large constellations and multiple satellite launches is the initiation sequence. Some Launch vehicles are currently not equipped with more than a few initiation signals, limiting the possible number of separations.

The solution lies in optimizing the dispenser, the separation system as well as the initiation system for mass and cost. However the solution will be different for large constellations where the S/C is generally homogeneous, whereas the situation for small satellites in general will be a heterogeneous launch where there is a main payload and smaller S/C that share the upper Launch Vehicle compartment using an adapter and dispenser configuration.

Smaller platforms generally do not have a standard interface, but assume that there can be an adaptation to their specific demands in the launch vehicle integration. This is however often not the case and hence one should look into standardizing separation interfaces.

2.1 Separation systems

The separation system traditionally uses solutions with discrete point hold down mechanisms or circular separation interfaces e.g. clamp bands or similar.

Using Clamp bands will reduce the number of initiators needed compared to discrete points and allows for a circular interface that provides a more uniform load distribution. Larger clamp bands are normally optimized for a low weight horizontal mounting. The RUAG PAS 381S, PAS 432S and PAS 610S Systems are adapted to the small satellite market requirements for mounting in both vertical as well as horizontal position. These clamp band systems are all designed with flight heritage components and has a very high reliability.

The discrete separation systems have a tendency to deliver a comparatively high shock during separation. This is a concern for a lot of small satellite manufacturers using more COTS components or other sensitive equipment. The 3 RUAG separation addresses these requirements by using a downsized version of the RUAG Clamp Band Opening Device (CBOD), a CBOD-LT release mechanism that combines a high load capability and low shock release.

The PAS 318 S, see Figure 1, has a lightweight passive ring with a mass less than 3 kg. Both the passive and active rings are adapted to the ESPA 15" I/F with 25 x1/4 inch bolts placed around the 381 mm ring. The release energy can be adapted by adjusting the number of springs from 4 up to 24 with the same ring. Typical use would be a 200 kg S/C with a CoG at 0.5 m.

Figure 1 RUAG PAS 381 S Separation System – with active and passive rings.

The PAS 432 S, see Figure 2, is originally designed for small satellites for the VEGA launch vehicle, but will suit slightly larger satellites and can handle a higher clamp band pre-tension 15kN compared to the PAS 381 S. The nominal setup has an active ring with a mass of 5.1 kg and 6 separation springs with total release energy of 28.2 J. Typical use would be a 300 kg S/C with a CoG at 0.5 m.

Figure 2 RUAG PAS 432 S Active ring with umbilicals – clamp band in resting position

The PAS 610 S, see

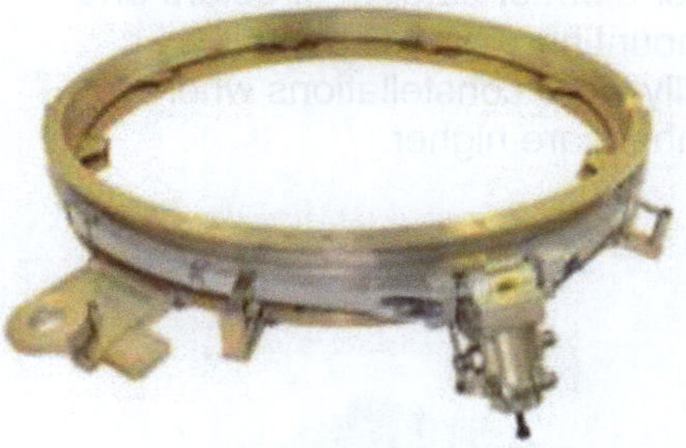

Figure 3, has been designed with a low ring profile to save weight and optimize the used height. The system has a height including both passive and active rings of only 73 mm. The height does not include the separation springs, which will protrude down into the underlying adaptor ring.

Both the passive and active rings are adapted to the ESPA Grande 24" I/F with 36 x1/4 inch bolts placed around the ring. The release energy can be adapted by adjusting the number of springs from 4 up to 10 up to a total of 47 J. Typical use would be a 400 kg S/C with a CoG at 0.5 m.

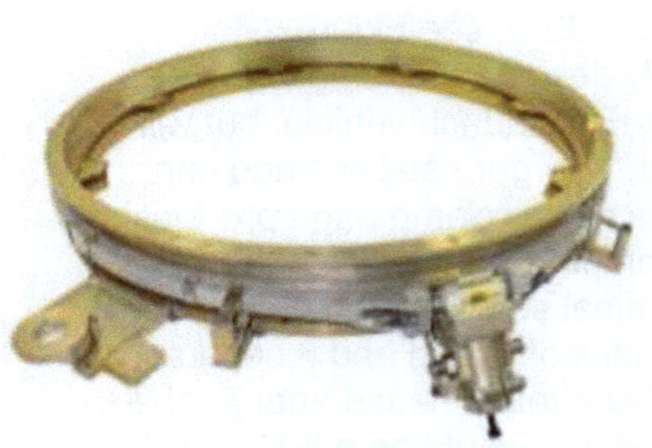

Figure 3 RUAG PAS 610 S Separation System – here shown in the low ring version without separation springs

To bring down cost, mass and the flyaway mass, which goes with the satellites, RUAG has also developed a new system consisting of 4 hold down mechanisms and a plate structure, but with only 1 release point making the system an almost shock free design. Still building on RUAG systems with flight heritage (this system allows for a very tight mounting of the S/C towards a dispenser structure lowering the distance from dispenser to CoG of the S/C. The simplicity of the design also allows for a better satellite structure and faster mounting. This will benefit especially large constellations where the numbers are higher.

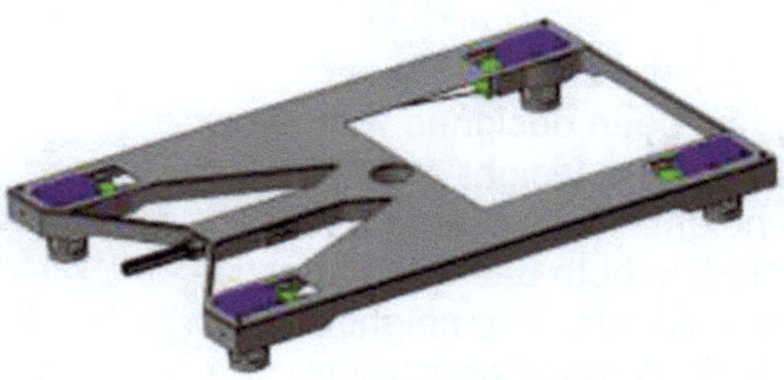

Figure 4 RUAG 4-point low profile separation system.

2.2 Dispenser Structures

Dispenser structures can be divided into three main categories. Small constellation dispensers where a few (typically 2-4) small to medium sized satellites of the same type are mounted. Secondary payload systems where the main passenger is accompanied by a different and smaller satellite. Finally one has to consider large constellations as a special case where multiple satellites of the same type are launched.

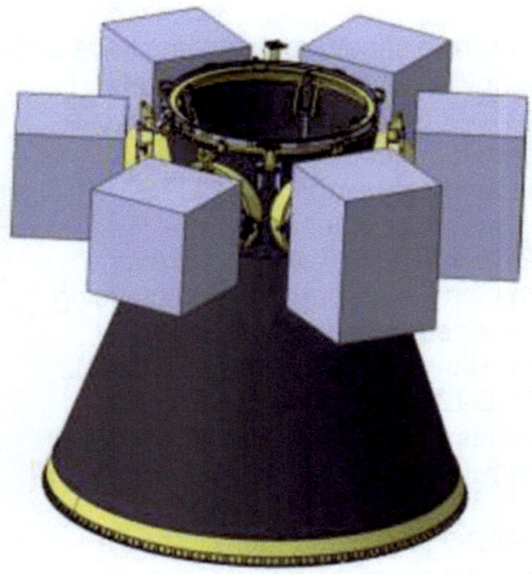

Figure 5 RUAG secondary payload adapter.

2.2.1 Secondary payload solutions

For a long time heterogeneous solutions, has been a good option for the satellite community to make use of the spare capacity on an already scheduled launch.

To make more effective use of this capacity, a solution where the adaptor ring is integrated with extra satellite dispensers and extended has been developed.

Since the system is built as a single structure it removes the need of the extra interface between adaptor and payload ring. In addition to this, the main structure is made up of CFRP in order to optimize mass. The system has an integrated upper separation ring with a standard 937 mm separation interface.

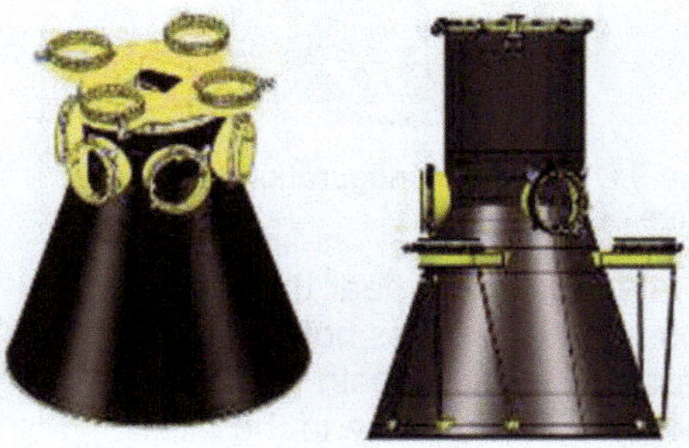

Figure 6 Secondary payload adapter configurations - to the left an additional platform on top for more small sats and to the right an extended version with struts to allow for S/C mounted in a axial position

The payload adapter is very versatile and can be configured to handle different S/C sizes and mounting configurations – see Figure 6

2.2.2 Small constellation solutions

Small to medium sized constellation dispenser structures are usually optimized to the specific program and launch vehicle used. Due to the small size of the constellation the development cost tied to this type of dispenser can become significant and has to be covered by the launch campaign.

There is a possibility to save cost in using already defined dispenser from previous programs, but they may not be optimized for the S/C intended. This is however a path that should be explored from the S/C manufacturers side. For small constellation of medium sized satellites RUAG has made quite a few designs that are already flown and others scheduled for flight, see Figure 7.

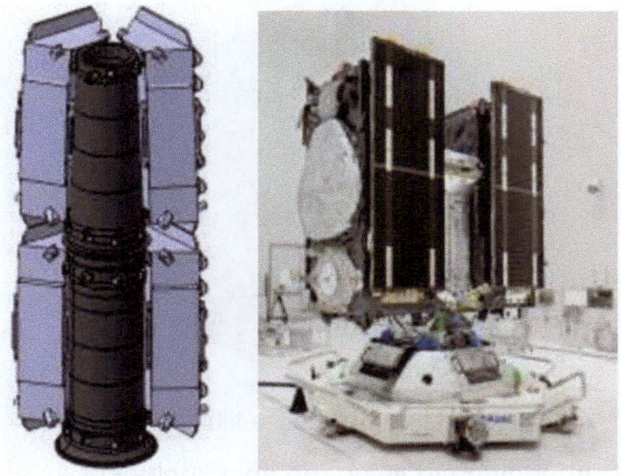

Figure 7 Example of dispenser systems. To the left S/C stacked on 2 structures on top of each other to accomodate 6 S/C. To the right the Gallileo dispenser shown with 2 S/C attached.

Improvements in the dispenser design has yielded a new dispenser version. This design reduces cost and weight however maintaining the same structural performance.

In some cases the size of the satellites compared to the launch vehicle fairing results in a very tight packaging. In these cases special measures has to be considered. In order to address these issues and increase the separation margins as well as reducing risk, a tilting mechanism has been invented. The tilting mechanism will deploy prior to separation in order to increase distance between satellites and also adjusting the direction of the separation. In order not to introduce any additional risk the system builds on already flight proven technology.

The structure developed by RUAG will smoothly tilt a S/C up to 1500 kg.

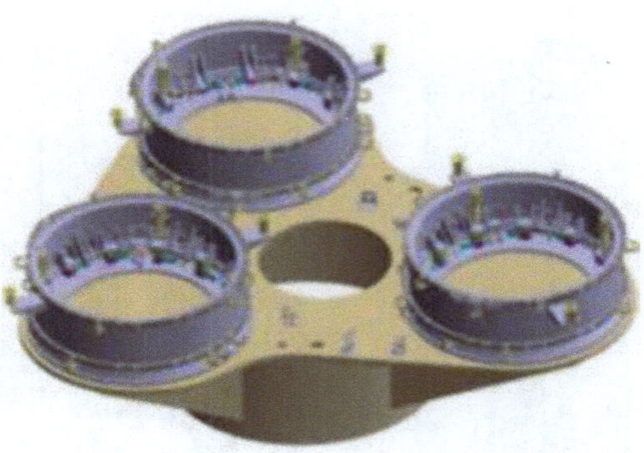

Figure 8 RUAG 3-Satellite tilting dispenser

In addition to tilting the spin, including transverse spin, can also be achieved. This becomes a factor especially in dispenser solutions not situated around the center axis of the launcher or if the launcher is unable to achieve spin on its own accord.

2.2.3 Large constellation Solutions

For large homogeneous spacecraft constellations the dispenser structure can be more optimized with regards to mass and stiffness towards the specific S/C.

2.2.3.1 Large constellation Dispenser structure

RUAG has developed alternative dispenser structure solutions. These solutions can be adapted to a large number of launchers. The baseline solution is capable of carrying up to 32 satellites with an individual mass of 150 kg. Satellites are placed on the outside of the dispenser structure dispenser in rows of 8 satellites in a circular fashion.

The first dispenser solution is a circular central structure similar to a central cylinder used in some of the larger communication satellites and a rail system to hold the harness and satellites. The CFRP Cylinder Structure is a monolithic shell.

Figure 9 Dispenser configuration for 32 satellite dispenser.

The interface areas near the ends and areas near the access holes are reinforced. The load path from the launch vehicle interface to the cylindrical dispenser structure is carried out by an aluminium structure. This structure can be adapted to different L/V interfaces.

Figure 10 Circular dispenser structure for 32 satellites

When considering a large number of S/C one also has to consider the logistics around getting 32 S/C onto the dispenser and in the end onto the launch vehicle. In order to handle this in a more effective way a rail system has been invented. Four satellites or one column of the entire S/C stack is mounted on the rail. This can be done with all the S/C prior to mounting rail onto the dispenser. After this the rail (with the four satellites) is then attached to the dispenser core structure in a simple and time saving operation.

The alternative design to attach the satellites to the launch vehicle is using the dispenser structure as the rail. In other words the dispenser structure is totally built up by a number of flat panels used as rails for satellite handling. The only remaining part of the structure beside these panels is the lower Interface Structure interfacing the launcher.

Based on 8 spacecraft at each level, an octagonal shaped structure can be beneficial and with 8 flat attached panels as the load carrying structure the S/C interface could be performed without intermediate load carrying structures, rails, as in the above circular design.

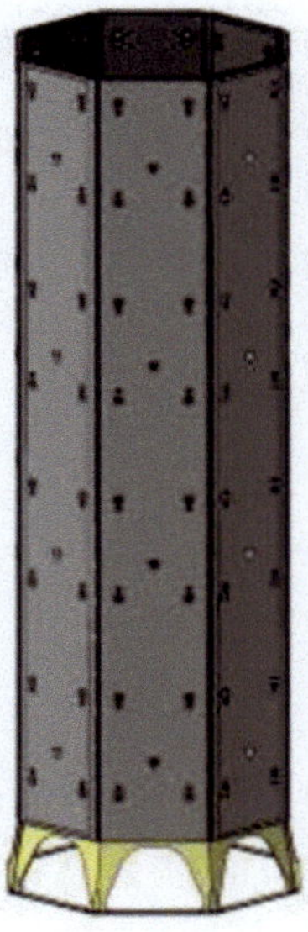

Figure 11 Octagonal dispenser structure for 32 satellites

This removal of intermediate structural components saves cost and liberates some mass that can be allocated to the panels themselves. Taking benefit of this, a sandwich panel has been designed within the same mass envelope, making the panel self-supporting also when carrying up to 4 S/C per panel.

An additional benefit of a panel with ample out of plane stiffness is the properties of the octagonal structure that is created when the panels are joined together.
With panel connections capable of transferring shear loads and bending moments, the need for stiffener rings is avoided. The octagonal shaped "tube" is fully self-supporting.

3 Challenges and other considerations

For large constellation and multiple S/C dispenser one has to of course consider the stiffness and mass of the system. In doing so also the harness must be accounted for. Although not covered here the harness mass will be significant if not using a more novel approach than classical pyro signals. Solutions are available, but have to be taken into account at dispenser design time.

Many different interfaces exist and have to be dealt with in order to keep a generic design. The launch vehicle interfaces can be addressed by changing the lower ring in the adapter interface. The S/C interfaces can be handled through the various separation interfaces already available.

Integration on the dispenser of both main passenger and secondaries has to be handled smoothly to avoid delays. Vertical and horizontal mounting processes can be utilized.

Last but not least the integration at launch site of the finished dispenser stack has to be considered. There are concepts and solutions for the integration sequence. And the dispenser and separation systems must be designed in a way that enables these in order to get a low effort and swift integration.

4 SUMMARY

There are many challenges when launching a multiple satellite constellation. For cost effectiveness and lead time reasons standards needs to be implemented in the interfaces between S/C and LV. If such standards are established it will benefit the community as a whole and small satellite launchers in particular.

Developing products that enables easy integration as well as ensures reliability requires experience from the environment at hand. Effective solutions for multiple satellite launches require knowledge about the launcher conditions as well as understanding the specificities on the dispenser side.

RUAG has the experience from many launch projects to be able to design effective solutions for the growing small satellite market.

Reinventing Space Conference
BIS-RS-2015-36

Development of Low Cost Propulsion Systems for Launch- and In Space Applications

Peter H. Weuta
WEPA-Technologies GmbH

Neil Jaschinski
WEPA-Technologies GmbH

13th Reinventing Space Conference
9-12 November 2015
Oxford, UK

S. Hatton (ed.), *Proceedings of the 13th Reinventing Space Conference*,
https://doi.org/10.1007/978-3-319-32817-1_25

Development of Low Cost Propulsion Systems for Launch- and InSpace Applications

Dr.-Ing. Peter H. Weuta, Dipl.-Ing. Neil Jaschinski
WEPA-Technologies GmbH

Jägerstrasse 30 / 51375 Leverkusen / Germany
Phone: +49-151-11657345
peter.weuta@wepa-technologies.de

The development of low cost propulsion systems for launch-, in-space and space tourism applications are a key component to enable private space flight.

An overview of the technology under development at WEPA-Technologies will be given. All propulsion systems focus on simple, cost effective design, high reliability and use of ‚green propellants'. To reduce development efforts, the basic configuration is partially based on proven designs used in the rocket programs of the former Soviet Union and USA.

Low cost propulsion will be realized by:
1) simplified design of rocket engines and turbo pumps
2) low-level operational parameters (< 60 bar chamber pressure)
3) use of low-cost materials and manufacturing technologies
4) unification of design of propulsion systems for the first and second stages of launch systems via clustering
5) environmentally benign and easy to handle propellant components as LOX / H_2O_2 / Ethanol / Kerosene – no NO_2 / N_2O_4 or hydrazines

The development of LPRE does encompass thrust chambers and turbo pumps.
At present the activities are focussed on a turbo pump fed, 3.5 to thrust demonstration unit (50 bar chamber pressure) using LOX respective H_2O_2 and Ethanol.
<u>H_2O_2-based propellants significantly facilitate development and reliable operation of propulsion- and overall systems architecture – key advantages can be summarized as follows:</u>

- simplified system design and increased responsiveness due to non-cryogenic characteristics of the storable propellants used
- reliable ignition / operation due to (quasi) hypergolic ignition
- facilitated reusability of propulsion systems
- environmentally begnin oxidizer

<u>Taking into account these advantages, WEPA-Technologies considers the use of Hydrogen Peroxide to be one very attractive option to enable fast track development of propulsion systems.</u>

Therefore the development acivities in this area will be discussed in detail and are focussed on the following areas:

- non-cryogenic turbo pump (~ 75 bar exit pressure)
- thrust chamber (regenerativly cooled)
- injector systems (pre-decomposition resp. liquid injection)
- concentration systems for Hydrogen Peroxide production (50 % => up to 97 %)

As ready and low leadtime availabilty of Hydrogen Peroxide even at a concentration level of 87,5 % is not always given and concentrations in the range of 90 – 97 % are - if at all - available under severe legal restrictions only, WEPA-Technologies does offer custom designed concentration plants.
Stationary plants delivering up to 90 % are available on a commercial basis at present and can be visited at a customers site (capacity: ~ 50 kg / d). Process technology to deliver up to 97 % is under development and will be available by late 2016. Safe and fully automatic, 24 h operability are key features of the plants.

In-flight qualification of propulsion systems are a key point of WEPA-Technologies' development strategy. The design of a sounding rocket under construction is discussed.

WEPA-Technologies GmbH has been founded in 2011 to provide development and manufacturing solutions in the field of Automation, Engineering and Aerospace. An extensive range of manufacturing technologies is available at the company owned workshop facility. Fast track prototyping therefore can be realized in house.

Development of Low Cost Propulsion Systems for Launch- and In Space Applications

2015 Reinventing Space Conference (RIspace 2015)

Oxford (UK), 2015-11-10

Dr.-Ing. Peter H. Weuta
Dipl.-Ing. Neil Jaschinski

WEPA-Technologies GmbH (Germany)

Introduction: WEPA-Technologies GmbH

Development of Low Cost Propulsion Systems for Launch- and In Space Applications
Dr.-Ing. P. Weuta, Dipl.-Ing. N. Jaschinski
Seite 2; 2015-11-04

Introduction: WEPA-Technologies GmbH

- **Background**
  - **Founded in 2011 via spin-off (origin: mechanical engineering company)**
- **Company focus**
  - **Engineering-, Automation- and Aerospace-Solutions**
- **Business premises**
  - **700m² work shop area**
  - **150 m² office space**

=> R&D focussed engineering office and manufacturing company

Development of Low Cost Propulsion Systems for Launch- and In Space Applications
Dr.-Ing. P. Weuta, Dipl.-Ing. N. Jaschinski
Seite 3; 2015-11-04

Business Activities

Development of Low Cost Propulsion Systems for Launch- and In Space Applications
Dr.-Ing. P. Weuta, Dipl.-Ing. N. Jaschinski
Seite 4; 2015-11-04

WEPA Technologies

Business Activities (Manufacturing) 1

Generell

- **<u>Planning, development and realization of non-standard solutions</u>**
- **Manufacturing of prototypes and small lots (company owned workshop)**
- **Broad range of manufacturing technologies**
  - **CNC-machining**
    - **Turning (max. 1.4 m diameter x 4 m length) (up to 4 axis)**
    - **Milling (max. 3.0 m x 0.8 m x 0.8 m) (up to 5 axis)**
    - **Metal spinning**
    - **Wire eroding**
  - **Conventional machining**
    - **Grinding, welding, sheet metal work**

- **Public references include…**
  - **CASSIDIAN GmbH (Airbus Defence & Space)**
  - **Dynamit Nobel Defence GmbH**
  - **EU-customer (H_2O_2 - concentration plant)**

Development of Low Cost Propulsion Systems for Launch- and In Space Applications
Dr.-Ing. P. Weuta, Dipl.-Ing. N. Jaschinski
Seite 5; 2015-11-04

Business Activities (Rocket Technology) 2

Business and development segments

- **Rocket technology (development)**
  - **Propulsion**
    - **Liquid propellant rocket engines (LPRE)**
    - **Turbo pumps for LPRE**
    - **Solid rocket motors (SRM)**
  - **(Complete systems)**
    - **Suborbital sounding rockets (propulsion unit)**
    - **H_2O_2 - concentration plants (max. 98 %)**
- **Engineering (business)**
  - **Construction and manufacturing of mechanical parts**
- **Automation (business)**
  - **Focus on control retrofits of CNC-machine tools**

CASSIDIAN contract development
(Airbus Defence & Space)
solid rocket motor test (thrust: 20 kN)

Development of Low Cost Propulsion Systems for La
Dr.-Ing. P. Weuta, Dipl.-Ing. N. Jaschinski
Seite 6; 2015-11-04

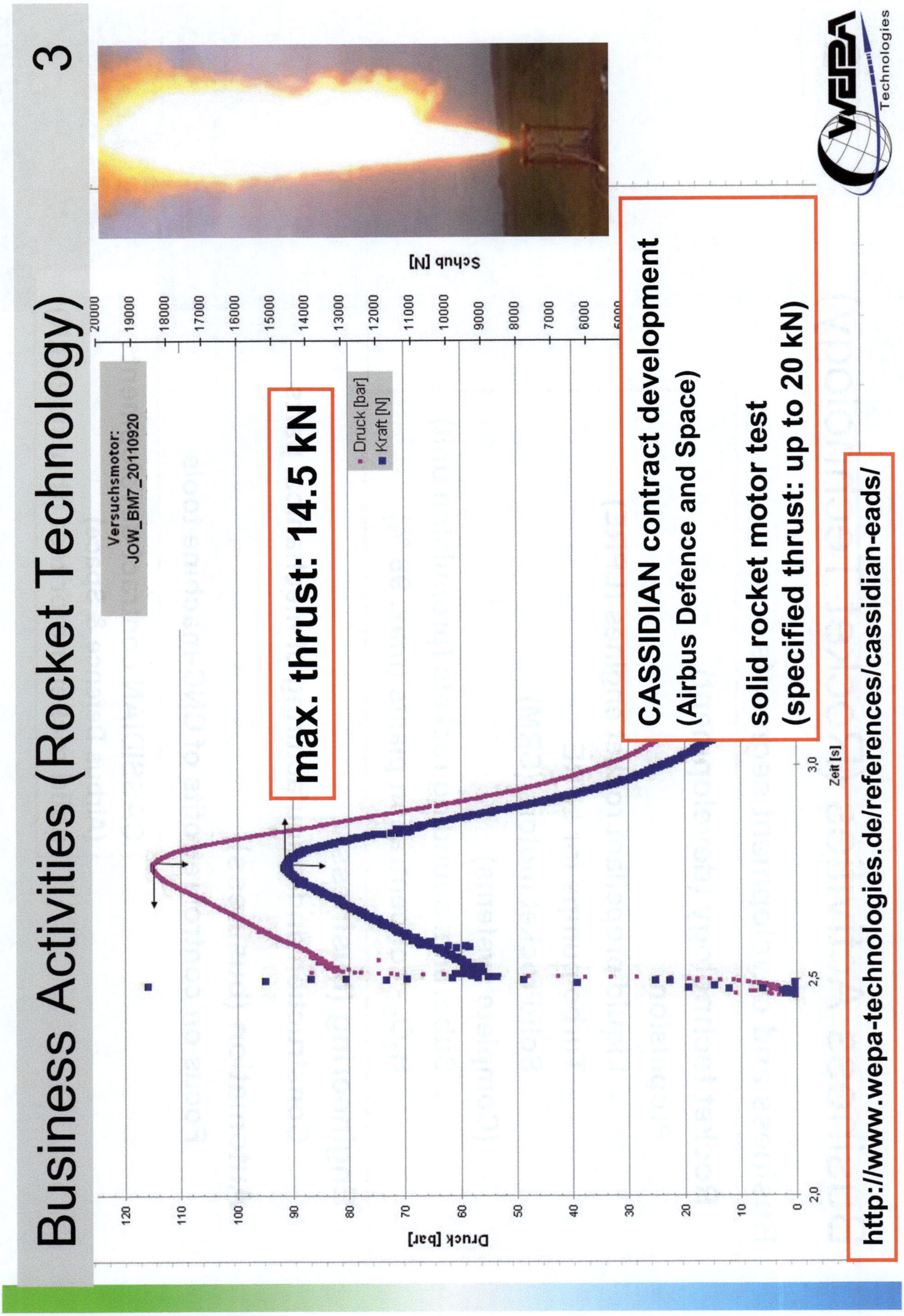
Business Activities (Rocket Technology)
3
Versuchsmotor:
JOW_BM7_20110920
max. thrust: 14.5 kN
Druck [bar]
Kraft [N]
Druck [bar]
Schub [N]
Zeit [s]
CASSIDIAN contract development
(Airbus Defence and Space)
solid rocket motor test
(specified thrust: up to 20 kN)
http://www.wepa-technologies.de/references/cassidian-eads/
WEPA Technologies

General Development Strategy: Rocket Technology

Development of Low Cost Propulsion Systems for Launch- and In Space Applications
Dr.-Ing. P. Weuta, Dipl.-Ing. N. Jaschinski
Seite 8; 2015-11-04

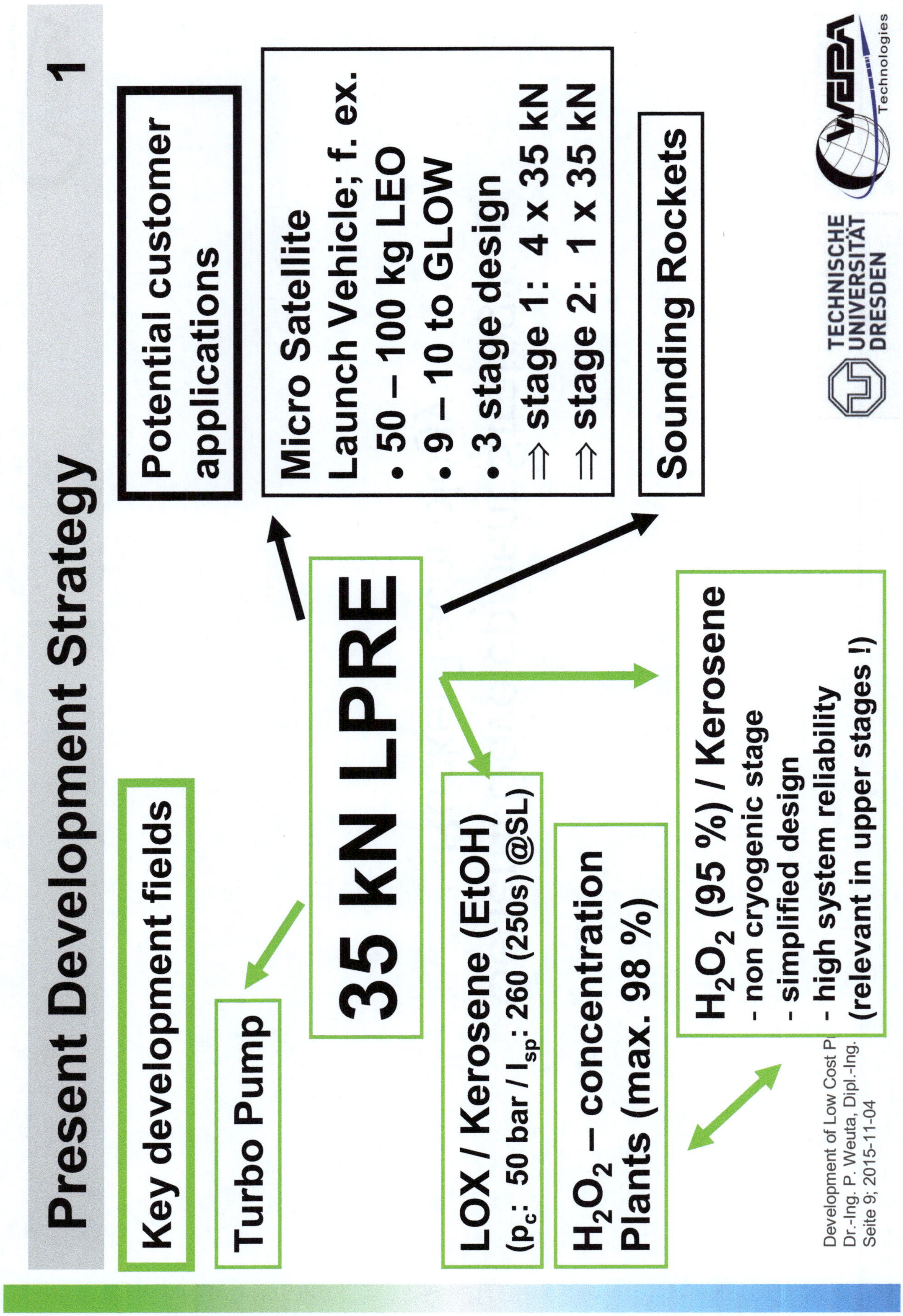
Present Development Strategy
1
Key development fields
Turbo Pump
35 kN LPRE
LOX / Kerosene (EtOH)
(p_c: 50 bar / I_{sp}: 260 (250s) @SL)
H_2O_2 – concentration Plants (max. 98 %)
H_2O_2 (95 %) / Kerosene
- non cryogenic stage
- simplified design
- high system reliability
(relevant in upper stages !)
Potential customer applications
Micro Satellite
Launch Vehicle; f. ex.
• 50 – 100 kg LEO
• 9 – 10 to GLOW
• 3 stage design
⇒ stage 1: 4 x 35 kN
⇒ stage 2: 1 x 35 kN
Sounding Rockets
Development of Low Cost P
Dr.-Ing. P. Weuta, Dipl.-Ing.
Seite 9; 2015-11-04
TECHNISCHE UNIVERSITÄT DRESDEN
Technologies

Present Development Strategy 2

Low cost propulsion system considered (one) key component to realize low cost launch- and in space applications !

How to achieve low cost propulsion ?

- Simplified design of rocket engines and turbo pumps
- Low-level operational parameter (chamber pressure, temperature)
- Use of low cost materials and manufacturing technologies
- Unification of propulsion system design for first and second stages via clustering (identical propellants assumed !)
- Prefer numbering-up instead of scale-up
- Environmentally benign and easy to handle propellant components (LOX resp. H_2O_2, EtOH, Kerosene, LCH_4)
  - Avoid NO_2 / N_2O_4 and hydrazine !

Development of Low Cost Propulsion Systems for Launch- and In Space Applications
Dr.-Ing. P. Weuta, Dipl.-Ing. N. Jaschinski
Seite 10; 2015-11-04

TECHNISCHE UNIVERSITÄT DRESDEN

WEPA Technologies

Development of Liquid Propellant Engines

Development of Low Cost Propulsion Systems for Launch- and In Space Applications
Dr.-Ing. P. Weuta, Dipl.-Ing. N. Jaschinski
Seite 11; 2015-11-04

TECHNISCHE UNIVERSITÄT DRESDEN

WEPA Technologies

Development of LPRE 1

Overview

- **Goal: construction of low cost engines**

  => Significant reduction of development and production costs required

- **Approach: improve designs based on proven technologies (USA / USSR / Europe 1960 – 1980)**

- **Use of 'green propellants'**
  - **Oxidizers: LOX / H_2O_2**
  - **Fuels: EtOH / Kerosene / LCH_4 / (LH_2)**

  => No significant environmental issues (test & launch area)

- **Present development: 35 kN technology demonstrator**
  - Chamber pressure: 5 MPa
  - Exit pressure: 0,5 MPa
  - Regenerative cooling

- **Increase to 100 – 200 kN thrust mid term goal** **(depending on market requirements)**

=> commercialisation of LPRE intended

Development of Low Cost Propulsion Systems for Launch- and In Space Applications
Dr.-Ing. P. Weuta, Dipl.-Ing. N. Jaschinski
Seite 12; 2015-11-04

TECHNISCHE UNIVERSITÄT DRESDEN

WEPA Technologies

Development of Liquid Propellant Engines 2

Current development (technology demonstrator)

- **Operating parameter**
  - Thrust
    - 35 kN
  - Chamber pressure:
    - 5 MPa
  - Nozzle exit pressure:
    - 0.05 MPa (=> 1. stage engine)
  - Propellants:
    - LOX / EtOH
    - H_2O_2 / EtOH
- Construction overview
  - Regenerative cooling: LOX / EtOH, H_2O_2 / EtOH
  - Thrust chamber: use series production enabling technologies (welding)
- Injector:
  - coaxial type (LOX and H_2O_2-based propellants)

Development of Low Cost Propulsion Systems for Launch- and In Space Applications
Dr.-Ing. P. Weuta, Dipl.-Ing. N. Jaschinski
Seite 13; 2015-11-04

TECHNISCHE UNIVERSITÄT DRESDEN

WEPA Technologies

Development of Liquid Propellant Engines 3

Current development (technology demonstrator): specifics of H_2O_2-based propulsion systems

- Use of highly stabilized H_2O_2 preferred
  - Significantly increased safety level during H_2O_2 production, storage and use
  - ⇒ Lower cost systems !
- Injector
  - Coaxial type
  - Catalytic pre-decomposition of H_2O_2 not mandatory
    - Direct liquid injection preferred
      - No catalyst required !
    - Hypergolic ignition by initial injection of organic or inorganic starters !
- Thrust chamber
  - Significantly lower combustion temperature compared to LOX systems
    - Facilitated reusability of propulsion systems ! (space planes / space tourism)
  - Regenerative cooling by H_2O_2

Development of Low Cost Propulsion Systems for Launch- and In Space Applications
Dr.-Ing. P. Weuta, Dipl.-Ing. N. Jaschinski
Seite 14; 2015-11-04

Development of Liquid Propellant Engines 3

Current development (technology demonstrator): specifics of H_2O_2-based propulsion systems

- Advantages especially within upper stages:
  - Storability / no evaporative losses during pre-operation time
  - Simplified, non cryogenic feed system (turbo pump resp. pressure feeding)
  - No chill down of system prior to ignition required
  - Reliable, “hypergolic” ignition process
  - Multiple burns possible (=> advantage to achieve precise orbital insertion !)
  - No safety / toxicity issues compared to N_2O_4 / UDMH (standard hypergol !)

=> Reduced system complexity / increased operational reliability !

Development of Low Cost Propulsion Systems for Launch- and In Space Applications
Dr.-Ing. P. Weuta, Dipl.-Ing. N. Jaschinski
Seite 15; 2015-11-04

Turbo Pump Units

Development of Low Cost Propulsion Systems for Launch- and In Space Applications
Dr.-Ing. P. Weuta, Dipl.-Ing. N. Jaschinski
Seite 16; 2015-11-04

TECHNISCHE UNIVERSITÄT DRESDEN

WEPA Technologies

Current Development: Turbo Pump Unit – overview 1

- Goal: minimize engineering, testing + manufacturing effort by low level operational parameter
  - exit pressure: max. 75 bar
  - max. 30,000 RPM; single shaft design
  - open gas generator cycle (LOX / EtOH or H_2O_2 + catalyst)

- Propellant systems: LOX / EtOH (H_2O_2 / EtOH)

- Mass flow rate: ~ 14 – 14.5 kg/s (35 kN engine)

- Weight: max.35 kg (incl. gas generator + control unit)

- Arrangement: turbine – EtOH – oxidizer

Development of Low Cost Propulsion Systems for Launch- and In Space Applications
Dr.-Ing. P. Weuta, Dipl.-Ing. N. Jaschinski
Seite 17; 2015-11-04

Current Development: Turbo Pump Unit – overview 2

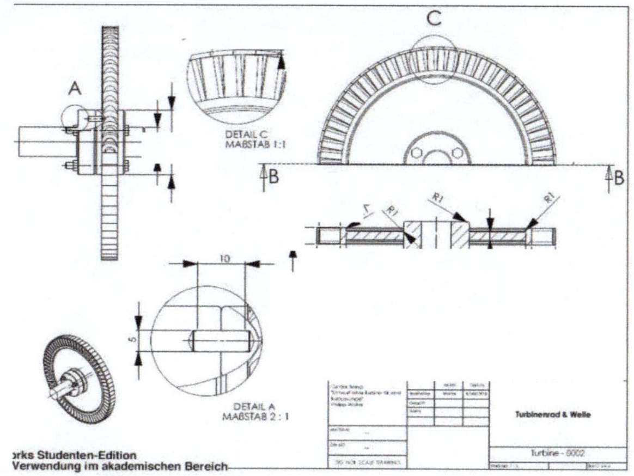

- Turbine
  - single or double axial stage, impulse type
  - partial admission of drive gas
  - inlet temperature: < 850 K
- Pump
  - single radial stage
- Seals
  - dynamic type
- Weight: max.35 kg (incl. gas generator + control unit)
- Arrangement: turbine – fuel – oxidizer

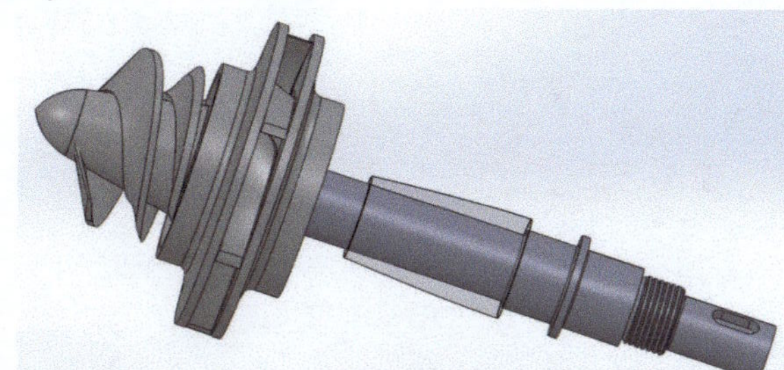

- <u>Status</u>
  - First hot firing test of LPRE and TPU to commence in Q3 2016

Development of Low Cost Propulsion Systems for Launch- and In Space Applications
Dr.-Ing. P. Weuta, Dipl.-Ing. N. Jaschinski
Seite 18; 2015-11-04

TECHNISCHE UNIVERSITÄT DRESDEN

WEPA Technologies

H_2O_2-Concentration Technology

Development of Low Cost Propulsion Systems for Launch- and In Space Applications
Dr.-Ing. P. Weuta, Dipl.-Ing. N. Jaschinski
Seite 19; 2015-11-04

WEPA Technologies

Supply of H_2O_2 (c > 88 - 98 %) 1

- **Motivation**
  - Due to non-cryogenic nature of H_2O_2 overall system architecture is significantly reduced (no isolation required, no formation of ice, less complicated TPU)
  - H_2O_2-based propulsion systems show very high operational reliability
  - Very high strength H_2O_2 required for high performance propulsion systems
  - Increase of H_2O_2 concentration (85 => 95 %): identical payload capacity compared to LOX (outer envelope kept constant !)

Development of Low Cost Propulsion Systems for Launch- and In Space Applications
Dr.-Ing. P. Weuta, Dipl.-Ing. N. Jaschinski
Seite 20; 2015-11-04

Supply of H_2O_2 (c > 88 - 98 %) 1

- **Commercial supply situation (present)**
  - Very limited availability at c > 88 %
  - Transport via public ground prohibited by law
    => on site production in specialized plants required !
  - Small production plants cannot be rented, only bought
    (> 1,8 Mio EUR, ~ 1 kg H_2O_2 / h)

=> not a very attractive situation for developing / using H_2O_2 - based propulsion processes….

Development of Low Cost Propulsion Systems for Launch- and In Space Applications
Dr.-Ing. P. Weuta, Dipl.-Ing. N. Jaschinski
Seite 21; 2015-11-04

WEPA Technologies

Supply of H_2O_2 (c > 88 - 98 %) 2

- **H_2O_2 concentration plant developed by WEPA-Technologies (current EU-customer)**
  - **Capacity: up to ~ 50 kg / d (90 %)**
  - **Feed: 50 % - 70 % H_2O_2**
  - **Fully automatic, 24 / 7 operability**
- **Working packages supplied by WEPA-Technologies**
  - **Conceptional process design incl. safety concept**
  - **Detail Engineering (process-, control- and electrical diagrams)**
  - **Equipment purchase**
  - **Erection and commissioning**
  - **Trouble shooting / maintenance**

Reference plant open to customer visits (commissioning complete: 10/2015)

- **Very safe production process up to 98 % concentration under development (10 kg / h): fully mobile set-up in 20 – 40 ft container**

Development of Low Cost Propulsion Systems for Launch- and In Space Applications
Dr.-Ing. P. Weuta, Dipl.-Ing. N. Jaschinski
Seite 22; 2015-11-04

Supply of H_2O_2 (90 %) : Reference Plant 1

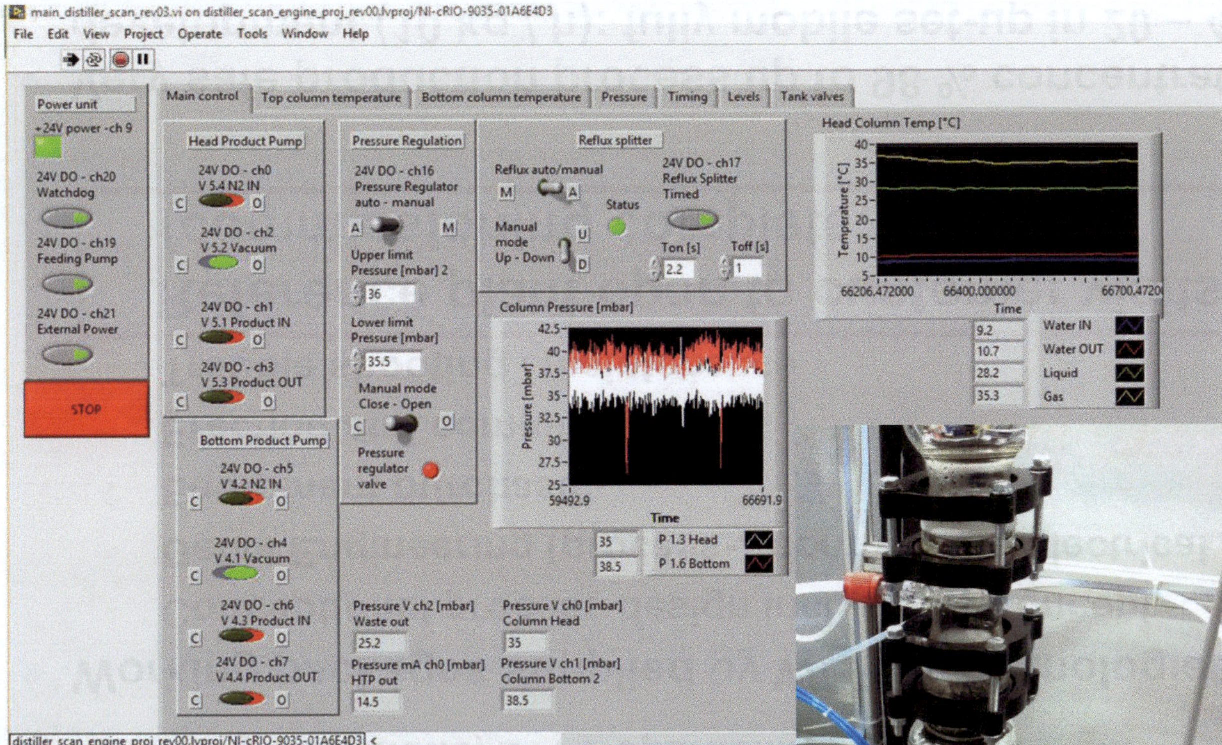

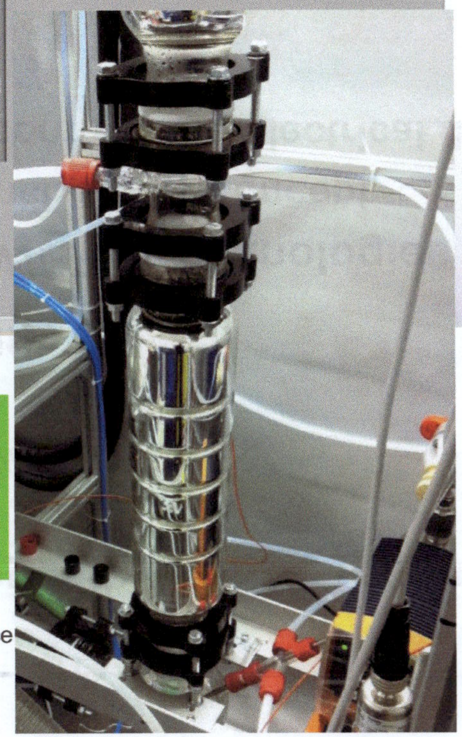

=> general commercialisation of H_2O_2 supply intended (90 – 98 %)
=> customer requests welcome !

Control by PLC:
LabVIEW RT
(alternative: TWINCAT)

Development of Low Cost Propulsion Systems for Launch- and In Space
Dr.-Ing. P. Weuta, Dipl.-Ing. N. Jaschinski
Seite 23; 2015-11-04

WEPA Technologies

Summary

Development of Low Cost Propulsion Systems for Launch- and In Space Applications
Dr.-Ing. P. Weuta, Dipl.-Ing. N. Jaschinski
Seite 24; 2015-11-04

TECHNISCHE UNIVERSITÄT DRESDEN

WEPA Technologies

Summary - Present development at WEPA-Technologies

- **Liquid propellant rocket engines incl. turbo pump units**
  - **Present: 35 kN LPRE technology demonstrator (LOX/EtOH; H_2O_2 / EtOH + TPU)**
    - **Use of LOX / Kerosene and LCH_4 projected !**
  - **Low cost focus**
  - **Possible application: booster for micro satellite launch vehicles or sounding rockets**
  - **First hot firing test of LPRE and TPU to commence in late Summer 2016**
- **H_2O_2**
  - **H_2O_2 significantly facilitates development and reliable operation of propulsion systems – however: very difficult supply situation at concentration > 88 % (sometimes even valid below 88 % !)**
  - **WEPA solution: development of mobile concentration unit**
    - **Key features:**
      - **Safe and fully automatic, 24 / 7 operability**
    - **Production technology of 90 % H_2O_2 fully developed (reference plant available : ~ 50 kg / day capacity)**
    - **Process to yield 98% H_2O_2 available by late 2016**

=> customer requests welcome !

WEPA Technologies

Thank you for your attention !

Pico- and Nano-Satellite Conference (09/2015)

Dr.- Ing. P. Weuta

Dipl.- Ing. N. Jaschinski

Develop
Dr.-Ing.
Seite 2

WEPA
Technologies

peter.weuta@wepa-technologies.de
www.wepa-technologies.de

Development of Low Cost Propulsion Systems for Launch- and In Space Applications
Dr.-Ing. P. Weuta, Dipl.-Ing. N. Jaschinski
Seite 27; 2015-11-04

Back-Up

Development of Low Cost Propulsion Systems for Launch- and In Space Applications
Dr.-Ing. P. Weuta, Dipl.-Ing. N. Jaschinski
Seite 28; 2015-11-04

TECHNISCHE UNIVERSITÄT DRESDEN

Application of Propulsion Technology: Conceptional Design of Sounding Rocket „SILBERPFEIL“ („Silver Arrow“)

Development of Low Cost Propulsion Systems for Launch- and In Space Applications
Dr.-Ing. P. Weuta, Dipl.-Ing. N. Jaschinski
Seite 29; 2015-11-04

TECHNISCHE UNIVERSITÄT DRESDEN

Technologies

Central Design Decision:
Liquid- or Solid Propellant Rocket Engines ? 1

By far most sounding rockets use Solid Rocket Motor propulsion systems

- Surplus military motors
  - ready availability not always given
- Very high acceleration of vehicle
  - significant stress on payload
- Thrust / time profile and total impulse cannot be modified
- Safety and cost issues using solid propellants
  - regulations for "explosives" becoming even more stringent:
    - transport
    - storage
    - handling / on site integration

Conclusion: Solid Rocket Motors show significant disadvantages for frequent low cost launches !

Development of Low Cost Propulsion Systems for Launch- and In Space Applications
Dr.-Ing. P. Weuta, Dipl.-Ing. N. Jaschinski
Seite 30; 2015-11-04

TECHNISCHE UNIVERSITÄT DRESDEN

WEPA Technologies

Central Design Decision: Liquid- or Solid Propellant Rocket Engines ?

2

Advantages of Liquid Propellant Rocket Engines

- Completely safe handling of rocket during payload integration, handling and transport (=> fuel tanks empty)
  - no stringent safety regulations to be followed
- Low peak acceleration possible to realize
  - low stress on payload
- Launch readiness can be kept up for many weeks: responsive, very low lead time launch possible (while using storable, H_2O_2 oxidizer)
- Environmentally friendly ("green") propellants (while using H_2O_2 or O_2 oxidizer and Kerosene fuel)

Conclusion:

Liquid propellant rocket engines show significant advantages for frequent launches ...but have to be made low-priced !

Development of Low Cost Propulsion Systems for Launch- and In Space Applications
Dr.-Ing. P. Weuta, Dipl.-Ing. N. Jaschinski
Seite 31; 2015-11-04

TECHNISCHE UNIVERSITÄT DRESDEN

Technologies

Central Design Goal: Low Cost !

Low cost characteristics of sounding rockets can be achieved by multiple, parallel approaches (focus: propulsion system):

- Significantly reduced safety regulations due to avoidance of explosives (solid propellants)
- Simplified design of propulsion system (rocket engine and turbo pumps)
- Low level operational parameters (chamber pressure)
- Environmentally benign and easy to handle propellant components (H_2O_2 / Kerosene)
- Simple tank structures / no thermal isolation; common bulkhead
- Low-cost materials and manufacturing technologies
  - avoid typical aerospace grade materials and manufacturing processes
- Simple guidance systems / thrust vector control for ballistic flight required
- Goal: 1900 – 3800 EUR / kg @ 400 kg (300 km) payload (0,75 – 1,5 Mio EUR)
  - Depending on flight rate and depreciation of development costs
  - Ground support not included

TECHNISCHE UNIVERSITÄT DRESDEN

WEPA Technologies

Preliminary Design of Sounding Rocket: Definition of Payload Section

- Payload section is very specific to mission requirements
  - Can be adapted to customers needs: length, diameter, total mass
- Choose representative (commercial) payload size: TEXUS module (DLR, ~ 400 kg)
  - Advantages: qualified equipment could be re-used (data acquisition + downlink, power supply, telemetry, recovery systems…)
- Use 35 kN technology demonstrator engine
  - Thrust / time profile could be adapted to mission's needs

Development of Low Cost Propulsion Systems for Launch- and In Space Applications
Dr.-Ing. P. Weuta, Dipl.-Ing. N. Jaschinski
Seite 33; 2015-11-04

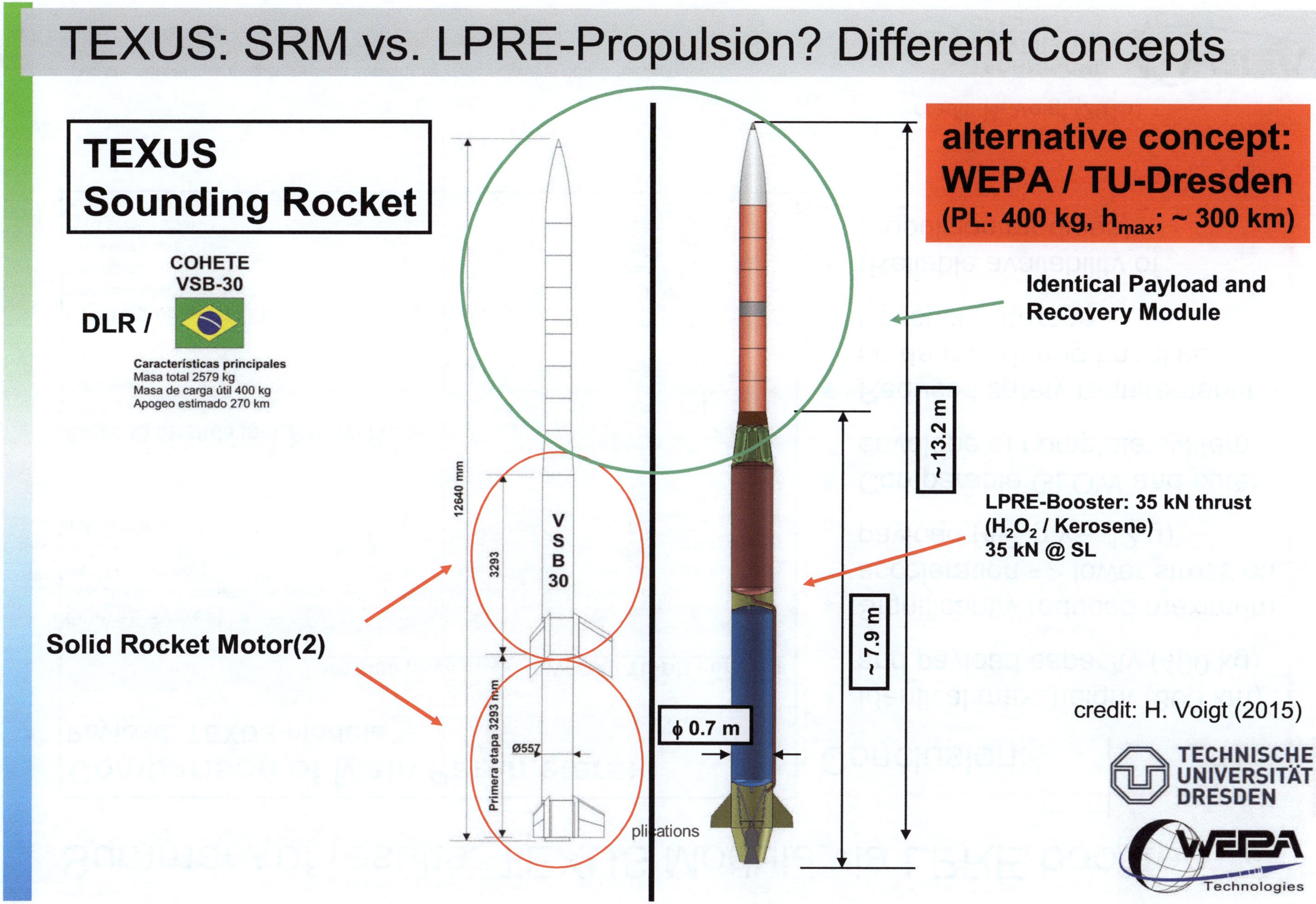
TEXUS: SRM vs. LPRE-Propulsion? Different Concepts
TEXUS
Sounding Rocket
DLR /
COHETE
VSB-30
Características principales
Masa total 2579 kg
Masa de carga útil 400 kg
Apogeo estimado 270 km
Solid Rocket Motor(2)
12640 mm
3293
V S B 30
Primera etapa 3293 mm
Ø557
alternative concept:
WEPA / TU-Dresden
(PL: 400 kg, h_{max}: ~ 300 km)
Identical Payload and
Recovery Module
~ 13.2 m
~ 7.9 m
LPRE-Booster: 35 kN thrust
(H_2O_2 / Kerosene)
35 kN @ SL
ϕ 0.7 m
credit: H. Voigt (2015)
TECHNISCHE
UNIVERSITÄT
DRESDEN
WEPA
Technologies

Summary of results: TEXUS Module via LPRE booster

Comparison of Main Parameters:

Payload: TEXUS-module

	VSB-30 base case	WEPA / TU-DD concept	
Overall System			
payload / communication / recovery	DLR-Standard	DLR-Standard	
max. height of flight	~ 270	~ 300	[km]
GLOW	2600	2790	[kg]
max. diameter	0,57	0,7	[m]
total length	12,6	13,2	[m]
Payload Module incl. Recovery			
length	4,5 - 5,5	4,5 - 5,5	[m]
diameter	0,44	0,44	[m]
mass	max. 400	max. 400	[kg]
Propulsion System			
number of stages	2	1	[-]
propellants	solid	liquid	[-]
propallant mass	~ 1575	2050	[kg]
max. accelaration	~ 12	4,65	[g]
burn time	31 (11 + 20)	125	[s]

Conclusion:

- Identical max. height (300 km) and payload capacity (400 kg)
- Significantly reduced maximum acceleration => lower stress on payload (4.7 g vs. 12 g)
- Comparable GLOW and outer envelope of complete system
- Reduced safety requirements: no danger during handling, transport, storage
- (Reliable availability of propulsion modules)

credit: H. Voigt (2015)

Development of Low Cost Propulsion Systems for Launch- and In Space Applications
Dr.-Ing. P. Weuta, Dipl.-Ing. N. Jaschinski
Seite 35; 2015-11-04

TECHNISCHE UNIVERSITÄT DRESDEN

WEPA Technologies

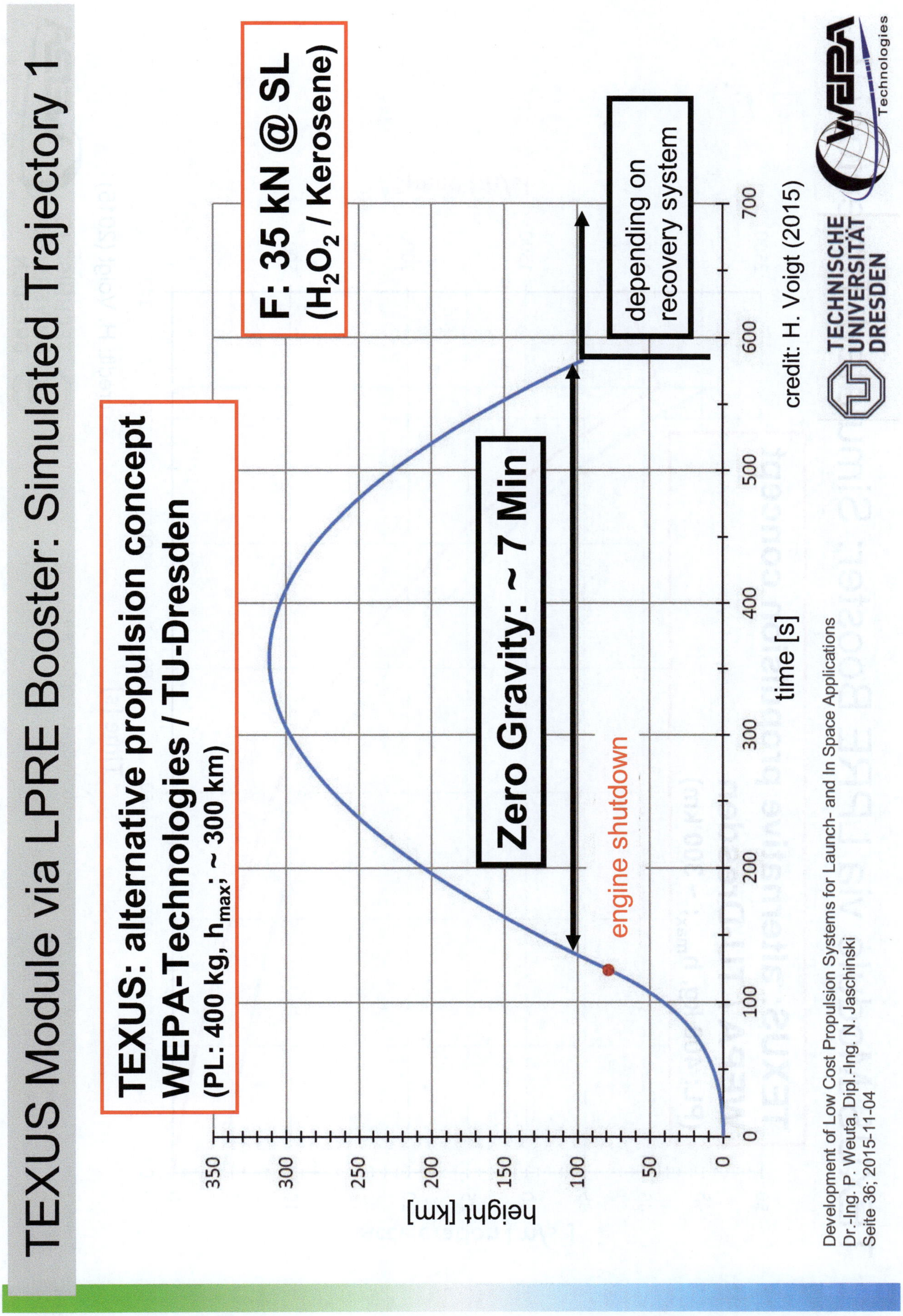

TEXUS Module via LPRE Booster: Simulated Trajectory 1
TEXUS: alternative propulsion concept
WEPA-Technologies / TU-Dresden
(PL: 400 kg, h_{max}: ~ 300 km)
F: 35 kN @ SL
(H_2O_2 / Kerosene)
Zero Gravity: ~ 7 Min
engine shutdown
depending on recovery system
height [km]
0
50
100
150
200
250
300
350
time [s]
0
100
200
300
400
500
600
700
credit: H. Voigt (2015)
TECHNISCHE UNIVERSITÄT DRESDEN
Technologies
Development of Low Cost Propulsion Systems for Launch- and In Space Applications
Dr.-Ing. P. Weuta, Dipl.-Ing. N. Jaschinski
Seite 36; 2015-11-04

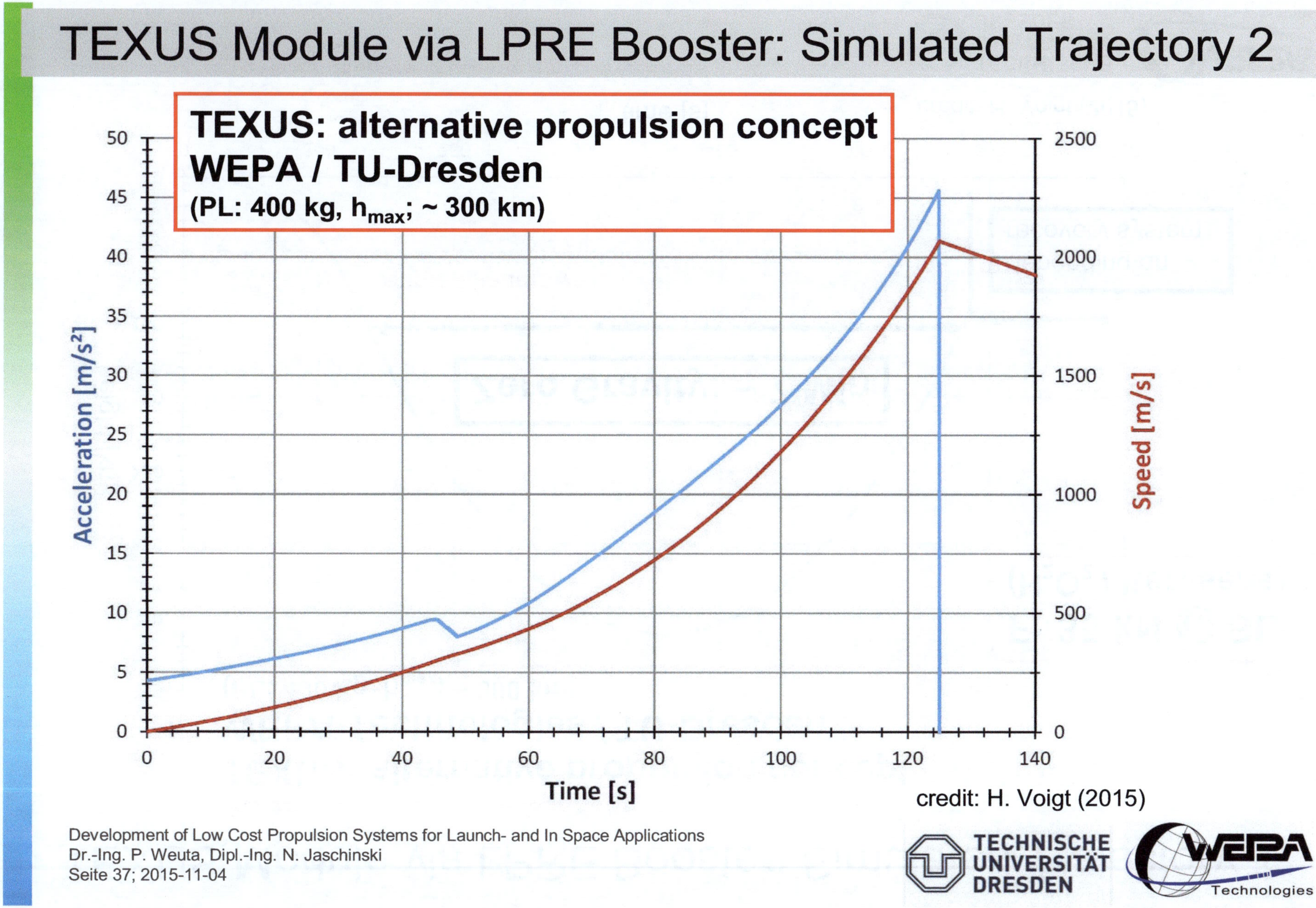
TEXUS Module via LPRE Booster: Simulated Trajectory 2
TEXUS: alternative propulsion concept
WEPA / TU-Dresden
(PL: 400 kg, h_{max}; ~ 300 km)
Acceleration [m/s²]
Speed [m/s]
Time [s]
0
5
10
15
20
25
30
35
40
45
50
0
500
1000
1500
2000
2500
0
20
40
60
80
100
120
140
credit: H. Voigt (2015)
Development of Low Cost Propulsion Systems for Launch- and In Space Applications
Dr.-Ing. P. Weuta, Dipl.-Ing. N. Jaschinski
Seite 37; 2015-11-04
TECHNISCHE UNIVERSITÄT DRESDEN
Technologies

Summary

- **Basic design parameter of a LPRE-propelled sounding rocket ("SILBERPFEIL") were described**
  - **Due to non-cryogenic nature of H_2O_2 overall system architecture is significantly reduced**
- **Possible applications of SILBERPFEIL**
  - **Zero-g experiments**
  - **Re-entry research**
- **TEXUS payload module (400 kg) has been chosen for reference**
  - **300 km height / ~ 7 min zero-g time**
  - **Other geometries / masses of payload section can be considered**

- **WEPA-Technologies is developing key propulsion-technologies (LPRE resp. turbo pumps) and H_2O_2 - concentration plants independent of the realization of sounding rocket projects**
- **To initiate development of the payload section and complete sounding rocket WEPA-Technologies is open to cooperations**

Development of Low Cost Propulsion Systems for Launch- and In Space Applications
Dr.-Ing. P. Weuta, Dipl.-Ing. N. Jaschinski
Seite 38; 2015-11-04

Reinventing Space Conference
BIS-RS-2015-29

Electromagnetic Launch to Space

Ian R. McNab
Emergent Power Solutions LLC, Austin, Texas.

Timothy R. Wolfe
L3-Com Applied Technologies, San Diego, California.

13th Reinventing Space Conference
9-12 November 2015
Oxford, UK

S. Hatton (ed.), *Proceedings of the 13th Reinventing Space Conference*,
https://doi.org/10.1007/978-3-319-32817-1_26

Electromagnetic Launch to Space

Ian R. McNab[1] and Timothy R. Wolfe[2]

[1]Emergent Power Solutions LLC, Austin, Texas. [2]L3-Com Applied Technologies, San Diego, California

ABSTRACT

A study was undertaken to determine if a ground-based electromagnetic (EM) acceleration system could provide a useful reduction in launch-to-orbit costs compared with current large chemical boosters, while increasing launch safety and reliability. The study evaluated the launch of a two-stage-to-low-Earth-orbit projectile, with the initial velocity being provided electromagnetically and the orbit insertion via a rocket motor. Several electromagnetic accelerator options are available but railguns were chosen for this study based on their demonstrated performance capabilities. The second stage of the system was assumed to be a chemical rocket that would carry a payload into low-Earth orbit.

Electromagnetic launch systems of this type will be governed by the same fundamental principles as tactical railguns with a major difference being that the EM accelerator track—which may be tens or hundreds of meters in length—cannot be powered only from the "breech" as in a tactical railgun, since electrical resistive losses will become unacceptably large. To overcome this, a distributed feed system will be required.

This study shows that the capital cost of the pulsed power system for the EM accelerator will dominate the system economics. With present pulsed power approaches, multiple launches will be required to offset the capital cost and provide low costs. The development of novel pulsed power concepts and/or low-cost manufacturing approaches will ensure that the EM system will be economically attractive and options for such approaches are discussed.

BACKGROUND

Figure 1 illustrates the basic concept of a railgun launcher in which a pulse of high current applied to the breech of the railgun causes an EM force to be generated that accelerates the launch package along the "barrel." For this simple configuration, high currents in the range of mega-amperes (MA) are required to launch masses of tens of kilograms. Unlike conventional powder propellant guns, the current pulse can be tailored to provide a relatively constant acceleration throughout the launch process. Taken with the absence of gas expansion limits, this allows hypervelocities (> 2000 m/s) to be achieved. For most tactical applications it is desirable to keep the barrel length to about ten meters or less, so that a hypervelocity launch necessarily requires the launch package to withstand very high accelerations. Typical parameters for a tactical railgun system are shown in Table 1.

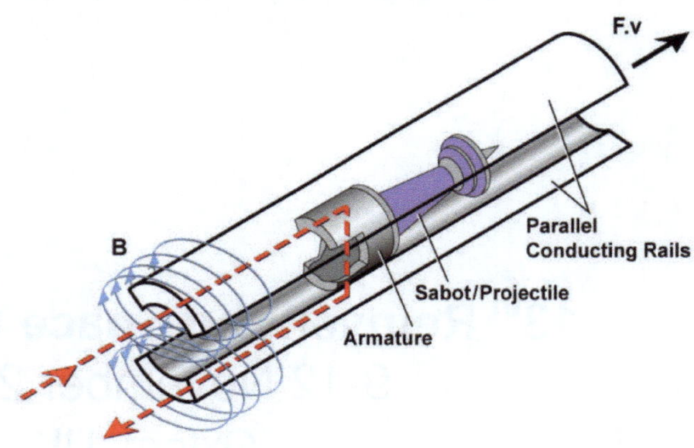

Figure 1. Basic railgun concept.

Table 1. Typical railgun parameters.

Parameter	Units	Value
Muzzle velocity	m/s	2000
Launch acceleration	kGees	30
Barrel length	m	12
Average current	MA	4.5
Launch mass	kg	16
Muzzle energy	MJ	32
Stored energy	MJ	105

Following earlier studies [1], the US Navy's Office of Naval Research (ONR) has publically stated that it is developing high energy tactical railguns for ship installation and deployment [2-6]. A large laboratory system is shown in Figure 2 and Figure 3 and an innovative naval prototype railgun undergoing testing is shown in Figure 4 and emplaced on a ship in Figure 5. The type of hypervelocity guided projectile planned for launch in this system would be similar to that shown in shown in Figure 6. This is essentially a small angle conical aerobody with fin stabilization.

Figure 2. Navy laboratory railgun [7].

Figure 3. Power feed to Navy laboratory railgun [7].

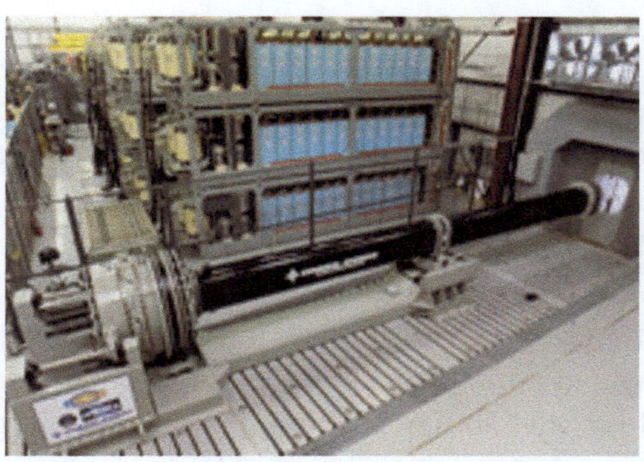

Figure 4. Laboratory tests of Navy railgun showing capacitor power supplies [6].

Figure 5. Railgun emplaced on ship deck [8].

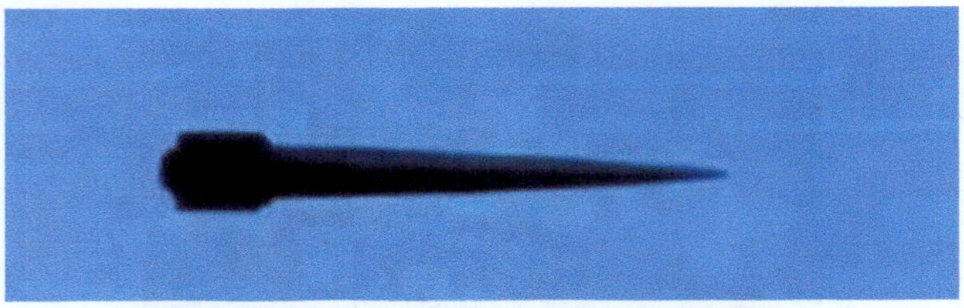

Figure 6. Hypervelocity projectile [9].

Although this Navy system is still under development, much of the technology demonstrated in the laboratory could serve as the basis for the launch to space system described here.

As mentioned, unlike conventional powder propellant guns, railguns can achieve very high muzzle velocities using controlled pulses of electric current. Tactical guns that operate at sea level will be designed for velocities of 2 to 3 km/s but laboratory experiments have operated to over 5 km/s and small experiments to over 6 km/s. Figure 7 shows a 4.4 km/s non-aerodynamic 0.6 kg projectile shortly after launch from a railgun while the 20 m long distributed railgun barrel shown in Figure 8 achieved 5.5 km/s with a 0.1 kg projectile.

Figure 7. 4.4 km/s railgun experiment.

LAUNCH TO SPACE CONCEPT

Earlier EM launch to space studies have evaluated a range of different concepts, from 1000 kg flight bodies [10], airborne launch [11-20], to augmentation of large two-stage chemical boosters [21]. Studies have been done by several groups, including the German Aerospace Research Center, e.g. [22].

This preliminary study has built on these earlier studies and recent Navy developments to evaluate whether railguns could launch small payloads into low Earth orbit (LEO). The EM launcher required for this application will be similar to tactical railguns but the EM accelerator track may need to be tens to hundreds of meters long to reduce acceleration loads on the payload during launch. For this reason it cannot be powered from the breech as in a tactical railgun, since electrical resistive losses become unacceptably large. A distributed power fed system similar to – but longer than – that shown in Figure 8 will be required.

Figure 8. 20-meter long distributed railgun.

The launch package will consist of two parts – an aerobody similar to Figure 6 to withstand the aerothermal heating after launch and transit through the densest portion of the atmosphere and a lighter "second stage" containing the rocket motor and fuel required for orbit insertion and circularization, together with the payload. Depending on the launch velocity and angle, an apogee close to LEO can be achieved. At apogee, the aerobody will separate from the second stage containing the rocket motor, fuel and payload. This rocket-powered section is required to circularize the orbit from the horizontal component of the launch velocity. Some parameter examples are given in Table 2.

Table 2. Notional launch to space railgun parameters.

Parameter	Units	Value
Muzzle velocity	m/s	5500
Launch acceleration	kGees	10
Barrel length	m	154
Average current	MA	3.2
Launch mass	kg	25.4
Muzzle energy	MJ	384
Stored energy	MJ	512

Comparison with Table 1 shows similarities and differences. The launch to space system clearly requires a much higher muzzle velocity and, with an assumed lower payload acceleration limit, the "barrel" length is much longer. However, the average current is less than that considered for a tactical railgun and the launch mass is comparable (about 50% larger). The muzzle energy is ten times larger than the tactical railgun but the use of a more efficient distributed power feed system means that the required stored energy is only about five times larger.

Given the different mission requirements, a much lower firing rate has been assumed for the launch to space system than would be necessary for a tactical system. This impacts the rating of the power supply needed to recharge the energy storage system between launches. Assuming five launches per hour and that half of the time between launches is available for recharging the energy storage system (i.e., about 6 minutes), a diesel generator rated at about 1500 kWe could provide the required power.

To reduce aerothermal heating on the projectile nosetip, the launcher should be located at the highest altitude possible and near the Equator to benefit from the Earth's rotational contribution (Figure 9).

The notional design of the second stage was based on a launch angle of 40 degrees, yielding an apogee (without allowance for drag) of about 640 km and a horizontal velocity component of 4200 m/s. To achieve the total estimated delta-V of 9447 m/s (orbital velocity 7558 m/s plus an estimated aero and drag loss of 25%) a further ΔV = 4890 m/s will be required. To achieve this a fuel I_{sp} = 300 sec was assumed, yielding a fuel mass of 14.9 kg which, with an aerobody mass of 7 kg, a second stage structure mass of 10% and a payload of 2 kg yielded a total launch mass of 25.4 kg.

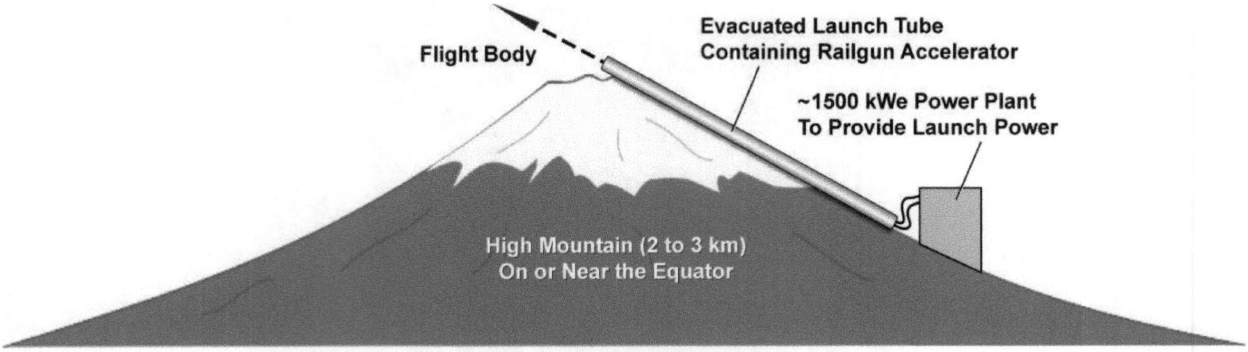

Figure 9. Schematic concept layout.

LAUNCH ECONOMICS

An initial estimate of the launch economics can be obtained by estimating the cost of the energy storage and launcher: range operating costs have not been estimated here but should be less than for chemical boosters given the likely improved reliability and safety.

The system investment cost will be dominated by the capital cost of the pulsed energy storage system. With larger-scale manufacturing and "lessons-learned" from the US Navy developments, it is conservatively estimated that the cost of the pulsed power can be reduced to \$0.25 per Joule. Together with an estimated launcher cost of \$100K/m, an approximate total EM system cost neglecting site preparation would be about \$144M.

The purpose of developing a system of this kind is to create a different paradigm for the launch of nanosatellites than exists at present with large chemical boosters. Thus, the facility would be expected to operate frequently and on a daily basis to place many nanosatellites into LEO. Over a five year period operating with 5 shots per hour for 8 hours per day and five days per week, about 50,000 launches could be achieved even including 10% down time for maintenance and repair. With a 2 kg payload per launch this would be 100 tonnes placed into LEO at an average cost of \$1440/kg. Future reductions in the cost of pulsed power can be expected to further reduce this cost.

CONCLUSIONS

This preliminary study shows that even with present pulsed power approaches, attractive low launch costs appear possible for an operating scenario based on the multiple launch capability for a system of this type. The capital cost of present-generation pulsed power technology dominates the capital cost but improved pulsed power concepts and/or low-cost manufacturing are expected to ensure that future EM systems are economically viable.

Many technical challenges remain to be overcome before a system of this type can be built with confidence. Nevertheless, this preliminary study indicates that further more detailed assessments should be worthwhile and could lead to a new, different and low cost launch system for the future.

ACKNOWLEDGEMENTS

The support provided by L3Com Applied Technologies Division in this study is gratefully acknowledged.

REFERENCES

1. I.R. McNab, S. Fish and F. Stefani. "Parameters for an Electromagnetic Naval Railgun." IEEE Trans. Magnetics. 37(1), 223-228, January 2001.
2. "Navy sets new world record with latest EM railgun demonstration." ONR Media Release, December 10, 2010.
3. "Navy researchers fire 1000th shot in laboratory railgun." ONR Media Release, October 31, 2011.
4. "Navy to begin tests on EM railgun prototype launcher." ONR Media Release, February 6, 2012.
5. "With a Bang Navy begins tests on EM railgun prototype launcher." ONR Media Release, February 28, 2012.
6. "Navy evaluating Second EM railgun Innovative Naval Prototype." ONR Media Release, October 9, 2012.
7. ONR Facebook information page, December 10, 2012.

8. ONR Facebook information page, July 14, 2014.
9. GA Blitzer projectile.
10. I.R. McNab. "Launch to Space with an Electromagnetic Railgun." IEEE Trans. Magnetics. 39(1), 255-304, January 2003.
11. I.R. McNab. "A Research Program to study Airborne Launch to Space." IEEE Trans. Magnetics. 43(1), 486-490, January 2007.
12. F. Stefani, I.R. McNab, J.V. Parker, M. Alonso and T. Klatt. "A Plasma Railgun Experiment addressing Launch to Space Issues." IEEE Trans. Magnetics. 43(1), 194-197, January 2007.
13. I.R. McNab. Electromagnetic Launch to Space." JBIS, Vol. 60. pp.54-62. 2007.
14. R. Gosse and G. Candler. "CFD Study of an Electromagnetic Launched projectile for Access to Space." 26th AIAA Applied Aerodynamics Conference. Paper AIAA 2008-6741. 18-21 August 2008, Honolulu, Hawaii, USA.
15. R. Gosse and G. Candler. "Evaluation of Carbon-Carbon Ablation Models using a Fully-Coupled CFD Server." 40th AIAA Thermophysics Conference. Paper AIAA 2008-3908. 23-26 June 2008, Seattle, Washington, USA.
16. M.J. Guillot and I.R. McNab. "Aerothermal Analysis of Small Payloads delivered into Low Earth Orbit from an Airborne Launch Platform." AIAA Summer Heat Transfer Conference, Paper HT2008-56438. August 10-14, 2008. Jacksonville, Florida, USA.
17. D.A. Wetz, F. Stefani, J.V. Parker and I.R. McNab. "Advancements in the Development of a Plasma-Driven Electromagnetic Launcher." IEEE Trans. Magnetics. 45(1), 495-500, January 2009.
18. I.R. McNab. "Progress on Hypervelocity Railgun Research for Launch to Space." IEEE Trans. Magnetics. 45(1), 381-388, January 2009.
19. D.A. Wetz, F. Stefani and I.R. McNab. "Experimental results on a 7-m long Plasma Driven Electromagnetic launcher." IEEE Trans. Plasma Science. 39(1), 180-185, January 2011.
20. M.J. Guillot and I.R. McNab. "Analysis of Miniature Payloads Launched from an Airborne Gun Launch Platform." J. Spacecraft and Rockets, Vol. 50(1), Jan-Feb 2013.
21. I.R. McNab. "Electromagnetic Augmentation can reduce Space Launch Costs." IEEE Trans. Plasma Science. 41(1), 1047-1054, May 2013.
22. O. Bozic. "Possibilities of Lorentz Electromagnetic Accelerators for Launch of Payloads in Low Earth Orbit." 59th International Astronautical Congress. Paper IAC-08-C.4.6.6. 29 September – 3 October 2008. Glasgow, Scotland.

Bristol Spaceplanes Limited

SPEARHEADING THE SPACEFLIGHT REVOLUTION

The prospect for orbital airliners

David Ashford

Spacebus - Designed to be the first orbital airliner

Airline service to orbit (passengers and cargo)
· Possible within 15 years
· Cost per seat to orbit 1000 times less than today
England - Australia in 75 minutes flying time

	Current Launch Vehicles (Derived from Ballistic Missiles, One Flight Only)	Airliner

Typical Cost per seat, £	20 million	500 (Long distance flight)

Saunders Roe SR.53 rocket fighter

- The Saunders Roe SR.53 rocket fighter first flew in 1957
- Both it and Black Knight used rocket engines with hydrogen peroxide (HTP) and kerosene as propellants
- This technology has never been bettered
- Saunders Roe proposed a viable suborbital space research conversion in 1958

Flight Paths (Trajectories)

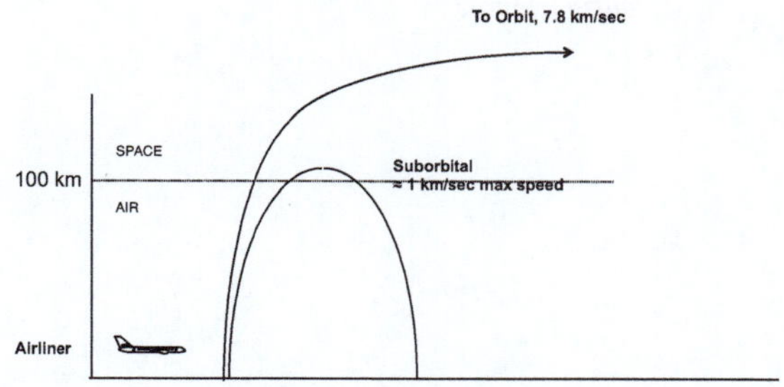

Orbital launch needs about eight times more speed than suborbital

Suborbital used for sounding rockets

Market for orbital launch far greater than for suborbital

Airliner 'Conversion'

- Change shape (for stability over wide speed range)
- Add rocket engines
- Higher propellant mass fraction
- Two stages (to enable use of existing technology
- Hydrogen fuel
- Additional systems (reaction controls, thermal protection)
- Extra attention to safety of rocket propulsion system

- ALL of these technologies have flown on other projects

EUROPEAN SPACEPLANE PROJECTS OF THE 1960s

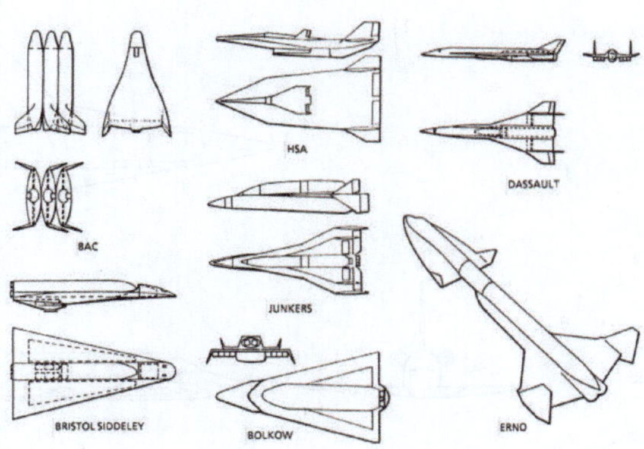

S. Hatton (ed.), *Proceedings of the 13th Reinventing Space Conference*,
https://doi.org/10.1007/978-3-319-32817-1_27

Leading Design Features of the First Orbital Spaceplane

- Design for 'safety soon'
  - One to two tonne payload
  - Piloted
  - Winged
  - (Horizontal take-off and landing)
  - Two stages
  - Separation at high supersonic speed
- **Most of the European 1960s spaceplane designs had these features!**

Spacebus does not need new technology

- Orbital spaceplanes were widely studied in the 1960s (Boeing, British Aircraft Corporation, Dassault, Douglas, Junkers, Lockheed, Martin, et al)
- The consensus was that they were the obvious next step and feasible with the technology of the time
- Not developed because the Cold War race to the Moon soaked up the available funding
- Spacebus uses technology comparable to that of the early designs of the Space Shuttle, which were fully reusable. A budget cut led to the eventual largely expendable design
- Prototypes could be built in about seven years
- A further eight or so years of operational service and product improvement are needed to achieve long life and rapid turnaround time

Spaceplane Development Strategy

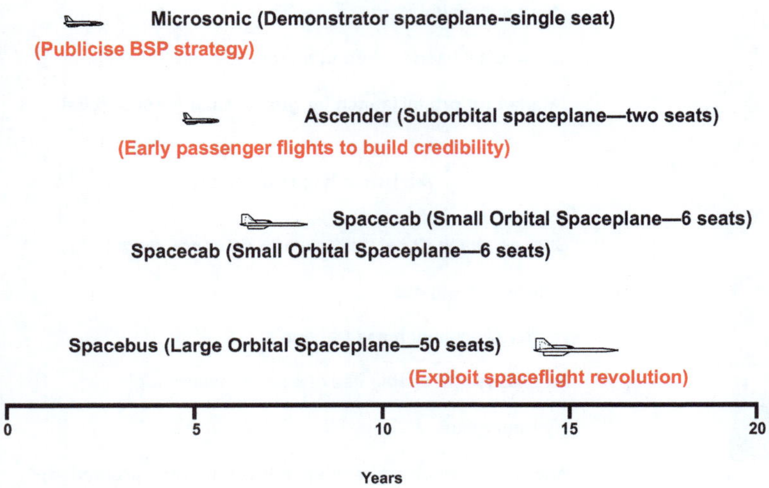

Step-by-step Development Programme

- Step-by-step development programme
- Starting with small demonstrator and ending with Spacebus
- Progressive build up of operational experience, safety man agement, and technical maturity

Microsonic

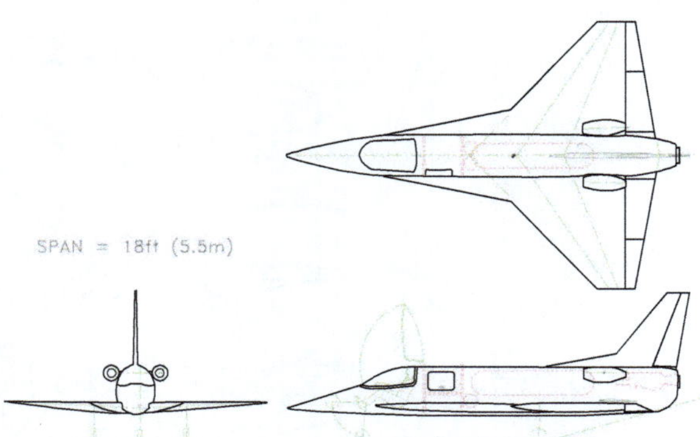

Microsonic

- Demonstrator rocketplane
- Single seat, small science payload to near-space height
- Least cost and risk rocketplane project
- Only project grounded in well-established heritage designs
  - Smaller and simpler than SR.53 rocket fighter of 1957
- Only project designed specifically as lead in to orbital space plane
  - Piloted, winged, horizontal take off and landing, jet and rocket engines

BIS-RS-2015-30

Dynamically Supported Launch Infrastructure

John Knapman
International Space Elevator Consortium
Chandlers Ford, SO53 2DU, UK; 02380254385
JMKnapman@aol.com

ABSTRACT

The space cable is a system of evacuated tubes held aloft by fast-travelling objects called *bolts* inside them using magnetic levitation to minimize friction. It is an extension of existing technology using permanent magnets stabilized electronically. No new materials are required. The proposed configuration uses five pairs of tubes and is suitable for accelerating spacecraft of 80 tonnes mass to 1.6 km/sec at 50 km altitude, thus replacing a first-stage rocket. It is a development of the Launch Loop or Lofstrom Loop, which could launch 5-tonne spacecraft to orbital velocity many times a day. The structure is stabilized to deal with gusting high winds. Mainly for this reason, it has been proposed as the bottom stage – *high stage one* – of the space elevator.

KEYWORDS: Launch Loop, Space Cable, Space Elevator, Atmospheric Turbulence, Magnetic Levitation, Evacuated Tube

1. INTRODUCTION

There have been many proposals to reduce the cost of reaching space by creating fixed infrastructure. The best-known is the space elevator, which is a simple idea that as yet lacks a suitably strong lightweight material. Edwards, B. and Westling, E., The Space Elevator, BC Edwards, Houston, Tx, 2003 Other ideas are the orbital ring and the skyhook, which can be built with today's materials. Birch, P., "Orbital Ring System and Jacob's Ladders," Journal of the British Interplanetary Society, Vol. 35 (1982), pp. 475-497·Moravec, H. P. "A Non-Synchronous Orbital Skyhook," Journal of the Astronautical Sciences 25 (1977), 307–322

Ground-based solutions include electromagnetic launch assistance and the rail gun. Jayawant, B V, Edwards, J D, Wickramaratne, L S, Dawson, W R C and Yang, T C, "Electromagnetic launch assistance for space vehicles," IET Science Measurement and Technology, 2 (1) pp. 45-52 (2008)·McNab, I. R., "Electromagnetic Launch to Space", Journal of the British Interplanetary Society, Vol. 58, No. 2, pp.54-62, 2007 They have the advantage that they can be constructed on Earth and so are much less expensive to develop and test than space-based infrastructure. One of their drawbacks is that payloads have to survive high speeds in dense parts of the atmosphere.

The Launch Loop can be constructed on Earth but lifts its payloads to high altitude so that air resistance and friction are minimized. Lofstrom, K., The Launch Loop, AIAA Paper 85-1368, July 1985 It can accelerate vehicles directly to orbit. A smaller version of the Launch Loop is called the *space cable*. Experimental versions less than a metre high can be constructed. Knapman, J., "Space Elevator Stage I," 62nd International Astronautical Congress, Cape Town, South Africa, 3-7 October 2011 A useful version could lift a ramjet to 20 km altitude while accelerating it to Mach 2 or above, so that it can take off without needing conventional jet engines. Knapman, J., "Diverse Configurations of the Space Cable," 61st International Astronautical Congress, Prague, Czech Republic, September 27-October 1, 2010 For space launch, a version reaching 50 km altitude is proposed that accelerates vehicles to 1.6 km/sec, replacing the first stage of a rocket. This makes it suitable for launching a rocket-powered spaceplane – a reusable, single stage to orbit craft that could reach orbit and return to Earth, landing on a runway. Such a spaceplane would use conventional rockets; it would not need an innovative propulsion system. Its mass could be as high as 80 tonnes.

The work is mostly theoretical, although prototypes of a few small-scale elements have been built.

S. Hatton (ed.), *Proceedings of the 13th Reinventing Space Conference*,
https://doi.org/10.1007/978-3-319-32817-1_28

2. DESIGN PRINCIPLES

The structure is supported by the forces generated as the bolts change the direction of their momentum vectors.Knapman, J., "Dynamically Supported Launcher," Journal of the British Interplanetary Society, Vol. 58, No. 3/4, pp. 90-102, 2005 The weight of the tubes does not change the magnitude of the bolts' momentum. Gravity causes them to lose momentum, and hence kinetic energy, while they ascend, but this is regained during their descent. By using magnetic levitation in evacuated tubes, friction is kept to a bare minimum, so that only a small amount of power is required to replace such losses when the bolts return to the surface stations.

At the surface stations, the bolts proceed to a semicircular track called the ambit, where they are turned around and directed up the ramp to continue their indefinite loop to the other surface station and back again (Figure 1).

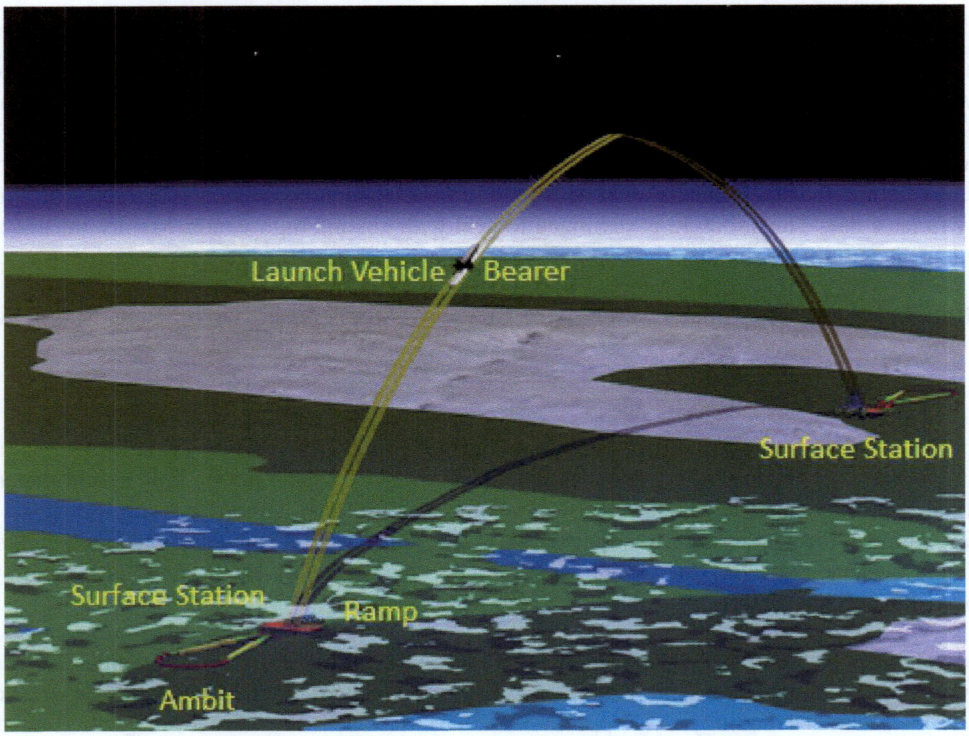

Figure 1 Overall structure of space cable

Magnetic Levitation

In the tubes, permanent magnets are sufficient to provide the forces needed (Figure 2). Non-conducting magnets – either ceramic ferrites or bonded neodymium iron boron (NIB) – are used to avoid eddy currents. Sintered NIB has twice the magnetization of bonded NIB and three times that of ceramic ferrites, but it is unsuitable because it is a conducting material and thus subject to eddy currents. In the ramps and ambits at the surface, superconducting magnets are used.

Figure 2 Prototype section of tube with permanent magnets (the cup shows the scale)

As the bolts travel round a closed loop, they transfer the structure's weight to the ramps at the surface; the ramps are either on the ground or floating on a platform at sea. The weight at each surface station is about 27,000 tonnes for the proposed structure, which rises to a height of 50 km.Knapman, J., "High Altitude Electromagnetic Launcher Feasibility," AIAA Journal of Propulsion and Power, Vol. 22, pp.757-763, 2006 Other dimensions are possible, but it is desirable to keep the structure below the altitude at which space debris becomes a problem.

Each bolt is equipped with permanent non-conducting magnets. These are aligned to be attracted to magnets of opposite polarity on each side of a tube. All the magnets are configured in Halbach arrays to maximize the magnetic field at the closest points. If the bolt is pushed up or down, there is a strong restoring force, and this is the main levitation force for the tubes. However, the bolt is in unstable equilibrium laterally, and sensors are required to control electromagnets at the front and rear of each bolt. Light-sensitive diodes sense the gap between the bolt and the tube, and electronic circuitry provides the moderate currents needed to maintain stability (Figure 3). The faster the electronic response, the less is the power consumption in the bolt, which is powered by capacitors backed up by lithium-ion batteries. Recharging takes place each time the bolts pass through the surface stations.

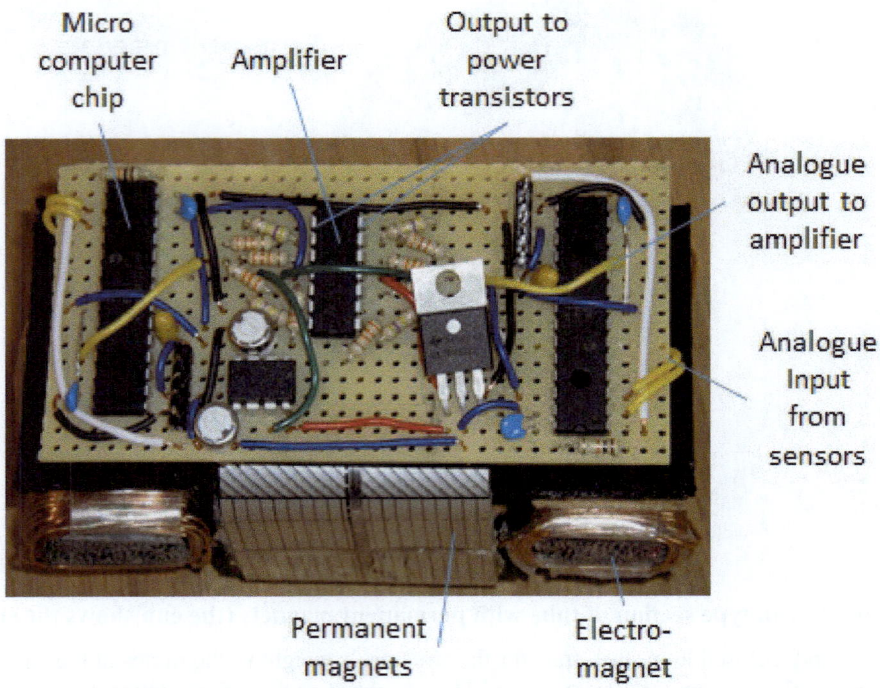

Figure 3 Prototype bolt

Tension and Other Forces

The tubes experience a levitation force from the bolts orthogonal to their direction of travel (Figure 4). The bolts support that component of the tube's weight that is normal to the tube. The component of the tube's weight parallel to the tube is transmitted further up as increasing tension. The tension itself creates a force on the bolts, which is also orthogonal to their direction of travel. At the top, the tube is horizontal, and the bolts there support the tube's weight at the top plus the weight transmitted from lower down in the form of tension.

Figure 5 shows three curves. If the bolts were travelling freely without the tube's weight but with no friction, they would follow a suborbital ellipse. The effect of the tube's weight is to change the shape to the form of an inverted catenary. The second-order differential equation governing that shape has been published.Knapman, J., "Dynamically Supported Launcher," Journal of the British Interplanetary Society, Vol. 58, No. 3/4, pp. 90-102, 2005 If we add another weight, such as a platform, the tubes require a sharper bend of several degrees at that point to sustain it.

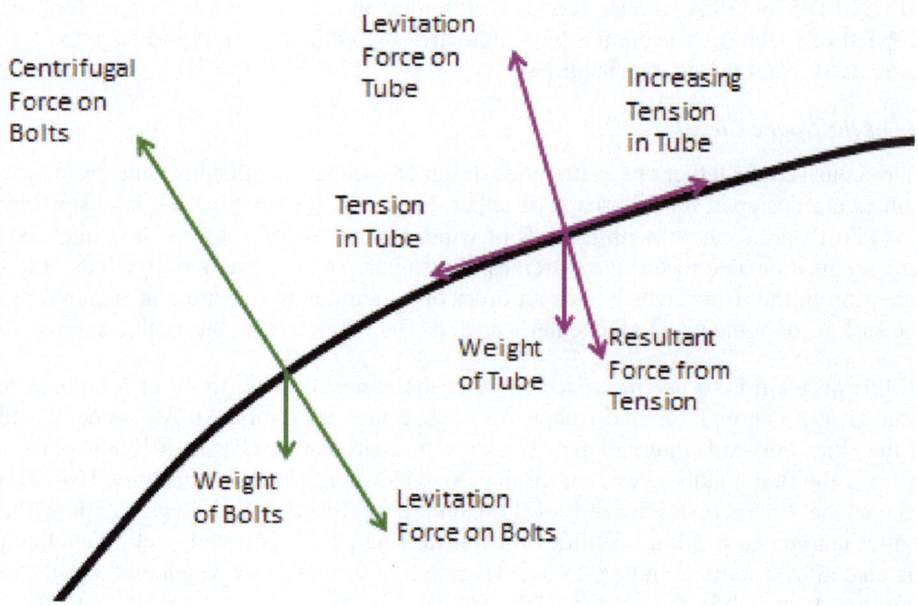

Figure 4 Forces on tube and bolts

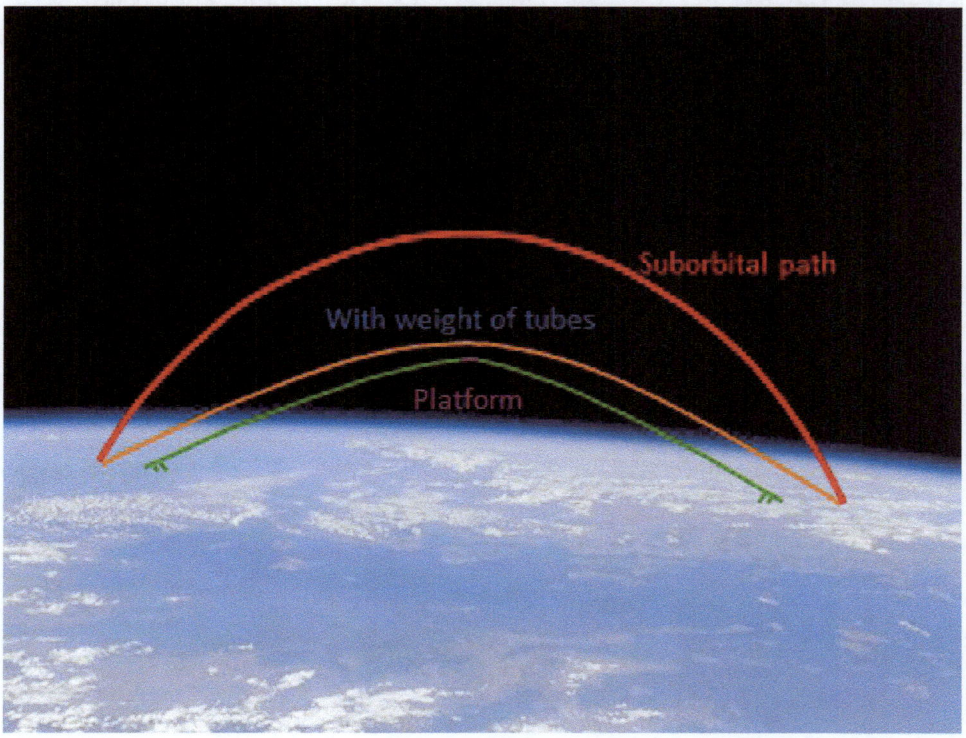

Figure 5 Effect of the tube weight and a platform on the shape of the curve

3. CONFIGURATIONS

The Launch Loop is 2000 km long and 80 km high. It can accelerate a spacecraft to orbital velocity (8 km/sec). The space cable is smaller and therefore less expensive. A version 150 km long and 50 km high is suitable for replacing a first-stage rocket, accelerating an upper-stage rocket or a single-stage-to-orbit spacecraft to 1.6 km/sec. A simpler version of similar size can be used with the space elevator to relieve its tether of potentially large forces due to wind and ice.Knapman, J. and Swan, P., "Design concepts for the first 40 km a key step for the space elevator," Acta

Astronautica 104 (2014) 526-530 A smaller version 55 km long and 20 km high is suitable for accelerating a ramjet to the required speed of Mach 2. A version a few kilometres high could support and astronomical telescope, other scientific instruments, or communications facilities.

High Stage One of the Space Elevator

The space elevator consists of a tether one metre wide designed to sustain collisions with space debris up to 10 cm in width. The climbers are designed for this shape of tether. However, the tether shape is unsuitable for dealing with the turbulence of Earth's atmosphere, with hazards of wind, ice and electric storms. It is intended to be permanent infrastructure and so must be able to survive extremes of weather. To deal with strong winds, it would be necessary to increase the tension in the tether, which adds an order of magnitude to the strength required of the tether, as the tension has to be sustained all the way to the tether's apex at 100,000 km from the earth's surface.

One solution to this problem is to use the space cable in the atmosphere up to 40 or 50 km altitude. Swan, P. A., Raitt, D. I., Swan, C. W., Penny, R. E. and Knapman, J. M., Space Elevators: An Assessment of the Technological Feasibility and the Way Forward, International Academy of Astronautics, Paris, 2013·Knapman, J. and Swan, P., "Design concepts for the first 40 km a key step for the space elevator," Acta Astronautica 104 (2014) 526-530·Knap man, J., High Stage One, International Space Elevator Conference, Seattle, WA, August 25th–27th, 2012 At the top, a platform provides automated handling facilities for transferring payloads to tether climbers that proceed upwards (Figure 6). This also allows tether climbers to use delicate lightweight solar panels without the added weight and complexity of folding them while they climb in the atmosphere. High stage one deals with the forces due to the atmosphere, transferring them back to the earth's surface.

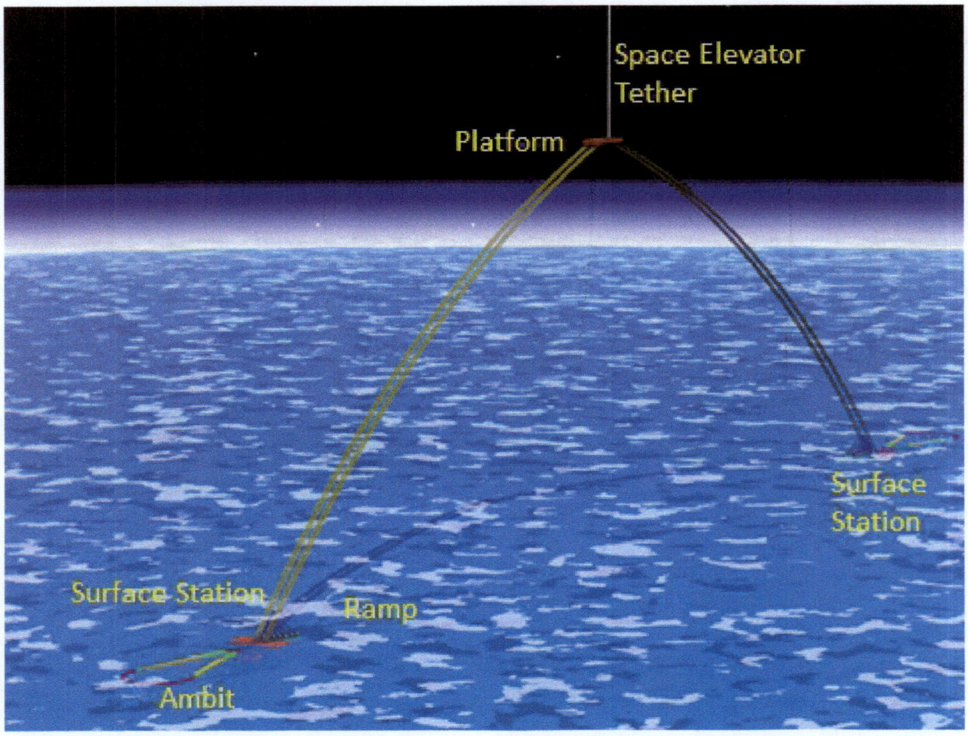

Figure 6 High stage one of the space elevator

The vehicles that travel on high stage one proceed as on a funicular or mountain railway at moderate speed. The structure is shown at sea because the preferred location for the space elevator's Earth terminus is near the equator west of the Galapagos Islands.

Launching Spacecraft

To accelerate a spacecraft to 1.6 km/sec, a component called a *bearer* (see Figure 1) travels along the tubes using magnetic levitation via permanent magnets in the tubes. It obtains thrust from the bolts ascending inside the tubes at 2.2 km/sec. The bolts, which are a metre long, contain electromagnets that are energized only in the vicinity of the bearer, and their force is transmitted to electromagnets in the bearer via passive ferrites in the tubes. The bearer is 150 metres long so that the weight and momentum extraction are spread over about 30 bolts at any time.

Because the forces involved are not conservative, the power (rate of change of kinetic energy) taken from the bolts to supply the necessary thrust (rate of change of momentum) is greater than that required to accelerate the spacecraft. Either this excess power can be disposed of as heat or it can be dumped by accelerating the descending bolts, which provides a little extra thrust.

At the desired altitude and speed, the spacecraft detaches from the bearer and ignites its engines. The bearer continues travelling along the tubes and decelerates towards the surface station. Five pairs of tubes are proposed, that is, five ascending and five descending. Together they are capable of supporting a bearer – about 15 tonnes – plus a launch vehicle of 80 to 85 tonnes, which causes a bend in the tubes (and the bolts' angle of inclination) of about 3°.

Many launches can be performed every day. Prior to a launch, the surface stations raise the bolts' speed by 10% so as to store up the needed energy. This has the effect of increasing the tension in the tubes. The structure remains vertically stable.

Launching a Ramjet

A commercial airliner using ramjets would be able to slash journey times to a third or less of those achieved today. To accommodate all passengers, including children and their grandparents, acceleration would have to be limited to 5 metres/sec^2 (0.5*g*). At that low acceleration, a distance of 36 km is required to reach Mach 2 (about 600 metres/sec). A space cable 20 km high is suitable. The tube length is 70 km, and it extends 55 km over the ground. The bolt speed is 1.2 km/sec.

Such a facility at high altitude would be ideal for the development and testing of ramjets and scramjets.

4. STABILIZATION

Figure 7 shows the estimated maximum wind pressure that the space cable may experience at various altitudes. It is based on wind speeds of four times the global average using the standard formula , where is the air density, and the drag coefficient is for a tube of circular cross section.

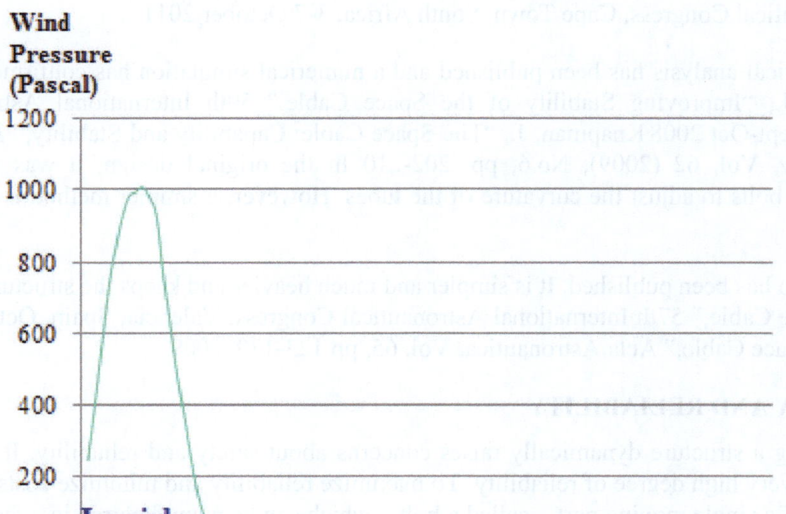

Figure 7 Estimated maximum wind pressure in Pascal

Careful mathematical analysis and computer simulation have shown that the structure can be kept stable in strong winds by a method known as *active curvature control*.Knapman, J., "The Space Cable: Capability and Stability," Journal of the British Interplanetary Society, Vol. 62 (2009), No.6, pp. 202-210 Consider a scenario in which a wind suddenly starts to blow. The natural reaction of the space cable is to bend at the boundary where the wind starts, as shown by the dotted line in Figure 8. Without control, it will continue to bend and become unstable. However, a small controlling force at the right moment can cause the tubes to bend until they reach the right curvature so that the bolts travelling inside them oppose the wind and no more. The wind provides the motive force. The control system moderates the force to limit the bending. As the bolts follow the bend, their centrifugal force balances the force of the wind.

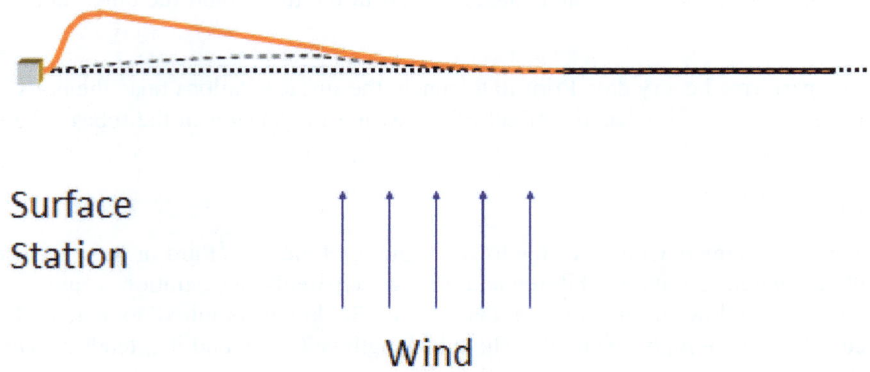

Figure 8 Effect of a sudden gust of wind

The control system works so that the tubes lower down are moved to support the lateral thrust caused by the wind, while the tubes higher up remain stable. After the manoeuvre, the curvature of the tubes matches the wind force. If there is no wind lower down, the tubes will be straight there but become realigned at a small angle to the vertical plane, as shown by the thick line in Figure 8. Movement at the surface station can be up to 160 metres in extreme cases, and an arrangement of tethers has to accommodate this.Knapman, J., "Space Elevator Stage I," 62nd International Astronautical Congress, Cape Town, South Africa, 3-7 October 2011

A complete mathematical analysis has been published and a numerical simulation has confirmed the validity of the approach.Knapman, J., "Improving Stability of the Space Cable," 59th International Astronautical Congress, Glasgow, Scotland, Sept-Oct 2008·Knapman, J., "The Space Cable: Capability and Stability," Journal of the British Interplanetary Society, Vol. 62 (2009), No.6, pp. 202-210 In the original design, it was proposed to use the electromagnets in the bolts to adjust the curvature of the tubes. However, a simpler method is to use piezo-electric actuators in the tubes.

An alternative solution has been published. It is simpler and much heavier and keeps the structure rigid.Knapman, J., "Stability of the Space Cable," 57th International Astronautical Congress, Valencia, Spain, October 2006·Knapman, J., "Stability of the Space Cable," Acta Astronautica, Vol. 65, pp.123-130, 2009

5. SAFETY AND RELIABILITY

The idea of supporting a structure dynamically raises concerns about safety and reliability. It will be necessary to design and build to a very high degree of reliability. To maximize reliability and minimize costs, the design employs thousands of copies of a single moving part – called a bolt – which can be manufactured in a factory. The other high altitude elements are kept simple – mainly permanent magnets and sheathing for the tubes. The technology employed is conventional: magnetic levitation in evacuated tubes to keep friction very low. However, the speeds and altitude involved are new and will require substantial engineering development and testing.

A rigorous, automated method of checking the continuing viability of bolts will be required. Five pairs of tubes operating in parallel are used so that one pair can be taken down for maintenance while the other four pairs continue in service. It is also possible to take down and raise the whole structure.

6. CONCLUSIONS

The space cable in its various forms has the potential to make access to space far cheaper and more reliable than conventional rockets. The capabilities required are a relatively straightforward extension of existing vacuum technology and magnetic levitation. Small versions can be built, from benchtop prototypes through stages up to several kilometres high. The author's work on the theory and limited prototypes pave the way for more professional groups to take the ideas forward.

Cost estimates have been given for constructing a full-scale version in the region of $2 billion to $3 billion.Knapman, J., "The Space Cable: Capability and Stability," Journal of the British Interplanetary Society, Vol. 62 (2009), No.6, pp. 202-210·Knapman, J., High Stage One, International Space Elevator Conference, Seattle, WA, August 25th–27th, 2012

Acknowledgments

The author is grateful to Keith Lofstrom for many useful discussions.

References

Edwards, B. and Westling, E., *The Space Elevator,* BC Edwards, Houston, Tx, 2003

Birch, P., "Orbital Ring System and Jacob's Ladders," *Journal of the British Interplanetary Society*, Vol. 35 (1982), pp. 475-497

Moravec, H. P. "A Non-Synchronous Orbital Skyhook," *Journal of the Astronautical Sciences* **25** (1977), 307–322

Lofstrom, K., *The Launch Loop*, AIAA Paper 85-1368, July 1985

Jayawant, B V, Edwards, J D, Wickramaratne, L S, Dawson, W R C and Yang, T C, "*Electromagnetic launch assistance for space vehicles,"* IET Science Measurement and Technology, 2 (1) pp. 45-52 (2008)

McNab, I. R., "Electromagnetic Launch to Space", *Journal of the British Interplanetary Society*, Vol. 58, No. 2, pp.54-62, 2007

Knapman, J., "Space Elevator Stage I," *62nd International Astronautical Congress*, Cape Town, South Africa, 3-7 October 2011

Knapman, J., "Diverse Configurations of the Space Cable," *61st International Astronautical Congress*, Prague, Czech Republic, September 27-October 1, 2010

Knapman, J., "Dynamically Supported Launcher," *Journal of the British Interplanetary Society*, Vol. 58, No. 3/4, pp. 90-102, 2005

Knapman, J., "High Altitude Electromagnetic Launcher Feasibility," *AIAA Journal of Propulsion and Power*, Vol. 22, pp.757-763, 2006

Swan, P. A., Raitt, D. I., Swan, C. W., Penny, R. E. and Knapman, J. M., *Space Elevators: An Assessment of the Technological Feasibility and the Way Forward*, International Academy of Astronautics, Paris, 2013

Knapman, J. and Swan, P., "Design concepts for the first 40 km a key step for the space elevator," *Acta Astronautica* **104** (2014) 526-530

Knapman, J., *High Stage One,* International Space Elevator Conference, Seattle, WA, August 25th–27th, 2012

Knapman, J., "The Space Cable: Capability and Stability," *Journal of the British Interplanetary Society*, Vol. 62 (2009), No.6, pp. 202-210

Knapman, J., "Stability of the Space Cable," 57th International Astronautical Congress, Valencia, Spain, October 2006

Knapman, J., "Stability of the Space Cable," *Acta Astronautica*, Vol. 65, pp.123-130, 2009

Knapman, J., "Improving Stability of the Space Cable," 59th International Astronautical Congress, Glasgow, Scotland, Sept-Oct 2008

Reinventing Space Conference
BIS-RS-2015-41

Fostering technology innovation in space through national activities: The Swiss example

Grégoire Bourban
Swiss Space Center

Prof. Volker Gass
Swiss Space Center

Johann Richard
Swiss Space Office

13th Reinventing Space Conference
9-12 November 2015
Oxford, UK

S. Hatton (ed.), *Proceedings of the 13th Reinventing Space Conference*,
https://doi.org/10.1007/978-3-319-32817-1_29

In 2010, Switzerland initiated national activities, in order "To foster and promote Swiss scientific and technological competences related to space activities". These measures where implemented by the Swiss Space Office of the State Secretariat for Education, Research and Innovation (SERI/SSO) and the Swiss Space Center was entrusted to ensure the technical implementation and the follow-up. This consists in issuing a "Call for Proposals" every other year, open to all technologies with promising application in the space domain and more specifically in the upstream market in line with the priority axes of the Swiss space policy. Considered as a success, this first initiative was supported in 2013 by the implementation of a "Call for Ideas" targeting early stage technologies. These national measures will be completed in 2016 by the establishment of an ESA Business Incubation Center (BIC). The first results achieved after five years of national activities are very encouraging, success stories are emerging which makes the space industry *"one of the most research-intensive and innovative industries in Switzerland"* [1].

I. Introduction

Switzerland is a founding Member State (MS) of the European Space Agency (ESA or hereafter the Agency). In 2015, over the 22 MS of ESA, with 135 mio Euros, Swiss financial support was ranked at the sixth position behind France, Germany, Italy, UK and Belgium. Moreover, during the ESA Council at ministerial level held in December 2012 in Naples, Switzerland was elected at the co-presidency of the Agency together with Luxembourg until the end of 2016.

This position within the Agency, coupled with the reputation of excellence attributed to Swiss entities (industry and academia), makes Switzerland a reliable and privileged partner in the international space business landscape.

Based on this fact, the Swiss Space Office (SERI/SSO), which is the division of the State Secretariat for Education, Research and Innovation in charge of leading the Swiss delegation at ESA, and also responsible for preparing and implementing the Swiss space policy, has decided to implement national activities to foster space technology innovation. The objectives are to strengthen the current position and support the development of new niches.

II. Swiss involvement in space business

The Swiss space community is composed by one big company, RUAG Space, leading supplier of products for the space industry in Europe, several Small and Medium Entities (SME) with limited number of key products, research centres and laboratories. One of the particularities compared to other ESA MS is the absence of a Large Space Integrator (LSI). This situation explains the necessity to target niche products or competences with high added values, which have to be competitive at the international level.

To support this strategy, the Swiss space policy is articulated along five technology axes: High-precision mechanisms and structures, atomic clocks, electro-optical data transmission, technologies for scientific instruments and technologies for user-funded applications [2]. The payload fairings produced by RUAG Space which equip all the European launchers of Ariane programme and VEGA as well as the American Atlas V500 are one example of success stories. Another one is the atomic clock developed by the company Spectratime in Neuchâtel and flying on-board Galileo satellites.

The clear strategy and the correct positioning of the Swiss delegation at ESA facilitated these developments. The delegation fully supports the technology development programmes of the Agency and fosters cooperation with the other MS by avoiding unnecessary duplications.

III. National activities – the concept

In 2009, a pilot study was initiated between an industrial and an academic partner addressing the technology transfer of a state-of-the-art sensor for space application. This study was selected and funded by the SERI/SSO with the technical support of the Swiss Space Center. Based on this experience, it was decided to launch a "Call for Proposals" open to all Swiss entities with the following main rules: consortium of at least one academic and one industrial partner, 12 months studies, 250'000CHF maximum which can fully fund the industrial part. This last rule was probably the most important one to raise industrial interest. Indeed, instead of asking for co-funding, a clear roadmap for the future development of the product was requested already in the proposal. This roadmap would be discussed again during the final review.

In the meantime, to support the implementation of this initiative known in French as "Mesures de Positionnement" (MdP), the SERI/SSO extended the mandate of the former EPFL Space Center which was renamed in Swiss Space Center (SSC) to highlight its national importance.

The first "Call for Proposals" was issued in February 2010, 23 proposals were received, 9 of which were selected for funding. The process was renewed in 2012 and 2014 with an increasing interest from the Swiss space community but also from newcomers in the space business. Figure 1 shows the number of proposals submitted (light blue) and the number of studies selected (dark blue) for each of the three calls.

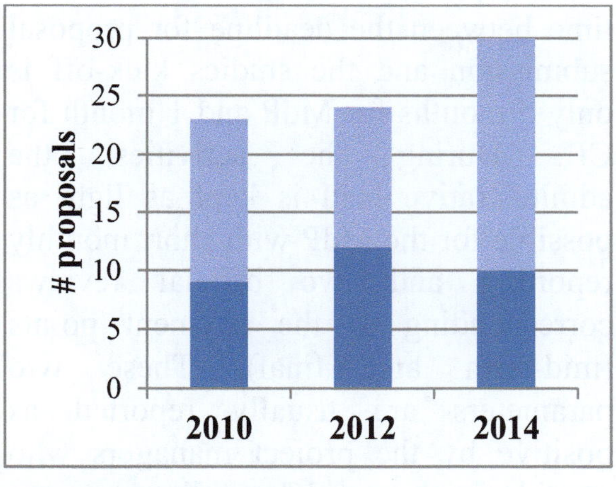

Figure 1 – MdP calls summary

Based on the success of the two first cycles and in order to identify disruptive space innovations based on ideas and concepts, it was decided in 2013 to launch a "Call for Ideas" (CfI). In this case, the target was to foster low TRL research and development studies (typically 1-2) over 6 months with a budget from 20'000 CHF up to

100'000CHF for small mission feasibility studies. Figure 2 shows the number of ideas submitted (light blue) and the number of studies selected (dark blue).

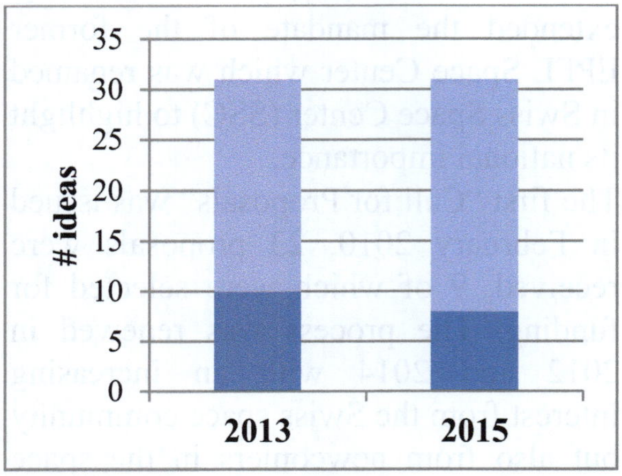

Figure 2 – Call for Ideas Summary

For both MdP and CfI, the key driver for the programme management (SERI/SSO and SSC) is the fast implementation. The time between the deadline for proposal submission and the studies kick-off is only 5 months for MdP and 1 month for CfI. During the activities, the administrative load is kept as light as possible for the MdP with short monthly reporting and two official reviews corresponding to the payment points (mid-term and final). These two parameters are usually reported as positive by the project managers who consider the SSC follow-up as beneficial.

The efficiency of the programme was recognised within the evaluation survey carried out by the University of Applied Sciences in Olten (FHNW) on behalf of the SERI/SSO [1] in April 2015. The opinion of the Swiss space community regarding the programme management organization and transparency of the national complementary measures was positive by 74% and the communication efficiency scored 79%. In comparison, the same questions were asked for ESA programmes and the answers were respectively 38% and 42%.

IV. First results

Out of the 140 proposals submitted in the frame of the three MdP and the two CfI calls, 47 were selected for funding after the assessment of a Tender Evaluation Board (TEB) composed of Swiss and international experts. This represents a 33% success rate. About 75% of the proposals were qualified as "good" to "very good" (e.g. an average mark above 65 over 100). In most of the cases, the limiting factor was the budget envelope allocated to the calls and not the quality of the proposals.

One of the most important conclusions was the number of different entities that submitted a proposal in the three MdP calls: 71 laboratories/institutes and 66 industries. This illustrates the broadness of the Swiss space community and the attractiveness of such calls.

Thanks to such activities (MdP and CfI), 8 new start-ups selected for funding were discovered and are now considering space as a new business model. One of these start-ups reported the signature of a contract in the automotive industry thanks to the image given by the space label. This "side-effect" illustrates the benefit for a newcomer to invest time in the proposal elaboration.

The quantity of proposals evaluated also allowed a better understanding of the Swiss interest in future technology development for space. The graph in figure 3 was created based on the proposals submitted. It shows the proportion of studies proposed according to technology domains.

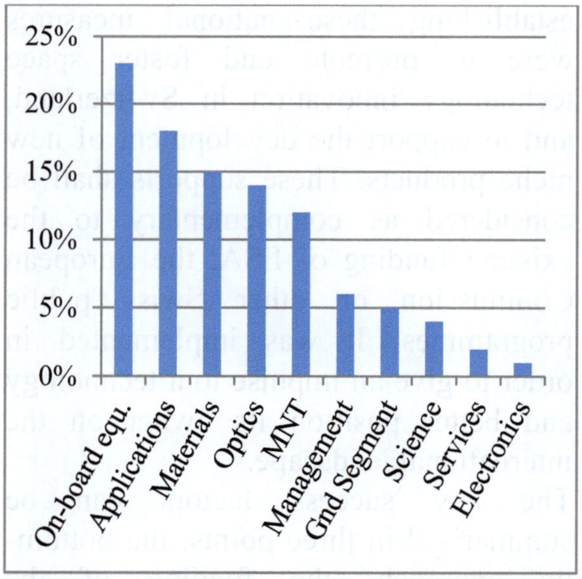

Figure 3 – Technology domains proposed (MNT=Micro Nano Technologies)

A survey was carried out during summer 2015 among the project managers or entities involved in the two first MdP calls. The objective was to determine how far the projects evolved one and three years after the end of the funding. The main results are in figure 4.

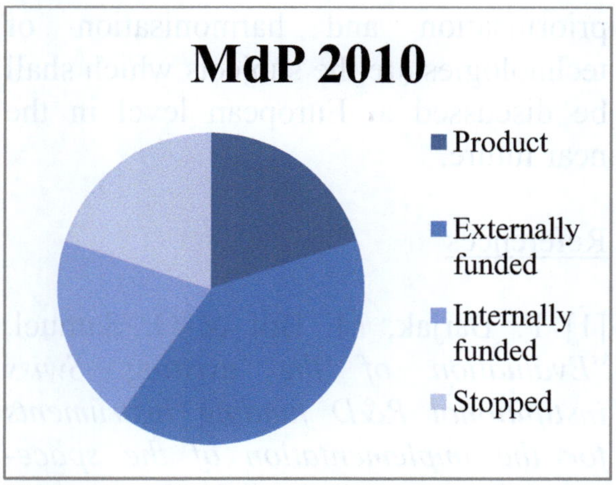

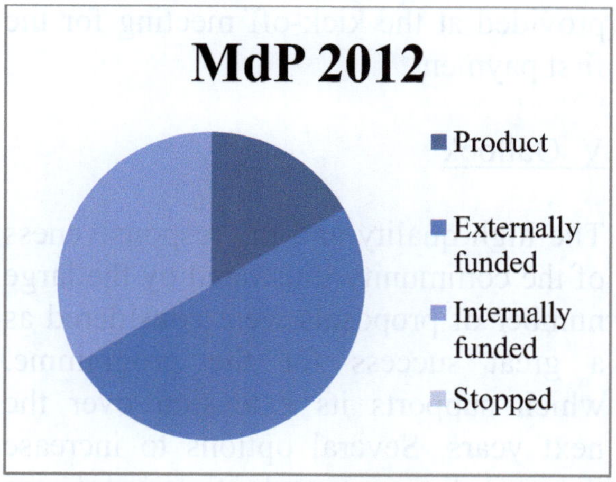

Figure 4 - Follow-up MdP 2010-2012

Over the 22 projects considered, 4 achieved a product maturity at the end of the funding (new software modules, terrestrial based instruments), 10 were pursued with external funding (ESA or the 7th Framework Programme of the European Commission), 6 were pursued with internal funding and only 2 were put on hold by the companies. In these 2 last cases, one of them could be explained by the low maturity reached at the end of the project which did not allow a transfer to the industry. For the other one, the interest is still there but the roadmap was not clear enough. It was developed by the largest consortium in the MdP programme in terms of entities number (2 industries, 1 research center, 1 laboratory). This induced discussions about intellectual property and slowed down technology development.

As this is a rather new programme and in order to continue its improvement, "lessons learned" are collected by the programme management and implemented if feasible. This was the case after the first MdP call, when it was decided to extend the project duration to 15 months and request an IP agreement between the partners to be signed and provided at the kick-off meeting for the first payment.

V. Outlook

The high quality and the responsiveness of the community illustrated by the large number of proposals were considered as a great success for the programme, which supports its extension over the next years. Several options to increase its attractiveness and its impact are considered but the importance to keep the same identity and continue to "build the brand" is also taken into account.

In parallel, together with the Integrated Applications Programme (IAP) and the Technology Transfer Office (TTO) of ESA, the SERI/SSO has decided to implement a Business Incubation Center (BIC) in Switzerland. The model, based on a Public Private Partnership (PPP), shall support the creation of a maximum of 10 start-ups per year with a funding of 500kEuros/start-up over two years of incubation. This initiative shall start in 2016 with the selection of the hosting entity among the existing business parks of Switzerland and the issue of the first call.

Although 12 other BICs are established in Europe, the Swiss one will be unique for two reasons: it is based on a PPP model, it targets not only the applications business (downstream) but also the space technology development (upstream).

This will make the ESA BIC Switzerland the third stage of the Swiss strategy to foster technology innovation in space after MdPs and CfIs.

VI. Conclusions

The objectives for the SERI/SSO when establishing these national measures were to promote and foster space technology innovation in Switzerland, and to support the development of new niche products. These supports shall be considered as complementary to the existing funding of ESA, the European Commission or other Swiss public programmes. It was implemented in order to give an impulse to a technology and better position its owner on the international landscape.

The key success factors can be summarised in three points: the bottom-up approach, the funding of the industrial partner and the fast implementation coupled with light administrative work.

The future challenge will be to facilitate the validation/qualification of the most promising developments by overcoming the "TRL valley of death" (range between TRL 4 and 7 in which many projects fail). Access to in-orbit demonstrators, revision of standards, prioritisation and harmonisation of technologies are the subjects which shall be discussed at European level in the near future.

References

[1] F. Barjak, M. Bill & O. Samuel, "*Evaluation of the existing Swiss institutional R&D funding instruments for the implementation of the space-related measures*", April 2015

[2] SERI/SSO "*Swiss Space Implementation Plan*", December 2013

MIX
Papier aus verantwortungsvollen Quellen
Paper from responsible sources
FSC® C105338

If you have any concerns about our products,
you can contact us on
ProductSafety@springernature.com

In case Publisher is established outside the EU,
the EU authorized representative is:
Springer Nature Customer Service Center GmbH
Europaplatz 3, 69115 Heidelberg, Germany

Printed by Libri Plureos GmbH
in Hamburg, Germany